ADHESIVES
HANDBOOK

Adhesives Handbook

J. Shields, B.Sc.

Senior Research Officer,
Engineering and Materials Technology Department
Sira Institute Limited

London
Newnes—Butterworths

THE BUTTERWORTH GROUP

ENGLAND
Butterworth & Co (Publishers) Ltd
London: 88 Kingsway, WC2B 6AB

AUSTRALIA
Butterworths Pty Ltd
Sydney: 586 Pacific Highway, NSW 2067
Melbourne: 343 Little Collins Street, 3000
Brisbane: 240 Queen Street, 4000

CANADA
Butterworth & Co (Canada) Ltd
Scarborough: 2265 Midland Avenue, Ontario M1P 4S1

NEW ZEALAND
Butterworths of New Zealand Ltd
Wellington: 26–28 Waring Taylor Street, 1

SOUTH AFRICA
Butterworth & Co (South Africa) (Pty) Ltd
Durban: 152–154 Gale Street

First published in 1970 by
the Butterworth Group

Second edition in 1976 by
Newnes–Butterworths

ISBN 0 408 00210 7

Filmset by Photoprint Plates Ltd
Rayleigh, Essex

Printed and bound in England by
The Pitman Press, Bath

CONTENTS

Preface to the Second Edition

The first edition of this book was sponsored jointly by Sira and the Ministry of Technology. In preparing the second edition I have attempted to bring the subject matter up to date without radical changes in the method of presentation.

Since the first edition was published there has been increased activity in the field of adhesives development. The multiplicity of adhesive products available today is such that there can be no possibility of comprehensive treatment within anything like the present compass. Increasingly, many industrial adhesives are highly specific and are designated for use in special situations, on particular equipment or specific tasks on one type of machine. The lifetime of many products is short, and typically, about a quarter of current sales are from products less than three years old. Clearly, the onus is on the user to approach the manufacturer where bonding processes involving specific application or processing equipment for mass production is concerned.

In this situation, attempts to compile a comprehensive trade product account became quite impracticable. Recognising this, care has been taken to select trade products which are of more general utility and not too restricted by equipment considerations and likely to remain on the market for at least five years. In making the selection of trade products, I would like to acknowledge with thanks the cooperation of Dr P. Bosworth and his colleagues of the British Adhesive Manufacturers Association (BAMA) and also those member companies who responded to my request for information.

It is hoped that the book can serve to introduce the newcomer to the technology of adhesive bonding and guide him to a useful understanding of the wide variety of adhesives which exists today.

Last but by no means least, sincere thanks are due to colleagues at Sira who have undertaken various tasks in the preparation of the manuscript. Particularly, I should like to acknowledge a major contribution to the preparation and typing of text material made by Mrs. B. Manderscheid, and the later assistance of Mrs. G. M. Jones with the typing of sections in Chapter 5.

J.S.

Preface to the First Edition

Recent years have seen the rapid development of adhesive bonding as an economic and effective method for the fabrication of components and assemblies. The recognition of this by Sira and the Ministry of Technology has led to the joint sponsoring of the present handbook. Past experience at Sira in connection with a large number of adhesive enquiries has shown that potential users are often deterred by a lack of reliable data or by unfortunate past experiences with adhesives. The book aims to assist in overcoming their difficulties in this field. No attempt has been made to deal comprehensively with the theoretical aspects of the subject which are discussed in greater detail in the books referred to elsewhere in the text.

A large part of the book describes adhesives which are representative of the wide spectrum of materials available today. More emphasis has been given to products based on recent synthetic polymers since many current bonding problems involving new materials or severe service conditions are solvable only in terms of these adhesives. A major objective of the handbook is to provide basic guidance for designers and technologists concerned with assembly processes with an outline of the basic concepts of adhesive bonding: proper design of the adhesive joint, adequate surface preparation of bonding materials, selection of a suitable adhesive and the specification of processing and testing techniques.

Many firms and organisations were kind enough to supply trade literature and other documents for the book and special thanks are due to the following: Ciba (A.R.L.) Ltd. Duxford; Imperial Metal Industries (Kynoch) Ltd. Kidderminster; Furniture Industries Research Association (FIRA), Stevenage; Shoe and Allied Trades Association (SATRA), Kettering; Forest Products Research Laboratory, Princes Risborough; B.A.C. (Operating) Ltd. Filton, Bristol. The last named provided valuable report material on adhesives evaluation work which has been used extensively and is acknowledged here and elsewhere.

I must express my appreciation to the following among the many colleagues at Sira who have assisted with the manuscript. Special credit is due to Mr. R. J. Wolfe for preparing the section on 'Joint Design', to Mr. J. Dracass for the contribution on 'Ceramic and Refractory Inorganic Adhesives', and to Mr. L. Holt and Mr. I. J. Humphreys for preparing the charts, diagrams and tables which have been so competently drawn by Mrs. E. J. Whiting and Miss C. Banks. Thanks are due to Mr. R. G. Burrows and my former assistant Gillian Collins (now Dix) for the collection and collation of the trade material and literature references used in the various sections. I am indebted to Dr. D. C. Cornish, Mr. B. Weight and Mr. M. Kenward who were kind enough to read and comment on the manuscript during its last stages and correct errors which must otherwise have escaped my attention. I also acknowledge the assistance given by the Ministry of Technology for this work; in particular, I thank Mr. D. G. Anderson for his encouragement and helpful advice and especially, Mr. L. Greenwood for his many comments and valuable suggestions following a close study of the script.

J.S.

CHAPTER 1 | # Introduction

1.1. USING THE HANDBOOK

While the handbook will be of value to those seeking general information on adhesives, it is primarily intended to provide basic guidance for the designer concerned with adhesive bonding as an assembly process. The non-specialist will readily locate those aspects of adhesives of particular interest to him by means of the contents. Where adhesive selection and usage with respect to a specific problem is the major consideration, the following procedure is suggested as a method of using the handbook to best advantage.

First read the introduction for background information on adhesives and note whether there are any particular advantages or disadvantages to adhesive bonding for the application in mind.

For certain applications joint design is important to adhesive selection and an early appreciation of the factors involved is essential and should be considered at this stage (p. 7).

Consider the section on Adhesive Selection (p. 23) and define the problem in sufficient detail with the aid of the Adhesive Checklist (p. 29).

Refer to Tables 9.1–9.7 to select candidate adhesive types for the materials to be joined. Tables 9.8–9.26, dealing with bulk physical properties, may indicate that some of these types have unsuitable properties and can be rejected.

Consult the section on Adhesive Materials and Properties (p. 30) for additional information on the selected materials. For many assembly problems commercial examples of these types described in the section Adhesive Products Directory (p. 80), will provide a ready answer or indicate potential adhesive products for comparative evaluation. At this point, refer to the section on adherend Surface Preparation (p. 219) before proceeding with the experimental evaluation of the selected adhesives.

The section on Physical Testing of Adhesives (p. 255) will also repay study at this stage. Other problems will require consultation with consultants or manufacturers.

A list of the basic adhesive types and their trade sources will be found on p. 320. Addresses of manufacturers of processing plant and equipment are listed on p. 251.

1.2. THE CLASSIFICATION OF ADHESIVES

A great many types of adhesives are currently in use and there is no adequate single system of classification for all products. The adhesives industry has generally employed classifications based on end-use, such as metal-to-metal adhesives, wood adhesives, general purpose adhesives, paper and packaging adhesives and so on. A limitation of this system is that a particular end-use adhesive may be useful in several other fields. Apart from end-use, adhesives may be classified according to physical form, chemical composition, method of application, various processing factors (e.g. setting action), and suitability for particular service requirements or environments. Some of these other classification schemes which have been adopted by adhesive technologists are briefly outlined below, together with the broad-based scheme used for this book.

Bonding temperature
This is based on the temperature required by the adhesive to establish a bond. Thus, the setting temperatures describe the classes of adhesive as follows: cold setting (below 20°C), room temperature setting (20–30°C), intermediate temperature setting (31–100°C), hot setting (above 100°C).

A variation of this system is found in the Ministry of Technology Aircraft Material Specification D.T.D. 5577 (Nov. 1965) *Heat Stable Structural Adhesives* which covers requirements for heat stable adhesives for use in bonding metallic and reinforced plastics airframe structures and honeycomb materials. They are classified according to mechanical performance in tensile shear and peel on exposure to various temperatures and aircraft fluids (Table 1.1).

Table 1.1.

Adhesive type	Classes	Temperature range (°C)
1	1P, 1H, 1PH	−60 to +70
2	2P, 2H, 2PH	−60 to +70
3	3P	−60 to +70
4	4P, 4H, 4PH	−60 to +150
5	5P	−60 to +220
6	6P	−60 to +350

The type number is determined by the ability of the adhesive to satisfy specified strength criteria (for tensile shear and peel stresses) for exposure periods up to 1 000 h at the maximum service temperature for the type.

The class letters are defined as follows:

Class P—an adhesive for (according to BS185, Section 3:1962) plate to plate bonding;

Class H—an adhesive for plate to honeycomb bonding;

Class PH—an adhesive suitable for class P and class H applications.

Origin
Adhesives may be broadly classified as being either natural (occurring naturally and requiring little change) or synthetic.

Method of bonding
This type of classification divides adhesives into categories that refer to the physical state or method of application of the adhesive. Typical groups are the pressure-sensitive adhesives, hot-melt adhesives, chemical-setting adhesives, solvent-release adhesives. etc.

Structural and non-structural adhesives
These classifications are somewhat arbitrary since there is no accepted specific definition of 'structural' in terms of bond strength. A structural adhesive is normally defined as one which is employed where joints or load-bearing assemblies are subjected to large stress loads. Non-structural adhesives cannot support heavy loads and are essentially employed to locate the components of an assembly or to provide temporary adhesion. A tensile bond strength exceeding 1 000 N/cm² at room temperature has been used, throughout the present text, as an

arbitrary criterion for designating an adhesive as a structural one.

Permanence
A durability classification is of particular importance to users of woodworking adhesives. The Draft British Standard Specification for Synthetic Resin Adhesives, Gap filling (Phenolic and Amino Plastic) for Constructional Work in Wood (BS1204:1964) gives durability ratings for urea, phenolic, resorcinol and melamine type adhesives. These four adhesive types are designated on the basis of the test used in the specification and partly known grades of durability as follows:

TYPE INT: INTERIOR joints with these adhesives (i.e. phenol and resorcinol formaldehyde) withstand cold water but need not resist attack by micro-organisms.

TYPE MR: MOISTURE-RESISTANT and MODERATELY WEATHER RESISTANT joints based on these adhesives (i.e. urea formaldehyde) survive full exposure to weather for a few years. They will withstand cold water for a long period and hot water for a limited time, but fail under a boiling water test.

TYPE BR: BOIL-RESISTANT. Joints made with these adhesives (i.e. melamine formaldehyde) withstand cold water for many years and have high resistance to attack by micro-organisms. Resistance to weather and boiling water is good and under prolonged exposure to weather, joints fail under weather conditions that Type WBP adhesives will withstand.

TYPE WBP: WEATHERPROOF AND BOILPROOF. This specification refers to adhesives which are known to have long term durability with respect to weather, boiling water, dry heat and micro-organisms. Phenolic resins are the only materials to have qualified as yet.

Other systems
The British Standards Institution is attempting to classify adhesives on the basis of the chemical type or major ingredient from which the adhesive is made. In addition to this, some consideration is being given to the use of 'tabulated' systems as means of completely describing adhesives in terms of the various factors of interest to the user. The employment of a 'tabulated' system (which could conceivably be extended to computer storage) would enable industrial users to retrieve information on those adhesive types satisfying requirements for form of material, properties, processing features, and performance, to economic advantage.

The following broad scheme (Table 1.2) is based on the origin, physical and chemical type of the main ingredient of the adhesive formulation. This basis for classification, together with other descriptive tables and charts, combines elements of all the aforementioned criteria and has been used to describe the adhesives in this handbook.

It is impossible to classify adhesive materials according to a single parameter (e.g. chemical constitution or end-use) without contradicting some principle of the particular classification as some adhesives may qualify for entry under a number of headings. Thus, natural rubber and cellulose could also be vegetable types in the scheme below. Many adhesives are compounded from basic materials belonging to different classes (e.g. casein-latex) and with some rubber-resin types of adhesive (e.g. polysulphide-epoxy) either component could be the

dominant one and the material classed as an elastomer (modified) or a thermosetting resin (modified).

1.3. BACKGROUND TO THE USE OF ADHESIVE BONDING

Up to the early part of this century, the only adhesives of major importance were the animal and vegetable glues which had been in use for thousands of years, and these materials are still widely employed for bonding porous materials such as paper. Casein glues were used structurally during World War I to construct the wooden main-frames of aircraft but these were found to have limited resistance to

Table 1.2.

Origin and basic type		Adhesive material	
Natural	Animal		Albumen, animal glue (inc. fish), casein, shellac, beeswax
	Vegetable	Natural resins	(gum arabic, tragacanth, colophony, Canada balsam, etc.); oils and waxes (carnauba wax, linseed oils); proteins (soyabean); carbohydrates (starch, dextrines)
	Mineral	Inorganic materials	(silicates, magnesia, phosphates, litharge, sulphur, etc.); mineral waxes (paraffin); mineral resins (copal, amber); bitumen (inc. asphalt)
Synthetic	Elastomers	Natural rubber	(and derivatives, chlorinated rubber, cyclised rubber, rubber hydrochloride)
		Synthetic rubbers and derivatives	(butyl, polyisobutylene, polybutadiene blends (inc. styrene and acrylonitrile), polyisoprenes, polychloroprene, polyurethane, silicone, polysulphide, polyolefins (ethylene vinyl chloride, ethylene polypropylene))
		Reclaim rubbers	
	Thermoplastic	Cellulose derivatives	(acetate, acetate-butyrate, caprate, nitrate, methyl cellulose, hydroxy ethyl cellulose, ethyl cellulose, carboxy methyl cellulose)
		Vinyl polymers and copolymers	(polyvinyl-acetate, alcohol, acetal, chloride, polyvinylidene chloride, polyvinyl alkyl ethers)
		Polyesters (saturated)	(Polystyrene, polyamides (nylons and modifications))
		Polyacrylates	(methacrylate and acrylate polymers, cyano-acrylates, acrylamide)
		Polyethers	(polyhydroxy ether, polyphenolic ethers)
		Polysulphones	
	Thermosetting	Amino plastics	(urea and melamine formaldehydes and modifications)
		Epoxies and modifications	(epoxy polyamide, epoxy bitumen, epoxy polysulphide, epoxy nylon)
		Phenolic resins and modifications	(phenol and resorcinol formaldehydes, phenolic-nitrile, phenolic-neoprene, phenolic-epoxy)
		Polyesters (unsaturated)	
		Polyaromatics	(polyimide, polybenzimidazole, polybenzothiazole, polyphenylene)
		Furanes	(phenol furfural)

moisture and to mould growth. These limitations of adhesives of natural origin have provided the stimulus responsible for the great expansion since the 1930s in the development of new adhesives which are based upon synthetic resins and other materials. The outstanding advantage of the new adhesives over the earlier types is their excellent resistance to moisture, mould growth and a variety of other hazardous service conditions.

Phenol formaldehyde was the first synthetic resin of importance to adhesive bonding, being mainly used for wood assembly and plywood manufacture. Later demands of the aircraft industry for materials suitable for metal bonding led to the employment of modified phenolic resins containing synthetic rubber components to produce adhesives displaying high shear and peel strengths. The 1950s saw the introduction of epoxy resin based adhesives offering equal strength properties and the processing advantages associated with 100% reactive solids systems.

Today, the number of applications for adhesives is large and ranges from industrial processes using considerable amounts to assembly jobs which depend on the use of small quantities of adhesive. Paper, packaging, footwear and woodworking still remain the major outlet for adhesives but usage has increased significantly in industrial equipment, building and construction, vehicle manufacture, instrumentation, electrical and optical assemblies, and for military and space applications. The last decade has seen the advent of many new synthetic resins and other components which have made possible the development of stronger, more durable and versatile adhesives for bonding surfaces which were difficult or impossible to bond before (e.g. recent thermosetting plastics and composites). These materials have been developed concomitantly with improved bonding equipment and techniques. As a result, adhesive bonding is now of considerable importance for joining metals to themselves and other materials in structural applications, and for a wide variety of other purposes.

A survey of all the applications, or even industries, that employ adhesives is not feasible; non-structural adhesives in particular have a limitless potential. The main applications of the various adhesive types are, however, considered in the following sections of the handbook:

Adhesive Materials and Processes deals with the major applications of each specific adhesive type described.
Adhesive Products Directory includes the main uses of commercially available adhesive products.
Adhesives Selection Charts. Table 9.1 refers to the adhesive types used to bond different materials. Table 9.3 gives an indication of the main industrial areas in which the adhesive types are

employed. The short list of recent survey papers (p. 345) will provide the reader with a more detailed consideration of adhesives applications pertaining to particular industries.

1.4. FACTORS DETERMINING WHETHER TO BOND WITH ADHESIVES

The basic function of an adhesive is to fasten the components of an assembly together and maintain the joined parts under the service conditions specified by the design requirements. In fulfilling this role, adhesive materials provide the answers to many joining problems, simplify and expedite assembly techniques, and reveal opportunities for design in new areas.

The consideration to use adhesives in the design of a product is generally called for where the following aspects are concerned:

The properties of the materials, or the special properties required of the finished assembly (or its behaviour in service), may indicate that adhesives are the only possible solution to a bonding problem. Often, the use of mechanical fastening methods (e.g. riveting, brazing, soldering, heat welding, screw or nail attachment) results in distortion, discoloration, corrosion or impairment of the assembly by virtue of other undesirable shortcomings.

Adhesive bonding may be preferred as a means of reducing production costs or improving performance even though alternative fastening methods such as riveting, welding and soldering, etc., are feasible. Adhesive bonding as an assembly method can offer cost advantages over alternative fastening methods, according to the requirements, dimensions and properties of the components. Unlike other methods, adhesives do not require substrates to be machined (e.g. drilling of rivet or bolt holes) so that overall costs tend to be reduced. However, adhesive bonding may involve other expense on equipment for application and curing the adhesive or jigs and fixtures for adherends so that, for some assemblies, mechanical fastening may be more economic.

Adhesives may be required to complement other fastening methods in an assembly.

Examples of application areas for which adhesive bonding is a practicable method of assembly include:

Dissimilar materials
combinations of metals, rubbers, plastics, foamed materials, fabrics, wood, ceramics, glass, etc.
Dissimilar metals which constitute a corrosion couple.
iron to copper or brass.

CHAPTER 2

Joint Design

Components that are to be connected by adhesive bonding must have specially designed joints. It is not sufficient to bond a joint that has previously been welded or riveted without first considering the loads and stresses that the joint must withstand. Thought should also be given to the means of clamping the joint during curing. The joint design chosen will usually depend on two main factors:

1. The direction of all the applied loads and forces that the joint will have to withstand in service

2. The ease with which the joint can be formed. This will depend upon the way in which the adherends are manufactured (cast, moulded, machined, etc.) and the material used.

2.1. TYPES OF STRESSES

There are four types of stresses that are important when considering adhesive bonded joints. These are shear, tension, cleavage and peel (Fig. 2.1).

Shear
A shear loading imposes an even stress across the whole bonded area. This uses the joint area to the best advantage, giving an economical joint that is most resistant to joint failure. Whenever possible, joints should be formed in such a way that most of the load is transmitted through the joint as a shear load.

Tension
The strengths of joints when loaded in tension or shear are comparable. Again, the stress is evenly distributed over the joint area, but it is not always possible to be sure that this is the only stress present. If the applied load is at all offset, then the benefit of an evenly distributed stress is lost and the joint will become more likely to fail.

It is also important that the adherends should be thick, and not liable to appreciable deflection under the applied load. If this is not so, then once again, the stress will be non-uniform.

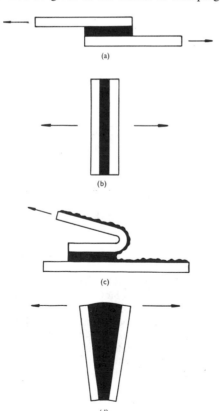

Fig. **2.1**. *The four important stresses:*
(a) shear, (b) tension, (c) peel, (d) cleavage

Cleavage

This type of loading is usually the result of an off-set tensile force or a moment. Unlike the previous cases, the stress is not evenly distributed, but concentrated at one side of the joint. A sufficiently large bonded area is needed to accommodate this stress in which case the joint will be less economical.

Peel

For this type of loading to be present, one or both of the adherends must be flexible. The effect of peel is to place a very high stress on the boundary line of the joint, and unless the joint is wide or the load small, failure of the bond will occur. This form of loading should be avoided whenever possible.

2.2. TYPES OF JOINTS

All bonded joints, however complex, can be reduced to the four basic types shown in Fig. 2.2.

Table 2.1 shows joint designs suitable for various loadings. If it is not possible to be sure in which directions the joined assembly will be loaded in service, choose the design marked 'MAY BE LOADED IN ANY DIRECTION'.

Each diagram carries one or more of the letters S, T, C and P. These denote the most important stresses that the joint will have to withstand, and are shear, tension, cleavage and peel respectively. Other types of stress may be present, but the values will be low and may be ignored.

It will be seen that for some loadings, a number of joint configurations are suitable. This is particularly so with butt joints. The type that is cheapest and easiest to produce should normally be chosen depending on the way the adherends are manufactured. If the joint must be machined (instead of cast, moulded, etc.) then a joint such as a single lap may be suitable. Sometimes, however, when very high strength is required, the joint involving the greatest bonded area may be chosen. In this case eccentrically loaded joints such as the simple lap, should be avoided. In the final analysis it is still possible to have more than one design and the choice will be a matter of taste.

2.3. SELECTION OF JOINT DETAIL

The joint or selection of joints required can be found by use of Table 2.2. It will be noted that after the primary selection of angle, tee, butt and surface joints already mentioned, the correct 'details of adherends' must be chosen. The table refers to these as homogeneous or laminated and rigid or flexible. Laminated adherends refers to all materials of an anisotropic nature where a force applied to the surface of the material might cause delamination. This applies principally to reinforced plastics and wood, laminated or otherwise. Homogeneous adherends is used to indicate the use of materials such as rubber, plastics, ceramic or metal.

The flexibility of an adherend depends upon the material used and the physical size of the component to be bonded. The most satisfactory way of deciding whether a component is rigid or flexible is to consider a force of the order to which the bonded

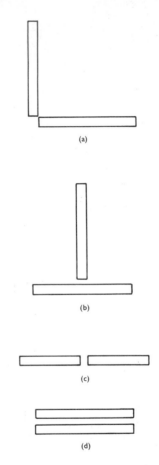

Fig. 2.2. The four basic types of joints: (a) angle, (b) tee, (c) butt, (d) surface

joint will eventually be subjected. If the component deflects appreciably when this force is applied at the end of a cantilever about 10 cm long, then the component may be considered flexible (Fig. 2.3). Thus, a soft rubber sheet 20 cm wide by 1 cm thick would be flexible when loaded with a 40N force as also would be a steel plate 1 mm thick by 4 cm wide. Examples of rigid adherends would be steel, hard rubber or plastics of similar proportions to the soft rubber under the same force.

The two adherends may be of dissimilar materials. The index table takes account of this possibility by providing two sets of columns, one vertical and one horizontal for selecting adherend details. If the

Table 2.1. RECOMMENDED JOINT CONFIGURATIONS AND STRESSES

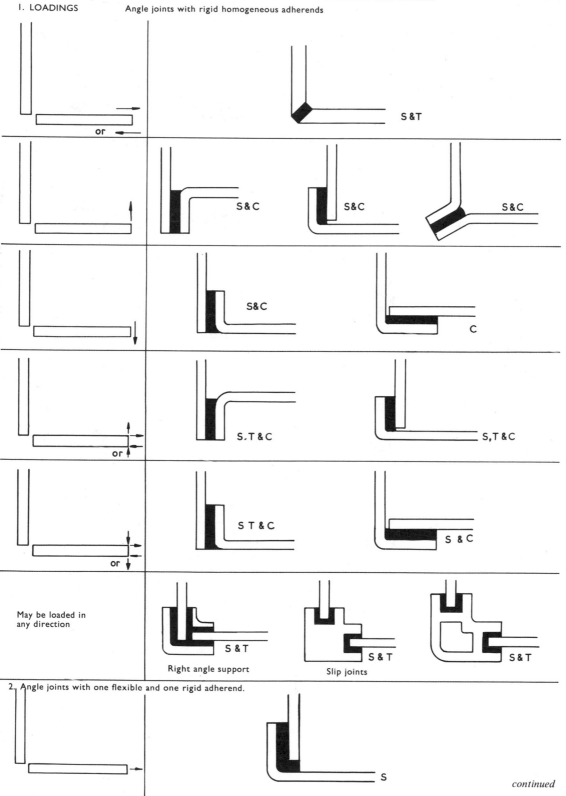

1. LOADINGS   Angle joints with rigid homogeneous adherends

2. Angle joints with one flexible and one rigid adherend.

continued

10

Table 2.1 continued

3.

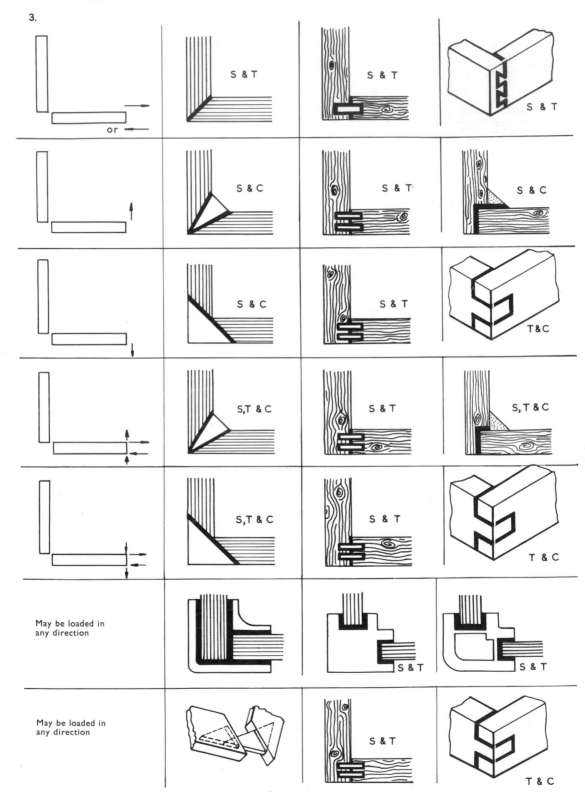

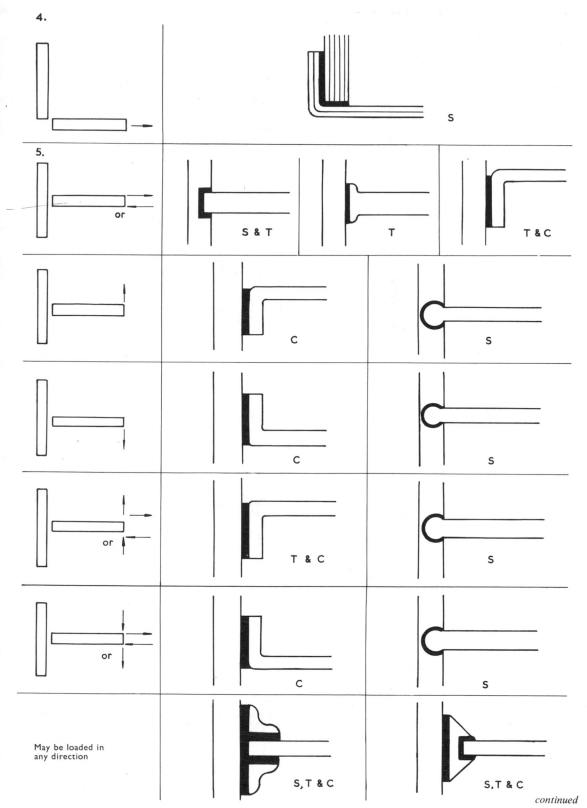

4.

S

5.

or

S & T

T

T & C

C

S

C

S

or

T & C

S

or

C

S

May be loaded in
any direction

S, T & C

S, T & C

continued

Table 2.1 *continued*

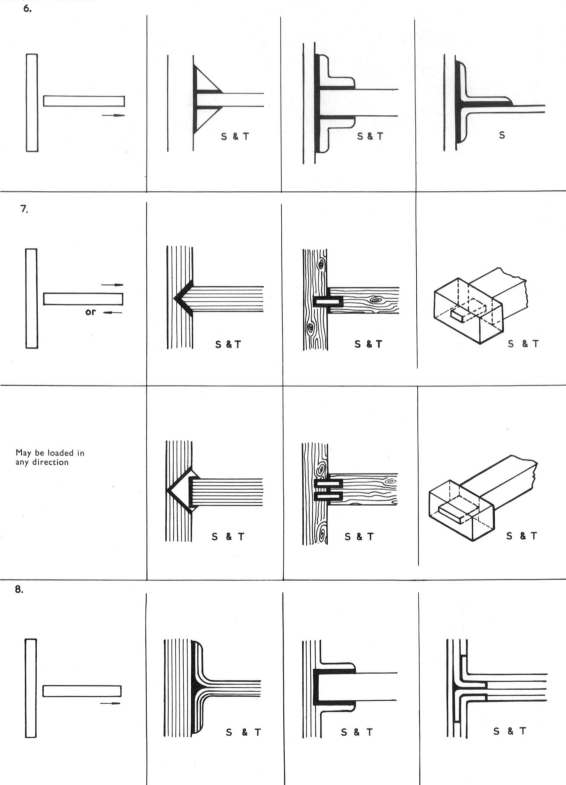

9.

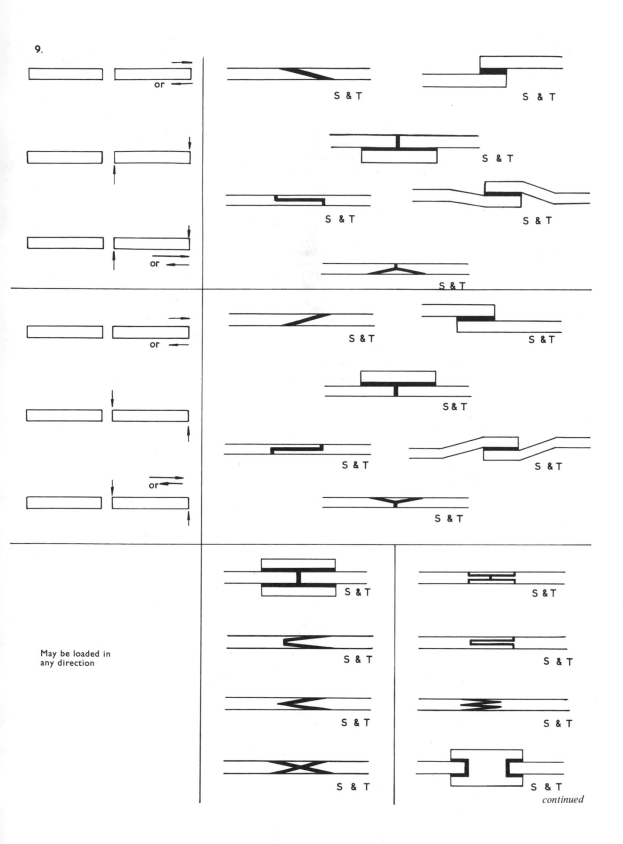

May be loaded in
any direction

continued

Table 2.1 continued
10.

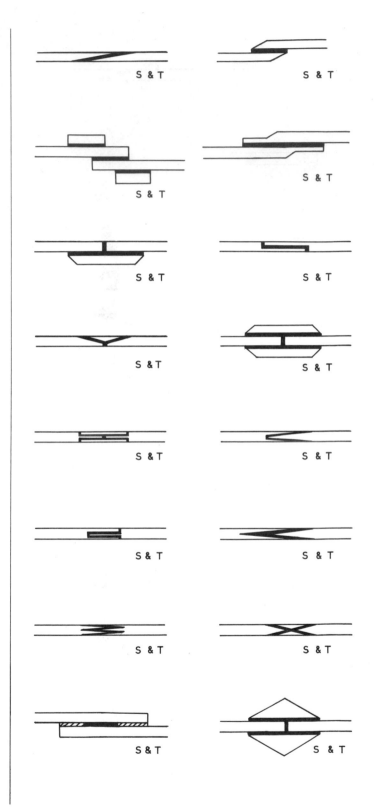

II.

May be loaded in
any direction

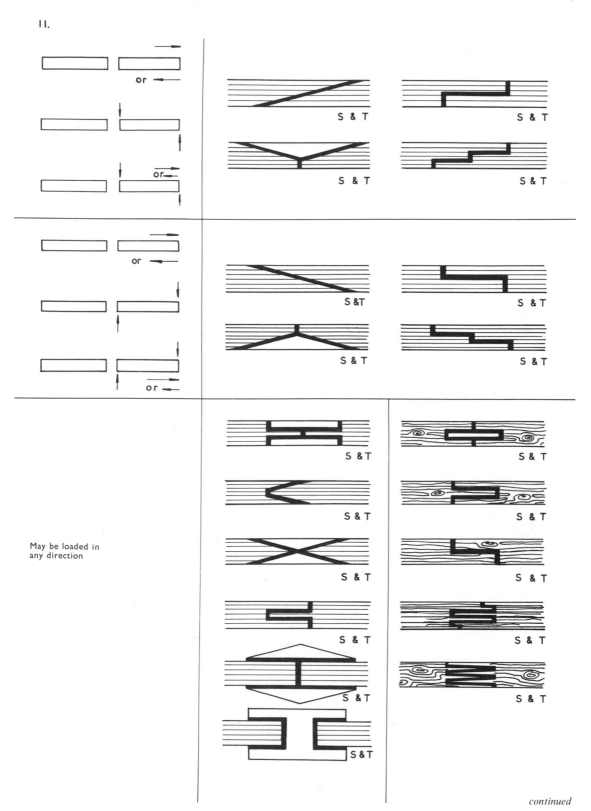

continued

Table 2.1 continued

12.

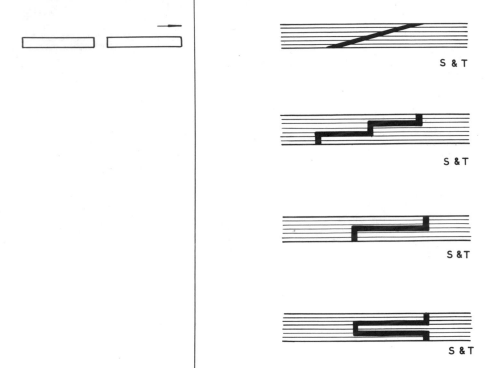

S & T

S & T

S & T

S & T

13.

May be loaded in
any direction

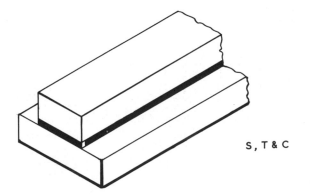

S, T & C

14

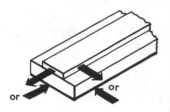

or — or

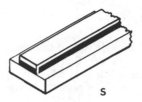

S

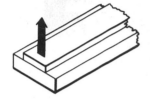

Rivet

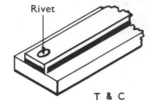

T & C

Bead end

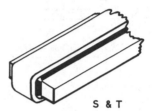

S & T

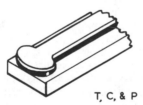

T, C, & P

Increase width

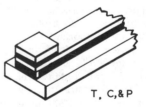

T, C, & P

Increase stiffness

These last two designs should only
be used where load is not
concentrated near the free end.

T, C, & P*

Thinned ends

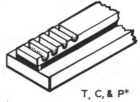

T, C, & P*

*of low magnitude

parts to be joined are of the same materials, then the *same* detail in each of the columns is selected, the table required is then indicated at the intersection.

2.4.   JOINT DESIGN CRITERIA

Table 2.1 attempted to provide joint details suitable for particular loadings on the four fundamental joint types. Applications will be met where the load on the joint is very low, perhaps only the weight of the joined parts. Under such conditions it is un-

designing for shear, reducing cleavage and peel stresses and preventing delamination.

Design for shear
Arrange the joint in such a way that the load carried by the components stresses the joint in shear. If this is not possible (and it quite often is not) try to arrange for *most* of the load to be taken in shear. The other type of stress that gives a uniform load across the bonded area is a tensile stress and is equally desirable but is not often possible to achieve in practice.

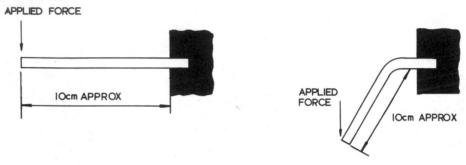

Fig. **2.3**. *Flexibility of adherend*

necessary to pay too much attention to joint design, and the shape most easily fabricated can be chosen. However, most applications require moderate, or even high stresses to be transmitted through the joints. This could be the result of a pressure difference across the bonded structure, the high self weight of the structure or external loads that the structure has been built to withstand. It then becomes very important to consider all the loads and forces that the structure must withstand and to design the joints accordingly.

The designs contained in Table 2.1 must not be regarded as restricting, they may be modified to meet specific needs, but it is important that the underlying principles are maintained. These are—

Reduce stresses when cleavage or peel have to be used
If it becomes clear that the bonded joint will have to withstand cleavage and/or peel stress, action must be taken to ensure that the maximum stress is sufficiently low. The maximum stress under cleavage loading occurs towards one side of the bonded area, either adjacent to the load or at the side stressed in tension by an applied moment. Under peel the stress is confined to a very thin line of adhesive at the edge of the bond. Stress can be reduced by increasing the joint area or by adding stiffeners or some mechanical connection such as a bolt or rivet. To increase bond area, it is preferable to increase the bond width rather than the amount of overlap, for the reasons described later.

Table 2.2.   SELECTION OF JOINT DETAIL

Type of joint		Angle joints				Tee joints				Butt joints				Surface joints			
Details of adherends		Homo-geneous		Laminated		Homo-geneous		Laminated		Homo-geneous		Laminated		Homo-geneous		Laminated	
		R*	F*	R	F	R	F	R	F	R	F	R	F	R	F	R	F
Homogeneous	Rigid	1	2	3	4	5	6	7	8	9	10	11	12	13	14	13	14
	Flexible	2	—	2	—	6	6	6	8	10	10	10	12	14	14	14	14
Laminated	Rigid	3	2	3	4	7	6	7	8	11	10	11	12	13	14	13	14
	Flexible	4	—	4	—	8	8	8	8	12	12	12	12	14	14	14	14

*R = rigid   F = flexible

Avoid delamination of anisotropic materials

Particular care must be taken when using materials such as wood, laminated plastics, etc., to ensure that applied loads do not pull the laminations apart. Situations that cause a tensile load on a localised area of the surface of the material should be avoided. The best approach is to arrange for the load to be transmitted to every layer in the material. In the case of wood this can be done by dowels, or by most of the traditional wood joints. Flexible laminated adherends can usually be 'stepped' by cutting through and peeling back each lamination. Rigid materials can either be bonded onto specially designed joining sections or edges can be machined to a shape that will distribute the load as required, e.g. the scarf joint.

2.4.1. DIMENSIONS OF ADHESIVE-BONDED JOINTS

It is important that the bonded area is large enough to resist the greatest force that the components are expected to withstand in service. Using the principles

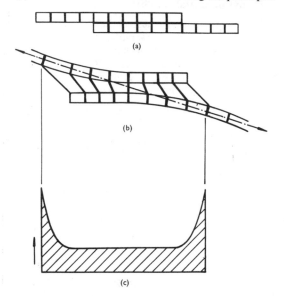

Fig. **2.4**. *Tensile force on lap joint showing: (a) unloaded joint, (b) joint under stress, and (c) stress distribution in adhesive*

above, this will normally be achieved, provided the load does not have a large mechanical advantage.

The calculation of stress in the adhesive joint is not a reliable method of determining the exact dimensions required. Firstly, it is not a simple matter to decide on an allowable stress. The adhesive strength is affected by environmental conditions, age, temperature of cure, material and size of adherends and the thickness of the adhesive layer.

The stress in the adhesive is rarely a pure one, but rather a combination of various stresses. The

relative flexibility of the adhesive to that of the adherends greatly affects the stress distribution. Fig. 2.4 is a typical example of a simple lap joint under tensile loading. Although a method does exist (Perry) for calculating the maximum stress in such a joint, it is of limited application. The method cannot easily be applied if the adherends are of unlike elasticity or unequal thickness, nor can it be applied to systems loaded other than in tension or compression.

It will be observed from the stress distribution

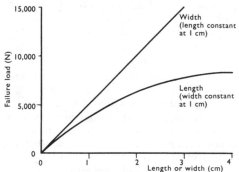

Fig. **2.5**. *Effect of overlap and width on the strength of a typical joint*

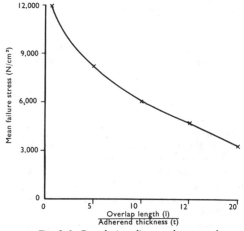

Fig. **2.6**. *Correlation diagram between shear strength and l/t ratio*

diagram (Fig. 2.4(c)) that most of the stress is concentrated at the ends of the lap. The greater part of the overlap (adjacent to the centre) carries a comparatively low stress. Hence if the overlap length is increased by 100%, the load-carrying capability of the joint is increased by a much lower percentage. The greater gain in strength is obtained by increasing the joint width.

The lap joint is typical of most adhesive joints. Increasing the width of the joint has the effect of giving a proportional increase in strength while increasing the overlap beyond a certain limit has very little effect at all. This is illustrated by Fig. 2.5.

In addition to overlap length and width, the

strength of the lap joint is dependent on the yield strength of the adherend. The modulus and thickness of adherend determine its yield strength which should not be greater than the joint strength. The yield strength of thin metal adherends can be exceeded where an adhesive with a high tensile shear strength is employed with a relatively small joint overlap. Figs. 2.5 and 2.6 typify the relationship between shear strength, adherend thickness, and overlap length.

The fall-off in the effective load-carrying capacity of the overlap joint is usually expressed as a correlation between shear strength and the l/t ratio (but sometimes t/l or $t^{\frac{1}{2}}/l$). The latter parameter is generally referred to as the 'joint factor' (De Bruyne). The use of 'joint factor' graphs to determine the dimensions of simple lap joints is discussed further in Section 2.4.2.

Another factor affecting joint shear strength is the thickness of glueline. For maximum strength and rigidity the glueline in a lap joint should be as thin as possible avoiding joint starvation. For thermosetting adhesives, the highest shear strengths are generally achieved with thin gluelines in the range 0·020–0·1 mm. The volume of adhesive in the joint must suffice to fill pores and capillaries, and level protrusions, of the adherend surfaces. Due allowance must be made for volume shrinkage during cure. This may result from changes in the molecular volume as a consequence of chemical reaction, or by loss of solvent. The possibility of loss of adhesive by diffusion into an absorptive substrate, such as wood, should not be overlooked and preliminary pore sealing with a suitable primer may be required.

If the adhesive is hard or rigid, as is the case with thermosetting materials, a thin glueline has more resistance to cracking on joint flexure. Larger forces are needed to deform a thin film than a thick one. The use of thick gluelines increases the probability that these will contain voids or air bubbles or other sources of joint weakness. Moreover, the population of 'frozen-in' stresses at the adhesive interfaces, and thermal stresses due to the use of adherends with mismatched expansion coefficients, tends to be proportional to glueline thickness. For soft, low modulus adhesives, the frozen-in stresses are less significant for strength reduction, and empirically it is well established that elastomeric adhesives should be applied to give thick gluelines where tensile loading is concerned. This is in accordance with the theory of elasticity, which describes stress concentration by the dimensionless coefficient Gl^2/Etd, where G is the shear modulus of the adhesive, l the overlap length, E Young's modulus for the adherend, t the adherend thickness, and d the adhesive thickness. For rigid, brittle adhesives, the discrepancy between theory and practical observation is ascribed to differences in the population of internal stresses.

With structural adhesives in which thermosetting resins are blended with a rubber or thermoplastic, the optimum glueline thickness is usually intermediate between the 'thin' and 'thick' gluelines preferred for the constituent materials.

2.4.2.   DETERMINATION OF JOINT DIMENSIONS

When the loads to be transmitted by the bond are high for the bonded area available, or it is necessary to keep the bonded area to a minimum, practical tests should be made to determine the actual area required. A very rough estimate of the bond area can be made by considering the approximate failing stress of the adhesive and the safety factor to be included. Test specimens can then be made, taking care to resemble the following seven conditions as they will occur in the final assembly:

1. adhesive used
2. material of adherends
3. preparation of adherends
4. cure temperature and pressure
5. thickness of adhesive layer
6. bonded joint
7. environmental conditions.

The specimens should be tested, the load being applied as it will be in the final assembly.

For the determination of the dimensions of simple overlap joints, a more satisfactory procedure is to construct a correlation diagram (Fig. 2.7) between shear strength and the joint factor, t/l, and employ this

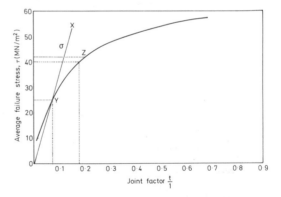

Fig. **2.7.** *Joint factor diagram for the design of overlap joints*

to calculate requirements for optimum strength. The stress condition of a particular joint is represented by a point value on this diagram which relates joint dimensions (on the X-axis) to mean shear stress, τ, in the adhesive (on the Y-axis) and the mean tensile stress in the adherend, σ (which is the slope of a line from the point to the origin).

The diagram is based on mean shear strength data derived from test specimens of various overlap lengths and adherend thicknesses. Sufficient tests must be conducted to plot the curve of shear strength against joint factor. It is important to remember that such a diagram is applicable to a particular set of conditions, as summarised in the list 1–7 above, and should any one condition change, the diagram will no longer be valid. In practice, allowance is made for joint strength reduction due to the adverse effects of, for instance, high temperature or humidity during service. Tests are made under actual service conditions to establish a series of curves each of which represents failure stresses at a certain percentage level of the initial strength. Usually, the mean shear stress data are divided by an appropriate safety factor to produce more reliable working curves for the intended service condition. The value of such curves is that they permit the designer to calculate the optimum overlap or optimum adherend thickness for a specified joint load. Procedures and examples of their use with the diagram (Fig. 2.7), are outlined below.

The relationship between the joint parameters is derived as follows:

$$\sigma = \frac{P}{t} \qquad \text{...(1)}$$

$$\tau = \frac{P}{l} \qquad \text{...(2)}$$

and from equations 1 and 2,

$$\tau = \frac{\sigma t}{l} \qquad \text{...(3)}$$

where σ = mean tensile stress in the adherend
τ = mean shear stress in the joint
P = load applied to unit width of joint
t = thickness of adherend (or thinnest adherend) in the lap joint
l = joint overlap length.

Equation (3) is used in conjunction with Fig. 2.7 to determine the optimum joint dimensions and mean failure stress.

Optimum thickness of adherend (t)
Procedure: given P and l, calculate τ from equation (2). Determine the corresponding value for t/l where τ value intersects the curve and from it calculate a value for t from the known overlap length, l.

Example
Required: Load to failure, $P = 480$ N/mm of joint width
Overlap length, $l = 12$ mm
Calculation
At failure, stress in adhesive $= \dfrac{480 \text{ N/mm}}{12 \text{ mm}}$
$= 40$ N/mm
For $\tau = 40$ N/mm, the intersection with the curve corresponds to a value for the joint factor, t/l of 0·177
Therefore optimum adherend thickness,
$t = 0·177 \times 12$ mm $= 2·12$ mm.

Optimum overlap length (l)
Procedure: given values of P and t, use equation (1)

to calculate σ. Construct a straight line of slope σ equal to $\tau/(t/l)$ on Fig. 2.7. Read off the value for τ at the intersection of the line and curve. Calculate the optimum overlap, l from t and the derived values of τ and σ, using equation (3).

Example
Required: Load to failure, $P = 420$ N/mm of joint width
Adherend thickness, $t = 1·2$ mm
Calculation
Tensile stress in adherend,

$$\sigma = \frac{P}{t} = \frac{420 \text{ N/mm}}{1·2 \text{ mm}} = 350 \text{ N/mm}^2$$

$\sigma = \tau/(t/l) = 350$ N/mm^2 which is the slope of the straight line OX on Fig. 2.7 drawn through the point whose co-ordinates are $\tau = 35$ N/mm^2 and $t/l = 0·1$
Line OX intersects the curve at point Y corresponding to,
(a) Mean failure stress, $= 25$ N/mm
(b) Joint factor, $t/l = 0·075$
therefore optimum overlap length,
$l = 1·2$ mm$/0·075 = 16$ mm.

Mean failure stress (τ)
Procedure: calculate the value of the joint factor t/l from given values of overlap length, l and adherend thickness, t. The point of intersection of the t/l ordinate with the curve determines the required mean failure stress, τ.

Example
Required: Adherend thickness, $t = 4$ mm
Overlap length, $l = 20$ mm
Calculation
Joint factor, $t/l = 4$ mm$/20$ mm $= 0·2$
Ordinate for 0·2 intersects curve at point Z corresponding to a mean failure stress,
$\tau = 42$ N/mm^2.

REFERENCES

PERRY, H. A., 'How to Calculate Stresses in Adhesive Joints', *Product Engineering·Design Manual*, McGraw-Hill, New York, (1959).
YUREK, D. A., 'Adhesive Bonded Joints', *Adhes. Age*, Dec. (1965).
SKOWRONEK, J., 'Design Principles of Adhesive-bonded Joints', *Engrs' Dig.*, **28**, No. 12, Dec. (1967).
EPSTEIN, G., 'Adhesive Bonds for Sandwich Construction', *Adhes. Age*, Aug. (1963).
KEIMEL, F. A., 'Design: The Keystone of Structural Bonded Equipment Enclosures', *Applied Polymer Symposia*, (Bodnar, Ed), Interscience, New York (1966).
BRYANT, R. W. and DUKES, W. A., 'The Effect of Joint Design and Dimensions on Adhesive Strength', *Applied Polymer Symposia*, (Bodnar, Ed), Interscience, New York, (1966)
BRYANT, R. W. and DUKES, W. A., 'Bonding Threaded Joints', *Engng Mater. Des.*, **7**, No. 3, 170, Mar. (1964).
TREITSCH, F. K., 'Joint Design in Adhesive Bonding, *Vortragsreihe über 'Metallkleben'*, 77, Jun. (1963).

HUDA, E. V., 'Bonding Friction Materials to Metals', *Adhes. Age.,* Apr. (1960).

BIKERMAN, J. J., *The Science of Adhesive Joints,* (2nd edn), Academic (1968).

DE BRUYNE, N.A., 'The measurement of strength of cohesive and adhesive joints', *Adhesion and Cohesion*, (Ed. P. Weiss), 46–64, Elsevier, Amsterdam (1967).

The following F.I.R.A. Technical Reports give useful information on the design and performance of wooden joints in the Furniture Industry.

WALTORS, R. A. and MERRICK, M. J., 'The Strength of Dowel Joints', No. 20, Jun. (1965).

SPARKES, A. J., 'The Strength of Dowel Joints', No. 24, Jun. (1966) and No. 28, Aug. (1967).

SPARKES, A. J., 'The Strength of Mortise and Tenon Joints', No. 33, Oct. (1968).

| # Adhesive Selection

3.1. INTRODUCTION

The basic function of adhesives is to hold materials together usefully by surface attachment. Generally, the first consideration in making a selection is the choice between the various types of adhesive which will adhere to the materials to be bonded. An indication of the adhesive types which have been used to bond various adherend materials is given in Table 9.1 (p. 274) which tabulates adhesives generally suitable for metals, glass, plastics, rubbers, wood, paper, etc. The tabulation, together with the more specific descriptions of adhesive products (p. 80), will assist in the selection of suitable adhesive types for simple bonding situations (i.e. in which the service conditions for the bond are not too severe or the mechanical performance demanded not too great).

Tabulations of adhesive properties are subject to limitations, and where particular designs are concerned it is undesirable to select adhesives solely on the basis of earlier similar applications. If the adhesive is selected on this basis, there is the possibility that the material chosen was originally intended for a specific purpose and will not give the best performance for the new application in mind. It should be appreciated that within a given chemical class there are wide variations of properties; adhesive types are continuously being re-formulated and modified, and new adhesives developed. It follows that, whenever possible, an adhesive selection should be made with the help of manufacturers or specialists who are expert in adhesives technology (see Section 'Sources of Further Information on Adhesives' p. 330).

Where outside advice is being sought, it is essential to provide complete information on the properties required of the adhesive and the final assembly. A specification should be as factual as possible so that the specialist can recommend potentially acceptable adhesive systems for further consideration by the user.

Many factors need to be considered in choosing an adhesive for a particular application and specified service conditions. Since no universal adhesive exists which will fulfil all the bonding requirements for all materials in every possible application, it is often necessary to compromise, bearing in mind the desired bond properties, and to decide which are the most important requirements and those which are less important for each application. The materials to be bonded, the strength and permanence requirements, the assembly requirements, and cost considerations are generally the major factors to be evaluated. Having satisfied the major requirements, the best possible solution must be effected for any other factors. It would be insufficient to specify a 'high strength adhesive for bonding rubber to metal for low temperature use'. The statement gives no indication of:

degree of strength required (it could vary from 10–1 000 N/cm²);

type of stress involved (i.e. tension, compression, shear, peel, etc.);

type of rubber (whether natural or synthetic, and which type of elastomer);

type of metal (whether ferrous or non-ferrous or an alloy);

low temperature range (above or below 0°C).

Also not mentioned are the dimensions of the materials and the assembly, the conditions of exposure (and whether continuous or intermittent), processing requirements for the adhesive, and many other important factors. It is important to realise too, that adhesives function as materials which affect, and are affected by, the materials with which they are in contact. Adhesive performance is always dependent on the environment.

The practice of specifying an adhesive for a given application by chemical type is fallacious, since the requirements for physical properties of the joint may be satisfied by several adhesives of quite different chemical composition. Thus, several thermosetting adhesives will bond wood or metal and will meet a specification for service temperatures exceeding 70°C. While some might offer extra strength characteristics in one respect or another, the gain is usually at the expense of another property or at an increase in cost. The designer must establish the specification requirements carefully and realise that over-specification is undesirable and can sometimes lead to a less reliable bonded assembly.

The purpose of the discussion which follows is to emphasise to the designer the several factors which

3.2. REQUIREMENTS OF BONDED ASSEMBLY

The type of assembly under consideration for bonding is frequently a determining factor in adhesive selection. Assemblies may be developed products, prototypes or mass-production units. Machine production items usually require the adhesive to be of a particular form which can be handled by processing equipment designed for fast assembly (e.g. laminating equipment is used in conjunction with liquid-adhesive roller coaters or solid film adhesives). Hand assembled units, such as toys, cameras and instruments, often employ adhesives in physical

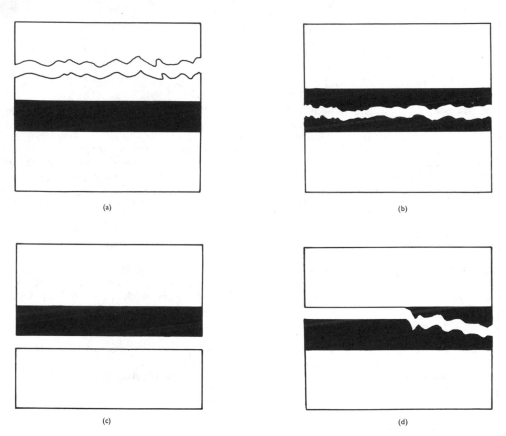

(a)

(b)

(c)

(d)

Fig. **3.1**. *Cohesive and adhesive bond failure: (a) and (b) cohesive, (c) adhesive, (d) 60% adhesive, 40% cohesive*

require consideration before selecting an adhesive, and to help him to communicate effectively with manufacturers or specialists when seeking advice. Information on the materials to be joined, bond stresses, processing requirements, service conditions and life, and other considerations are dealt with in the following paragraphs. Several of the aspects, included below in order to define the adhesive bonding problem, are given a more detailed treatment in other sections of the book.

forms unsuited to mass production machinery. The assembly of small, delicate parts is dependent on human skill and the relatively slow output involved makes it feasible to consider a wider range of adhesive types.

Adhesives are sometimes called upon to perform functions other than the adhesive one. Adhesives may be required to act as sealants against gases, moisture and solvents, or as thermal and/or electrical insulators. Resistance against corrosion of metal joints, vibration and fatigue are additional require-ments that may be needed in an adhesive assembly. The secondary functions of adhesives are some-times of major importance and adhesive selection may be considerably influenced by them. Some of these applications have been mentioned previously (see Advantages and Disadvantages of Adhesive Bonding, p. 5) and others are discussed in later paragraphs.

3.3. MATERIALS TO BE BONDED

The mechanical and physical properties of the adherends and the degree of surface preparation required prior to bonding are important factors to be considered in adhesive selection. Where several adhesives with different physical properties will adhere to a surface, the type of bond required limits the range of possible selections. Usually the objective is to develop adhesion in the joint to the extent that the adhesive will fail in cohesion when the joint is tested to destruction. Where ultimate adhesive failure rather than cohesive failure is essential, the adhesive may be selected or formulated to increase the level of cohesive strength so that joint failure occurs at the adhesive/adherend inter-face (Fig. 3.1).

Low strength materials such as fabrics, felt or some types of wood may be weaker than the adhesive so that joint failures occur cohesively within the material. For this type of application, the adhesive usually allows the assembly to be used under any physical conditions which the material will withstand without danger of bond failure. The use of a high-strength adhesive would constitute an expensive over-specification of bonding material.

The thickness and strength of the adherends are also important, particularly where the elastic constants of the adhesive are relevant to joint design. Flexible materials, such as rubbers or thin metal, plastic foils, etc. (which are subject to flexure in service), should not be joined with a rigid, brittle adhesive; a rigid bond may crack and cause a reduction in bond strength. Differences in flexibility or thermal expansion between adherends can introduce internal stresses into the glue-line. Such stresses can lead to the premature failure of a bond before the imposition of any external load, and present a particular hazard at sub-zero service temperatures. To a certain extent, stresses can be minimised by joint design but the characteristics of the cured adhesive are still affected by them. Where minimum stress between adherends of the same materials is the objective, it is desirable to choose an adhesive which is similar with respect to rheological properties, thermal expansion and chemical resistance. The swelling of one component of a joint in a chemical environment results in stresses adjacent to the adhesive-adherend interface.

For adherend materials of different composition (e.g. steel and plastics) it has been suggested[1] that an adhesive of modulus close to $\frac{1}{2}(E_1 + E_2)$ and a total relative elongation close to $\frac{1}{2}(L_1 + L_2)$, would satisfy minimum stress requirements for different adherends with elastic moduli E_1 and E_2 and total relative elongations L_1 and L_2 respectively ($L =$ change in length per unit length on application of stretching force). The relationship is a hypothetical one and the selection of an adhesive on the basis of mechanical properties intermediate to the adherends assumes that all other adhesive properties are satisfactory. Frequently, the other properties are more critical for the joint than its ultimate strength which, in many cases, is not always vital. Neverthe-less, where strength is a requirement, it is important

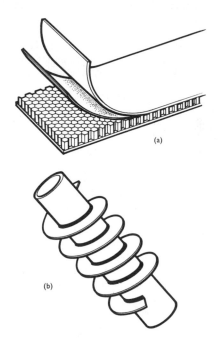

Fig. **3.2**.   *Joining (a) aluminium honeycomb structures to flat metal sheets, (b) spiral copper fins to aluminium tubes*

to have fore-knowledge of the thickness, elastic moduli and thermal expansion coefficients of adherend materials.

The possible distortion of flexible, critical shapes with solvent based adhesives is another factor which should not be disregarded. The crinkling of thin thermoplastic film laminates and edge joints, is often a consequence of bonding with solvent type adhesives. However, solvent adhesive action on rigid thermoplastic substrates often reduces the surface preparation required (*see* Chapter 6). The shape of the components often favours the use of a particular form of adhesive for effective bonding. The joining of aluminium honeycomb structures to flat metal sheets (Fig. 3.2 (a)) is best accomplished with thermosetting film adhesives (supported on glass cloth interliners) and liquid adhesive primers. On the other hand, it is more convenient to use a paste adhesive for the construction of heat exchangers formed from aluminium tubes and spiral copper fins (Fig. 3.2 (b)).

3.4.   COMPATIBILITY OF ADHERENDS AND ADHESIVES

Improper choice of an adhesive can lead to damage of an assembly where the adherend and adhesive (or one of its components) are incompatible.
Examples of this include:

the corrosion of metallic parts by acidic adhesives,

the migration of plasticisers from a flexible plastics material into the adhesive with consequent loss of adhesion at the interface,

the action of adhesive solvents and volatiles on plastics adherends (particularly thin plastics films).

The corrosion potential of some adhesives is often enhanced by poor process control over mixing and curing conditions. Whenever possible, it is advisable to submit adherend samples, together with their property specifications, to the adhesives manufacturer or technologist. Electronic components and printed circuits generally require adhesives which will not corrode copper conductors and other parts under storage or service conditions. Where explosives and similar pyrotechnic materials are bonded, the chemical reactions taking place can destroy adhesion and adversely affect the explosive (i.e. by sensitisation or desensitisation). This particular problem is a matter for specialists and is mentioned here simply to emphasise the point!

3.5.   BOND STRESSES

The cohesive strength properties of adhesives vary widely from soft tacky materials to tough, rigid substances with strengths approaching several thousands of Newtons per square centimetre. Those adhesives known to have a lower strength than that required need not be considered and those with a significantly greater strength can be disregarded unless selected for the importance of some other property. In some instances the adhesive may need to form only a temporary bonding function such as a locating and holding agent for component parts which are to be fastened later by alternative means. Thus, for applications where the adhesive has to satisfy particular strength requirements it is necessary to consider the stresses to which the bond will be subjected. Of importance are the nature and degree of stress and the conditions under which it is applied. The adhesive performance in a joint is dependent on many factors, the most important being joint design, the state of the surfaces being joined, the bonding technique employed, the glue-line thickness and the strength and thickness of the parts being bonded. Joint design determines the type and degree of stress to which the adhesive will be submitted. Bonds may be stressed in shear, tension or compression, cleavage or peel, or any combination of these stresses. Most adhesives display optimum strength properties in tension or compression. Some adhesives may have low peel strengths but high shear strengths, or vice versa, while other adhesive types give an acceptable performance in peel or shear. By increasing the bonding area sufficiently, it is frequently possible to achieve a required joint strength even with low strength adhesives. However, where it is not possible to design large area joints the use of a high-strength adhesive becomes necessary.

Film thickness of the adhesive in the joint is significant where the selection is being made to satisfy a required strength. The highest tensile and shear strengths are obtained with high modulus adhesives when the film thickness is minimum. (For thermosetting resins optimum strength is usually obtained for 0·06 to 0·12 mm glue-lines; below 0·03 mm strengths usually decrease, depending on the smoothness of adherends, and starvation of adhesive in the joint is a hazard). On the other hand, increased film thickness with elastic adhesives produces higher peel strengths (optimum strengths are generally achieved for glue-lines exceeding 0·13 mm).

The conditions under which the stress will be applied must be specified. Bond loading can be sustained, intermittent or vibratory in character and not all adhesives function equally well for all of these circumstances. Some adhesives form hard, brittle bonds which fail under vibratory conditions; other types are subject to creep and are unable to sustain continuous loads although they may withstand intermittent loading. An increase in the rate of loading is responsible for an apparent increase in bond strength (e.g. impact or shear) for many adhesives and is a factor to be considered.

3.6.   PROCESSING REQUIREMENTS

The conditions under which the adhesive is to be bonded are also important criteria for the selection of the correct adhesive. Certain assembly circumstances can restrict the choice of an adhesive to a product that will bond under factory or assembly line production. Often, it may be that the working properties of the adhesive are the factors of overriding interest to the potential user. This might be the case, for instance, when the adhesive bond is not to be exposed to significant stresses in service or to severe temperature or weather conditions.

Typical factors that are involved in assembly include: form of adhesive, method of preparation and use, shelf (storage) life, working life, method, or machinery necessary for bonding, and processing variables; the last mentioned includes time permitted between coating and bonding; drying time and temperature; temperature required for application and curing; bonding pressure required and time of application; rate of strength development at various temperatures and such properties as odour, inflammability and toxicity of the adhesive which may call for extra equipment or precautionary measures. Several of these factors are discussed at this stage as an indication of the appraisal required in selecting the most suitable adhesive for an application.

The method chosen for the application of an adhesive to the workpiece is determined by size and shape of the parts, the number of components to be coated and the dimensions of the part, as well as by the physical properties of the adhesive. Most adhesive types are available in forms which range in consistency from thin liquids to pastes and solids which require different application methods (e.g. thin liquids are sprayed, brushed or roller-coated, whereas pastes require spreaders or knife blade coaters). Insistence upon a particular method of adhesive application (perhaps with mass production assembly in mind) will point to the selection of an adhesive form (which is often a restriction on the type chosen) with the requisite handling and application properties.

Adhesive tack properties are frequently important for assembly processes. Tack or stickiness of an adhesive is responsible for bonding coated parts which are fitted together, and the tack range determines the time interval between the coating and assembly of components. The tack properties will thus dictate the conditions under which the adhesive must be employed (i.e. form, rate of delivery to substrate, mixing time, and application method). In contrast to the thermoplastic adhesives, the thermosetting types generally have little tack. Tack properties vary widely and are dependent on specific properties. Good tack is displayed by latex adhesives which become tacky only on removal of the bulk of the liquid carrier (water) and by solvent rubber adhesives which are tacky even with an appreciable content of solvent.

For some assemblies, the curing temperature influences the selection of an adhesive. A number of thermosetting adhesives require heat and pressure to form the bond so that where these processing conditions cannot be met, a cold setting adhesive would be employed. Temperature or pressure sensitive elements in an assembly often preclude the use of heat or bonding pressure to set the adhesive. The choice of adhesive may be determined by the geometry and disposition of the assembly parts. Gap-filling adhesives are generally required for loose-fitting parts, whereas low viscosity adhesives are called for where close-fitting components are concerned.

3.7. SERVICE CONDITIONS

The adhesive selected for a particular assembly must hold the component parts together throughout the expected service life and maintain its strength under the service conditions encountered. The designer must, therefore, be aware of all the circumstances likely to be met in order to specify a suitable adhesive. Most often the strength and permanence requirements are critical and the factors concerning these have been dealt with in Section 3.5.

Adhesive types display wide differences in their response to different stresses and stress rates. Thermoplastic adhesives are unsuitable for structural applications because they have a tendency to fail under low sustained loads; also they soften on heating. Thermoplastic adhesives are unable to withstand vibratory stresses for long periods although they may exhibit greater strengths than thermosetting types for short duration tests. Thermoplastic rubber types usually possess high peel strengths but relatively low tensile or shear strengths. The thermosetting resins, by contrast, are often used as the basic ingredients for structural adhesives. They give rigid bonds which retain much of their strength at elevated temperatures. Thus, in general, thermosetting adhesives are preferable for applications demanding high strength and good resistance to fatigue. Thermosetting resin or rubber-resin types also perform well under vibratory loads but are relatively poor under peel or cleavage stresses. For applications where impact loading is to be encountered, a resilient adhesive will prove more satisfactory than a brittle thermosetting one.

Another important factor to be considered is the temperature range over which the adhesive will be used and required to be effective. Specifications for the service temperature of a bond should refer to the actual glue-line temperature and not the ambient temperature. Bond temperatures are usually lower than ambient temperatures especially when intermittent or brief exposure to heat is involved. At high temperatures all adhesives lose some of their strength and some types soften or decompose to the extent that they become useless. Up to 70°C several types of thermosetting and thermoplastic adhesives may be used, but at 120°C while only a few thermoplastic adhesives (e.g. silicone rubbers) will withstand intermittent exposures at low stress rates, most thermosetting adhesives will function satisfactorily under continuous exposure. Above 120°C, only the more resistant thermosetting adhesives, such as phenolic-nitriles and epoxy-novolacs, will perform adequately. Although some other available adhesives can withstand 350°C, a service temperature limitation of 70 to 95°C is more usual for most types. Some polyaromatic adhesives (e.g. polybenzimidazoles) retain shear strengths exceeding 300 N/cm² after a 1 h exposure at 500°C.

Low temperatures cause embrittlement and internal stressing of many adhesives so that premature cohesive failure of the bond is not uncommon below —80°C. For sub-zero temperature conditions, it is preferable to choose flexible thermoplastic or elastomeric adhesive types which retain some resilience on exposure to cold. Flexible adhesives (e.g. polyurethanes and epoxy nylons) provide shear strengths in excess of 6 000 N/cm² at temperatures as low as —200°C.

In addition to temperature, many other conditions may be involved which may affect the strength and durability of an adhesive. Although performance varies considerably, most organic adhesives are adversely affected by moisture to some extent, especially when under conditions of stress; natural products are more vulnerable than the synthetic ones. Consideration must be given to such influences as chemical reagents, hot and cold water, oils, solvents, hydraulic fluids, chemical atmospheres (e.g. ozone, acidic gases, salt-spray), exterior weathering and ageing, the effects of biological agents, (e.g. mould growth, fungi, insect and rodent attack on adhesives of natural origin which are nutritious) radiation (e.g. sunlight, x-rays, radioactivity, infrared), hard vacuum conditions responsible for outgassing of adhesive resins. Which of these, and many other influences are most important must be determined before a realistic choice can be made. Service requirements must be specified accurately but without over specification. Thus, it would be improper to select an adhesive type resistant to steam when the atmosphere to be encountered has a very low relative humidity. However, some over specification may be inevitable where the adhesive has been selected to accommodate other environmental conditions of greater importance.

3.8. SERVICE LIFE

Adhesive bonds are usually required to have a service life which is equal to that of the assembly. For example, the adhesive used for bonding honing stones to metal holders must maintain a satisfactory bond until the stone has worn out. Footwear and automobile brake shoe provide other examples of a requirement for adequate service life.

Occasionally, short service life is an adhesive requirement. The parts of an assembly are sometimes temporarily attached with an adhesive to position them for assembly by other means. The use of adhesive binders for ceramic powders prior to sintering and for foundry mould materials (sand, etc.) are examples of temporary adhesion. Waxes and related materials function as short-term fixing agents and sealants for many applications.

3.9. ASSEMBLY STORAGE CONDITIONS

Some of the factors pertaining to storage have been discussed under the conditions of service. The possibility that abnormal extreme temperatures and impact (shock) loads might affect a bonded assembly in transit and in storage should not be overlooked when selecting an adhesive.

3.10. COST

Adequate properties should be more important to the selection of an adhesive than cost considerations. Where the selection indicates that several adhesives are valid, choice becomes a matter of economy for the bonding process considered as a whole. The use of the lowest cost adhesive may be outweighed by some of the processing factors:

efficiency of coverage in relation to bonding area or number of components;

ease of application and processing equipment needed (jigs, ovens, presses, applicators);

processing time (for assembly, for preparation of adherends, for curing, etc.);

cost of labour for assembly and inspection of the bonded parts;

amount of rejected material as compared with other methods of joining.

Often the use of a fast-setting expensive adhesive with a simple bonding technique, which avoids complex jigging of assembly parts, might be an economical choice.

3.11. SPECIAL CONSIDERATIONS

For the majority of applications, the conditions outlined above generally provide a sufficient basis on which to select suitable adhesives. There are, however, other circumstances which can influence the choice and in certain cases the following factors may need to be considered and should be made known when advice is being sought.

Electrical properties
Properties such as dielectric constant and loss, or insulation resistance often require specification for printed circuit manufacture. Adhesives can be formulated to be electrically and/or thermally conductive by compounding with suitable filler materials. It should be noted that the surface pretreatment given to adherends prior to bonding frequently affects the electrical and thermal properties of the bond.

Optical properties
Transparent, colourless glue-lines of a given refractive index are usually required for optical systems. For high quality optical systems it may be

important to avoid stressing the optical components (lenses, mirrors, etc.) as a result of the excessive shrinkage of an adhesive on curing. Where colour is important, the adhesives can be tinted with dyes and pigments. Some adhesives have permanent colour properties which prevent their application to many optical assemblies.

Hazards and safety precautions

For industrial applications such as food packaging, the adhesive must not be toxic or possess an offensive taste or odour which would be objectionable to the user of a finished product. The odour occurring during the application or mixing and curing of certain adhesives may militate against their use for large scale assembly processes. Adequate ventilation should be provided for adhesives known to evolve volatiles during processing.

Inflammability is another hazard that can prevent the employment of solvent-based adhesives for some manufacturing processes. A notable example of this is the fabrication of flock-coated substrates by deposition of electrostatically charged flock particles on a tacky adhesive coated fabric. The possibility of ignition of adhesive solvent vapour by electric spark discharge precludes the use of inflammable adhesives. Precautions against fire should always be taken where inflammable adhesive volatiles are concerned.

Adhesive formulations may contain ingredients which can cause skin irritations and dermatitis (e.g. certain epoxy-resin adhesives). Careless handling and exposure to vapours should be avoided.

3.12. CHECKLIST OF FACTORS FOR ADHESIVE SELECTION

Localised conditions often necessitate variations from 'standard' adhesive formulations to satisfy specific production problems. A check list of the factors discussed above is presented here. Discussion of these factors with adhesives manufacturers will help to secure the most suitable adhesive for a given application at the lowest overall cost for material and processing.

ASSEMBLY REQUIREMENTS
Bonding, sealing, insulating, flocking, encapsulation and joint reinforcement pertaining to development work, mass production, repair and maintenance.

SURFACES
Details of materials to be bonded, trade name or specification, finishes and base materials, chemical types, physical forms, alloys.

FORM OF ADHESIVE
Solvent base: water base, mastic, liquid, film, powder, paste, 1–2 part.

APPLICATION METHOD
Hand or machine: brush, knife, trowel, extruder, spray, roller-coater, dip.

PROCESSING REQUIREMENTS
Viscosity, tack time, time/temperature for drying/curing, pot-life, open/closed assembly time, bonding pressure, substrate pretreatments, preferred conditions and limitations of processing equipment, post-assembly processes—cleaning, painting, etc.

DESIGN FEATURES
Type of joint (overlap, butt, edge, etc.), bonding area, number of assemblies.

MECHANICAL REQUIREMENTS
Bond strength: temporary, moderate, low, structural.

JOINT LOADING
Sustained/intermittent/cycling loads, direction of loading, peel (N/cm), shear, compression, tension, cleavage, (N/cm^2).

SERVICE CONDITIONS
Continuous/intermittent/cycle exposure to heat, cold, temperature and pressure range, weather, humidity, water (hot or cold), chemicals (and atmosphere), sunlight and radiation, solvents and vapours, outdoor (location, exposed or sheltered), indoor (ambient or controlled atmosphere), biological influences, etc.

OTHER CONSIDERATIONS
Cost, shelf-life, toxicity, inflammability, odour, colour, corrosion properties, thermal, electrical, chemical, optical properties, etc.

SPECIFICATIONS
Government or other.

PRESENT ADHESIVE USED
Limitations, quantity used.

REFERENCE

BIKERMAN, J. J., *The Science of Adhesive Joints*, (2nd edn) Academic, (1968).

CHAPTER 4

Adhesive Materials and Properties

4.1. THE COMPONENTS OF AN ADHESIVE

Many of the older adhesives were based on single materials but, while some of these are still useful for certain applications, the majority of adhesives today are composed of mixtures of several complex materials which may be organic, inorganic, or hybrid (*see* Section 1.2). The components of the adhesive mixture are usually determined by the need to satisfy certain fabrication properties of the adhesive, or properties required in the final joint. The basic component is the binding substance which provides the adhesive and cohesive strength in the bond; it is usually an organic resin but can be a rubber, an inorganic compound or a natural product. Other constituents of the adhesive fulfil other functions.

Diluent
This is employed as a solvent vehicle for other adhesive components and also to provide the viscosity control which makes a uniformly thin adhesive coating possible. Occasionally, liquid resins are added to control viscosity.

Catalysts and hardeners
These are curing agents for adhesive systems. Hardeners effect curing by chemically combining with the binder material and are based on a variety of materials (monomeric, polymeric, or mixed compounds). The ratio of hardener to binder determines the physical properties of the adhesive and can usually be varied within a small range. Thus, polyamides combine with epoxy resins to produce a cured adhesive. Catalysts, which themselves remain unchanged, are also employed as curing agents for thermosetting resins to reduce cure time and increase the crosslinking of the synthetic polymer. Acids, bases, salts, sulphur compounds and peroxides are commonly used and, unlike hardeners, only small quantities are required to effect curing. The amount of catalyst is critical and poor bond strengths result where resins are over or under catalysed.

Accelerators, inhibitors and retarders
These substances control the curing rate. An accelerator is a substance that speeds up curing caused by a catalyst by combining with the binder (a catalyst may have the same effect but will not lose its chemical identity during the process). An inhibitor arrests the curing reaction entirely whereas a retarder slows it down and prolongs the storage and/or the working life of the adhesive.

Modifiers
There are many chemically inert ingredients which are added to adhesive compositions to alter their end-use or fabrication properties. Modifiers include fillers, extenders, thinners, plasticisers, stabilisers, or wetting agents, and each material is used for a special purpose.

Fillers are non-adhesive materials which improve the working properties, permanence, strength, or other qualities of the adhesive bond and those commonly used are; wood flour, silica, alumina, titanium oxide, metal powders, china clay and earths, slate dust, asbestos and glass fibres. Some fillers may act as extenders.

Extenders are substances which usually have some adhesive properties and are added as diluents to reduce the concentration of other adhesive components and thereby the cost of the adhesive. Extenders often have positive value in modifying the physical properties of the glue-line by providing reinforcement to resins which would otherwise craze. Common extenders are flours, soluble lignin and pulverised partly-cured synthetic resins. Thinners are generally volatile liquids which are added to an adhesive to modify the consistency of other properties. Plasticisers are incorporated in a formulation to provide the adhesive bond with flexibility or distensibility. Plasticisers may reduce the melt-viscosity of hot-melt adhesives or lower the elastic modulus of a solidified adhesive. Stabilisers are added to an adhesive to increase its resistance to adverse service conditions such as, light, heat, radiation, etc. Wetting agents promote interfacial contact between adhesive and adherends by improving the wetting and spreading qualities of the adhesive.

REFERENCES

Modern Plastics Encyclopedia, 45 (14A), McGraw-Hill, Inc. New York (1968).
Fillers for Araldite epoxy resins, CIBA (A.R.L.) Ltd., Duxford Publication C 40 B, August (1968).
Comparison of different fillers in epoxide resins. Tech. Bulletin 228, Croxton and Garry Ltd., Esher (1966).

4.2. ADHESIVES TYPES

One objective of this handbook is to indicate the basic properties of the different adhesives types. To a large extent the mechanical properties depend on the thermosetting or thermoplastic nature of the bond and the following general discussion of these differences provides background information to the detailed listing of the adhesives types (based on the major chemical ingredient) which follows.

4.2.1. THERMOPLASTIC ADHESIVES

The thermoplastic adhesives are classified under the general categories of thermoplastic resin and thermoplastic rubber adhesives. As a class, thermoplastic adhesives are fusible, soluble, soften when heated and are subject to creep under stress. Unlike the thermosetting resins, they do not change chemically in establishing a bond. The thermoplastic nature of these materials confines their application as adhesives to low load assemblies formed from metals, ceramics, glass, plastics and porous materials based on paper, wood, leather and fabrics and which are not subject to severe service conditions. Hot-melt adhesives which fall into this class are being increasingly employed for fast assembly of packaging materials and plastic film laminates.

Thermoplastic resin adhesives are based on various synthetic materials (typified by the polyamide, vinyl and acrylic polymers and cellulose derivatives) or on natural products such as rosin, shellac, oleoresins and the mineral waxes. The important hot-melt adhesives are invariably compounded from polyethylene, vinyl polymers and co-polymers, polystyrene, polycarbonates, polyamides and other polymers. Additives, including plasticisers, fillers, and reinforcing materials, are frequently compounded with the resins to confer particular properties on the adhesive. With the exception of pastes, these adhesives are available in the same forms as the thermosetting adhesives, i.e. liquid forms can be solutions, dispersions, or emulsions of the polymer and other modifying components in a volatile medium. Solid forms are also available as films (supported and unsupported), pellets, sticks, or extruded cord lengths, suitable for machine application. Other solvent-free liquid forms ('100% solids systems') contain the thermoplastic material as a monomer or as a pre-polymer which requires a catalyst to bring about polymerisation to a high molecular-weight solid.

Thermoplastic rubber adhesives are some of the most versatile industrial adhesives currently used. The rubber based adhesives discussed in the following pages include; natural and reclaim rubbers and synthetic elastomers such as polychloroprene (neoprene), butyl, styrene-butadiene and acrylonitrile-butadiene (nitrile). Most of the elastomers are available in solvent and latex forms or as water dispersions and other types are supplied with vulcanising agents. The thermoplastic rubber adhesives are generally modified with fillers, plasticisers and compounding ingredients. The types of rubber and solvent vehicle used partly determines the physical and chemical properties of the adhesives and the many compounding techniques employed result in widespread variations in strengths, tack ranges, drying rates, environmental resistance and other properties.

Heat or solvent activation is used to convert film adhesives into the fluid state prior to bonding. Solvent activation is applicable only to situations where an adherend is porous enough to permit solvent release by absorption and diffusion and heat activation is employed where adherends are impermeable and able to withstand the temperatures involved. Heating also has the effect of curing any thermosetting component which may be present in the adhesive. Both techniques are also used prior to bonding to activate substrates which have previously been coated with a solvent-base adhesive and dried to a tack-free state. Bonding is usually carried out under heat and pressure after joint assembly. Solid thermoplastic adhesives of the hot-melt type rely on heat to render them fluid and on cooling to bring about the setting action.

An unusual cure mechanism is displayed by the cyanoacrylates which are an example of chemically blocked materials. When confined between close-fitting parts, these one-component liquid monomers undergo polymerisation in a very short period (often 15 s). The thin moisture film which is usually present on exposed surfaces is sufficient to harden these materials if the glue-line is thin enough.

4.2.2. THERMOSETTING ADHESIVES

As a group, the thermosetting adhesives form bonds which are essentially infusible and insoluble through the action of heat, catalysts or combinations of these. In contrast to thermoplastics, the thermosetting resins display good creep resistance and provide the basis for many structural adhesives intended for high-load applications and exposure to severe environmental conditions such as heat, cold, radiation, humidity and chemical atmospheres. Thermosetting adhesives include materials of natural origin such as animal glues, soybean and vegetable proteins, casein and miscellaneous water-based adhesives as well as synthetic products based on epoxy, phenolic, polyester, polyaromatic and other thermosetting polymers.

Water-based adhesives prepared from low strength materials of animal or vegetable origin were the earliest adhesives used and are still important for furniture and plywood manufacture, paper and packaging materials, and similar applications where low strength and a limited durability to outdoor conditions are acceptable. In addition there are thermosetting rubber-resin adhesives (see Section 4.2.3) and other blends referred to as thermosetting-thermoplastic resin adhesives. These adhesives have increased toughness and strength while their improved resilience enhances stress-distribution properties. Examples of the latter class are the phenolic resins modified with nylon or various vinyl resins. The characteristics of these materials are, in general similar to those of thermosetting adhesives and hence many of these products are employed as structural bonding agents for metal to metal. Epoxy resins are modified with polysulphides to improve their flexibility and are thermosetting materials. However, polysulphide adhesives often function as sealing materials and may be thermoplastic or thermosetting according to formulation and cure.

Thermosetting adhesives are supplied as liquids, pastes and solids (which may be in film form). Liquid types are generally one- or two-component systems which are already non-solvent, containing 100% solids materials, or react to become so by catalytic action. Some liquid adhesives contain a volatile solvent which is non-reactive and which acts as a dispersant or improves the handling and processing properties of the system. The curing agent for a liquid system may be a powder which requires to be melted before mixing the components. As a result of added modifying agents, pastes are usually thixotropic and may be applied to vertical joints as non-sag adhesives which will not flow out during assembly and cure of a bonded structure. Film forms may be supported or unsupported and of various thicknesses. They have the advantages of easy, clean handling and can be cut to conform to the shape of the joint. The shelf-life of film and one-component types is increased by refrigeration but in the case of some film adhesives cold storage is essential to prevent room temperature curing.

Natural product thermosets, like animal-glue, set by loss of solvent. Many two-component liquid types, such as epoxy resins, cure by catalytic action with or without the aid of heat. Other two-part thermosetting rubber-resin adhesives can be vulcanised at room temperature but otherwise curing with heat and pressure is necessary. Some rubber-resin adhesives which are used to bond unvulcanised rubber to metal cure during subsequent vulcanisation of the rubber while other adhesives, employed to bond already vulcanised rubber to metal, are cured separately. Structural film adhesives invariably require heat and pressure to realise the maximum mechanical properties and curing temperatures ranging from 150–250°C with bonding pressures up to 100 N/cm^2 are not uncommon. Post-cures are often an additional processing requirement for structural adhesives where optimum strength is sought.

4.2.3. RUBBER-RESIN BLENDS

There are innumerable adhesives in which rubbers and resins, both natural and synthetic, are blended to obtain combinations of desired properties of both types of material. Blended adhesives may be employed for structural or general purpose bonding according to the type of resin and rubber used and their ratio in a formulation. Those consisting mainly of thermosetting resins modified with synthetic rubber are used for the structural bonding of metal and other rigid materials. Phenolic-nitrile and phenolic-neoprene adhesives are examples of this type, in which the rubber component serves to improve the flexibility of the cured bond and promote its resistance to impact or shock loading. Thermosetting resins alone tend to be brittle. Adhesives based on rubber, with a certain amount of natural or synthetic resin as a modifying component, represent the other end of the scale. In practice, the various types of rubber are rarely used alone as adhesives but are invariably modified with resins to improve such properties as tack, cohesive strength, specific adhesion to surfaces and heat resistance. Within these extremes are numerous formulations in which various ratios of resin to rubber are used. These adhesives have a wide range of applications which include: bonding of textiles; bonding of synthetic fabrics to wood and metal; affixing wallboards and tiles; lamination of paper, metal foil and plastic films; laying of flooring materials; and various other industrial or domestic applications.

The structural rubber-resin adhesives are available as films or tapes (supported or unsupported on fabric carrier cloths) and occasionally as solvent solutions. The films are cured at elevated temperatures up to 200°C and under bonding pressures ranging from 30–100 N/cm^2. Post-curing is often included to ensure optimum mechanical properties for the cured adhesive. Liquid types are dried to remove solvent and then processed as film adhesives. The non-structural rubber-resin adhesives are generally supplied as solutions in organic solvent mixtures and can be applied by brush, spray, dip or roller coater, spatula, or flow techniques. Because these adhesives rely on a loss of solvent before adhesive action can take place the shelf-life and working life are usually indefinitely long provided the solvent content is maintained. With porous adherends the assembly can be made with wet adhesive and time allowed for solvents to escape by diffusion through the material. Where solvents have a high volatility assembly times may be as short as 15 min by which time the substrates have lost the tackiness required for contact bonding. Impermeable materials are coated with adhesive and bonded together only after the bulk of the solvent has been dried off to leave the adhesive in a tacky state. Light assemblies can frequently be handled after a few hours but heavier assemblies require a setting period of at least 24 h. Maximum joint strength is not realised until after a few days following the removal of residual solvent traces. Wet-bonding generally produces joints having good strength and durability but poor solvent resistance. Optimum performance is given by heat curing (according to manufacturer's instructions) which has the effect of removing the trace solvents otherwise retained by these adhesives and which act as plasticisers and increase the thermoplasticity of the system. Heat also promotes cross-linking of the adhesive constituents and thereby increases the creep resistance of the joint. Processing conditions depend on the adhesive with bonding pressures ranging from 10–300 N/cm² according to joint factors such as rigidity, dimensions, closeness of fit and glue-line thickness. Low bonding pressures are more satisfactory for glue-lines exceeding 0·2 mm. Curing schedules range from 1 h at 80°C to 20–30 min at 140°C, with optimum properties resulting from the longer curing periods.

4.3. PROPERTIES OF BASIC ADHESIVES TYPES

This section has been prepared almost entirely from published material appearing in technical books and journals. The main information sources are listed in the literature references appearing after each adhesive type description, although other material has provided additional information on specific types of adhesive. The length of an entry is not indicative of the importance of the adhesive type under consideration since the amount of information available was found to vary considerably. Due regard has been paid to technical information in the trade literature received from the various adhesives manufacturers. Manufacturers' literature

was found particularly useful for confirmation of such adhesive properties as colour, available form, processing factors, and applications. These data were necessary to supplement material from published sources and effort has been made to keep the section free from any trade bias.

It has already been noted earlier that adhesives based on the same material may show considerable variation in their properties where modifying materials have been added to the formulation. Properties are dependent, not only on the adhesive composition, but also on the conditions under which it is prepared and used. Because of these possible variations any values given in this section should be regarded as representative of the probable behaviour of a basic type of material used under certain conditions. This is further complicated by the fact that a large number of commercial adhesives are blends of two or more basic adhesive types; in particular both natural and synthetic rubbers and resins are often used together and form the basis of the numerous 'resin-rubber' or 'rubber-resin' adhesives which are available. Consultation with the manufacturers concerned (*see* p. 320) is strongly recommended where detailed information is required on the behaviour of specific commercial adhesives under various service conditions.

ACRYLICS

Type
Thermoplastic resins based on acrylates (properties of polymethyl methacrylate types are discussed below) or derivatives (amides and esters).

Physical form
Available as emulsions, solvent solutions, and monomer-polymer mixtures (one or two components) with catalysts (liquid or powder). One component liquids which polymerise under ultra-violet radiation are available.

Shelf-life, 20°C
Up to 1 yr.

Working life, 20°C
30 min to 1 h according to setting action.

Closed assembly time, 20°C
As for working life period.

Setting action
Emulsion-solvent types set by evaporation and absorption of solvent. Polymer mixtures set through polymerisation by heat, ultra-violent radiation and/or chemical catalysts.

Processing conditions
Solvent types set over a period of 20 days at 20°C or 6 h at 80°C. Polymer mixture setting times

depend on polymerising method used: chemical action, 14 d at 20°C or 4 h at 80°C; ultra-violet action, 5 h exposure; heat action, 2 h at 55°C followed by 8 h at 80°C. Bonding pressures range from contact to 17 N/cm².

Permanence
Resistance to weathering and moisture varies from poor (solvent types) to excellent (polymer mixtures). Change from transparent to yellow colour may occur with time (over 1 yr period). Not affected by alkalies, non-oxidising acids, salt spray, petroleum fuels but attacked by alcohols, strong solvents and hydrocarbons (aromatic and chlorinated). Highly resistant to ultra-violet exposure. Service temperature range of acrylic resin adhesives is −60°C to 52°C.

Applications
Light structural assemblies based on acrylic plastics to themselves, wood, glass, metals, rubber, leather and fabrics; colourless jointing of decorative plastic laminates; production line assembly of components (ultra-violet effective here); outdoor applications such as plastic name-plates; aluminium foil work; windshields, instrument panels, lenses and optical components in aircraft, marine, and automotive industries. One-type, *n*-butyl methacrylate (alone or modified with Canada Balsam) is used as an optical cement. Ross-24 (modified type) is a heat-setting cement which is transparent ($N_D^{20°C} = 1.485$) and has good thermal shock resistance.

Remarks
Acrylic cements are compatible with pigments where coloured glue-lines are required. Gap-filling properties vary from 0·025–6·3 mm according to grade used. Bonding of acrylic plastics (e.g. Perspex) to other materials of lower expansion coefficient (e.g. steel) with acrylics, requires interlayer of rubber. The rubber is bonded to other materials with an epoxy resin adhesive. Shear strengths up to 4 500 N/cm² are realisable with heat-cured acrylic cements. For long service, tensile stresses should be limited to 1 000 N/cm² to avoid crazing of glue-line.

REFERENCES

WOODMAN, J. F., 'Acrylic resins', *Modern Plastics Encyclopaedia*, McGraw Hill, New York, (1968).
Perspex Acrylic Materials (Cementing), Tech. Bulletin I.C.I. (Plastics Division), May (1966).
SNYDER, W. H. et al., 'Investigation of fast setting acrylic adhesives for bonding attachments to human tooth surfaces', *J. appl. Polym. Sci.,* **11** (8), 1509, (1967).

ACRYLIC ACID DIESTERS

Type
Polyester-acrylic resins.

Physical form
One-component, solventless, low viscosity, polymerisable liquids or pastes.

Shelf life, 20°C
Up to 1 yr.

Working life, 20°C
Up to 1 yr.

Closed assembly time, 20°C
Within 5 min.

Settling process
Anaerobic curing materials which polymerise in the absence of oxygen (air).

Processing conditions
Cure automatically, without shrinkage or evaporation, into a rigid solid when placed between mating surfaces (which exclude the air). Cure in less than 10 min at 120°C to 24–72 h at 20°C under 3 N/cm² pressure only. Glue-line thickness in cured joint of 0·025–0·075 mm is recommended for optimum strength.

Coverage
6 cc per 0·1 m².

Permanence
Durability varies considerably with formulation. Resistance to chemical agents such as ketones, hydrocarbons, glycols, chlorinated hydrocarbons and water may be high or low. Prolonged exposure to moisture generally has an adverse effect on bond strength. Service temperature range is −55°C to 150°C.

Applications
Aluminium, steel, copper, (zinc, cadmium, gold, require priming with an organo-metallic activator in solvent, or a chromate rinse), glass, thermosetting plastics. Some thermoplastics may be attacked by uncured adhesive and will require evaluation (e.g. vinyls, acrylics, styrene, cellulosic types). Joining of cams, pulleys, screws and shafts subject to torsional stresses.

Remarks
Easy to apply and particularly suited to small assembly work. Surplus adhesive at the joint edge will remain uncured because of anaerobic nature of material, and should be removed with solvent (trichloroethylene), e.g. outgassing and condensation of volatiles in closed systems could affect sensitive devices. Low strengths preclude their use as conventional structural adhesives but resistance to torque shear is excellent. May be compounded to provide such properties as heat strength and high tensile or impact strength or plasticised to give low

strength. Metal joint strengths for these formulations at room temperature range from:

1 000 – 2 000 N/cm² Shear (ASTM D1002–64T)
1 500–4 100 N/cm² Tensile (ASTM D 897–49)
2—25 N/cm Peel (ASTM D1876–61T)

REFERENCES

DRAFFONE, A. P., 'How to use anaerobic adhesives', *Manuf. Eng.*, **66**, 46–48, (1971).
NAKANO, Y., 'Photosensitive adhesives and sealants', *Adhes. Age*, **16** (12), 28–33, (1974).
SHIELDS, J., and SUER, T. W., 'Evaluation of an ultra-violet curing structural adhesive—Loctite 353', *Report G388*, Sira Institute, (1974).

ALLYL DIGLYCOL CARBONATE

Type
Thermosetting unsaturated polyester.

Physical form
Low viscosity liquid of low volatility.

Shelf life
Catalysed solution, more than 3 months at 0°C. Unstable at room temperature.

Working life, 20°C
2 weeks at 20°C.

Closed assembly time, 20°C
Assemble within 30 min of preparation.

Setting process
Polymerisation on heat curing after catalyst addition.

Processing conditions
Benzoyl peroxide (3% w/w) is dissolved in the monomer (hastened by warming to 50°C). Cures after 2 h at 70°C to handling condition. Post curing by warming slowly to 80–85°C and retaining at this temperature (8 h) produces a cured cement in its hardest form. Alternative catalysts such as isopropyl percarbonate are also used, but handling is more difficult.

Permanence
Excellent resistance to virtually all solvents including ketones, aromatic hydrocarbons, petroleum and chemicals other than highly oxidising acids. Good gamma radiation resistance with only 5% transmittance loss after 10^8 roentgen exposure. Resistant to heat, cold and rapid temperature change and weathering conditions. Continuous service temperature is 100°C but with intermittent exposure (1 h duration) possible up to 150°C.

Applications
Optical cement for glass, cover-plates, windows, lenses, etc. where clarity properties comparable to optical glass are required ($N_D^{20°C} = 1·504$). Casting of sheets, rods, tubes, etc. Observation and instrument windows, plastic mirrors, projection slides, ophthalmic lenses, and electro-optical assemblies. X-ray equipment and biological test chambers.

Remarks
Available as CR–39 (Columbia Resin 39) from Arinor Ltd. London. Exceptional impact resistance (even at low temperatures) and resistance to abrasion. Radiation resistance level is fifty times that of acrylic resins before physical properties are reduced to the 50% level. Optical properties are maintained under difficult service conditions without the occurrence of internal cracks or surface crazing; photo-elasticity is excellent and there is no loss of optical qualities from strain.

REFERENCES

General properties and characteristics of Allymer C.R.39, Data Sheet No. 44–1*, *Cementing optical elements with Allymer C.R.39*, Data Sheet No. 44–7*, Pittsburgh Plate Glass Co. (U.S.A.) (*available from Arinor Ltd. 41, Duke Street, London, W.1).
'New interest in allyl diglycol carbonate', *Br. Plast.*, June (1966).
STARKWEATHER, H. W., ADICOFF, A., and EIRICH, F. R., 'Heat Resistant Allyl Resins', *Ind. Engng Chem.*, **47**, 302, (1955).

ANIMAL GLUES

Type
Based on collagen, or protein extracted from animal hides and tissues or bones.

Physical form
Liquid, jelly, or solids in the form of flakes, cubes, granules, powder, cakes, slabs, etc., for re-constitution with water.

Shelf life, 20°C
Liquids, jellies, limited to few months; solid glues, indefinite.

Working life, 20°C
Indefinite if protected from water, heat and fungus.

Closed assembly time, 20°C
Varies with type used. Assembly within 10 min where cooling of hot glue leads to premature setting.

Setting process
Loss of water and gelation.

Processing conditions
Dependent on type of glue. Set at room or high temperatures (80–90°C). Bonds form before glue cools and gels. Bonding pressures range from contact to 100–140 N/cm² (hardwoods) and 35–

70 N/cm² (softwoods) and are applied for 5 min to several hours. For wood, moisture content should not exceed 18%.

Coverage
0·1–0·15 kg/m².

Permanence
Generally resistant to organic solvents but do not withstand exposure to water, weather, heat or bacterial growths in humid conditions. Resistance to water and biodeterioration is slightly improved by additives such as paraformaldehyde, oxalic acid and chromates.

Applications
Furniture woodworking; leather; paper; textiles. Adhesive binder for abrasive papers and wheels.

Specifications
BS 745 : 1949
BS 647 : 1959

Remarks
Hide glues provide stronger joints than bone glues; bond strengths usually exceed the strength of the material for wood and fibrous adherends. High strength joints result where bonds are maintained under dry conditions, but structural applications are limited to interior service conditions. Gap-filling properties are useful where close fit joints are not feasible and filler powders are not required.

REFERENCES

CORNWALL, E. D., 'Animal Glues and their industrial application', *Adhesion and Adhesives*, (Houwink and Salomon, Eds), Elsevier (1967).
SUTTON, D. A., 'Modern natural and synthetic glues and adhesives,' *Aspects of Adhesion 2 Proc. Conf.*, **23**, 63 (1964).
Animal Glue in Industry, National Assocn. of Glue Manufacturers, Inc. New York, (1951).

BLOOD ALBUMEN

Type
Dried animal blood albumen.

Physical form
Light brown powder which is mixed with water, lime (hydrated) or sodium hydroxide.

Shelf life, 20°C
Dry powder, 1 yr.

Working life, 20°C
Several hours to 1 d.

Closed assembly time, 20°C
Within 15 min.

Setting process
Sets rapidly at 80°C by loss of water and coagulation of the blood.

Processing conditions
For plywood, 10–30 min at 70–120°C and 50–70 N/cm² bonding pressure. For porous materials, several minutes at 80°C.

Coverage
0·05–1·5 kg/m²

Permanence
Good resistance to boiling water but subject to biodeterioration and attack by insects. Ageing properties are moderate and these adhesives are suitable for interior use only.

Applications
Interior-grade plywood manufacture; porous materials such as cork, leather, textiles and paper. Packaging applications such as cork to metal in bottle caps.

Remarks
Glues based on blood albumen are now of minor importance in the United Kingdom but are still used extensively in other countries (e.g. Scandinavia) for medium grade plywood manufacture. A more recent use has concerned its use as a modifier for urea-formaldehyde adhesives.

REFERENCES

ROY, D. C., *et al.*, 'Animal blood as extender for U.F. resin adhesives', *Pop. Plast.* **12** (11), 11, (1968).
PERRY, T. D., 'Albumin glues', *Modern Wood Adhesives*, Pitman, London, (1947).

BUTADIENE—STYRENE RUBBERS

Type
Synthetic thermoplastic elastomers.

Physical form
Viscous liquids based on styrene-butadiene copolymers in solvent (usually hydrocarbons such as naphtha or petroleum) or latex and dispersion form. Usually compounded with tackifier and plasticisers.

Shelf life, 20°C
3 mth to 1 yr in sealed containers.

Working life, 20°C
Equals the shelf life where no solvent loss occurs.

Closed assembly time, 20°C
10–20 min for wet bonding.

Setting process
Solvent release from adhesive coated adherends and contact bonding of the tacky surface.

Processing conditions
Removal of bulk of solvent by evaporation for 20 min to several hours (may be assisted by warm air drying) followed by low pressure contact bonding.

Coverage
Dependent on viscosity of adhesive; about 0·2 kg/m² is typical range for 30% w/w solids content.

Permanence
Durability depends on the butadiene to styrene formulation ratio and like natural rubber, resistance to oils, solvents and oxidation is poor. Heat resistance varies widely with formulation and service temperature range is 5–70°C (special compounded types, 93°C). Heat resistance and ageing properties are superior to natural or reclaim rubber types; elevated temperature ageing increases rigidity, rather than softens, the adhesive. Water resistance of solvent types is good and slightly better than for latex types. Absorption of water is less than natural rubber.

Applications
Solvent types; vinyl plastics to metal, cork, paper, rubber; wood to chipboard bonding; upholstery fabrics. Manufacture of pressure sensitive adhesives where ageing properties are superior to natural rubber formulations. Latex types; lamination of metal foil, plastic films, paper and similar fibrous materials; rubber tyre cords (when compounded with resorcinol formaldehyde); fabric to metal.

Remarks
Limited application in adhesives bonding compared with nitrile and neoprene types. Strength characteristics of solvent and latex types are poor and there is a tendency to creep. Creep performance is better than for reclaim rubber adhesives but inferior to nitrile and neoprene rubbers. Vulcanising types perform better. Physical properties of latex types, such as wet strength, dry film strength are inferior to natural rubber latex adhesives. Lack of tack necessitates addition of tackifier to formulation.

REFERENCES

MCHUGH, J., 'Synthetic rubber latices in British carpet industry', *Rubb. Plast. Age,* **47** (10), 1087, (1966).
NICHOLS, M. J., 'How to utilise styrene-butadiene rubbers in adhesive formulations', *Adhes. Age,* **11** (3), 22, (1968).

BUTYL RUBBER AND POLYISOBUTYLENE

Type
Thermoplastic elastomers; polyisobutylene is a homopolymer, butyl rubber is a co-polymer of isobutylene and isoprene.

Physical form
Viscous liquids or mastics based on the elastomer in solvents (usually hydrocarbons such as naphtha, cyclohexane or chlorinated hydrocarbon) with additives. Solids content ranges from 5% (sprayable) to 70% (spatula).

Shelf life, 20°C
3 mth to 1 yr, in sealed containers.

Working life, 20°C
As for shelf life where solvent loss is avoided.

Closed assembly time, 20°C
Indefinite or up to 4 h for cured butyl rubber systems.

Setting process
Solvent evaporation from adhesive coated substrates, or vulcanisation (butyl rubber only) by addition of curing agents (sulphur and poly-nitroso compounds).

Processing conditions
Porous materials, wet bonded and solvent allowed to absorb into adherend. Non-porous materials, removal of bulk solvent by drying 5–20 min at 90°C followed by contact bonding of tacky substrates with minimum pressure. Vulcanisation by catalyst addition and heat curing (2 h at 115°C is typical) after drying.

Coverage
Dependent on viscosity of formulation.

Permanence
Polyisobutylene and butyl rubbers have good resistance to ageing, mineral oils, acids and chemicals (not hydrocarbons), ozone, water. Exceptionally low permeability to gases, vapours and moisture. Resilience, electrical resistivity and resistance of butyl to abrasion, acids and heat are improved by vulcanisation.

Applications
Not widely used in adhesive systems. Paper and stationery, fabrics and butyl rubbers to other elastomers (not silicone rubbers). Automobile tyre fabrication (tread to carcass bonding and latex tyre cord primer for rayon or nylon fibres). Metal to butyl rubber bonding. Mastics for sealing and caulking.

Remarks
Properties resemble those of reclaim rubbers. Polyisobutylene cannot be cured but may be used instead of butyl in other adhesive formulations. Poor strength characteristics and vulnerability to oils and solvents along with tendency to flow under low loads have restricted adhesive application. Cold flow reduced somewhat by compounding. Acid and chemical resistance improved by mica, talc,

graphite additives. Zinc oxide, alumina, carbon blacks are employed to promote tack and cohesive strength. Tack strengths exceed those of polymer of comparable molecular weight.

REFERENCES

HIGGINS, J. J. and STUCKER, N. E., 'Improved butyl intermediates assist adhesive processors', *Adhes. Age*, **9** (10), 29, (1966).

BALDWIN, F. P. *et al,* 'Elastomeric prepolymers for adhesives and sealants provide improved strength and versatility', *Adh. Age,* **10** (2), 22, (1967).

CASEIN

Physical form
Off-white powder which is mixed with water.

Shelf life, 20°C
1 yr

Working life, 20°C
1–24 h.

Closed assembly time, 20°C
Up to 1 h.

Setting process
Chemical reaction with loss of water by evaporation or absorption into adherend

Processing conditions
For general purposes, 2–6 h at 20°C. For structural applications, 16–24 h at 20°C. Bonding pressures range from 35–140 N/cm^2 for timber work. Moisture content of wood should not exceed 18%. It is feasible to reduce setting times by hot pressing some assemblies. Alkaline nature precludes use of copper or aluminium mixing vessels.

Coverage
0·1–0·15 kg/m^2.

Permanence
Casein adhesives are unsuitable for outdoor use although more resistant to temperature changes and moisture than other water based adhesives. Resistance to dry heat up to 70°C is good, but under damp conditions the adhesives lose strength and are prone to biodeterioration (chlorinated phenols inhibit this). Casein adhesives are often compounded with materials, such as latex and dialdehyde starch, to improve durability. Resistance to organic solvents is generally good.

Applications
Woodwork applications; furniture and joinery; interior grade plywood fabrication.

Specifications
BS 1444 : 1948.

Remarks
Strong alkaline nature of mixed casein adhesives often effects the bonding of timber with high resin or oil content by virtue of a saponification action on poorly wetting surface contamination. Resultant bonds may be stronger than those with synthetic resins. Hardwoods are subject to staining. Gap-filling properties are good. Natural or synthetic rubber latex-casein formulations age well and have good strength; resistance to weather, moisture and biodeterioration is comparable to unmodified casein adhesives. Casein-latex is used for bonding wood to metal panelling, laminated plastics to wood and metal, and linoleum to wood.

REFERENCES

WEAKLEY, F. P., ASHBY, M. L. and MEHLTRETTER, C. L., 'Casein-dialdehyde starch adhesive for wood', *Forest Prod. J.,* **13,** No. 2, (1963).

BROWNE, F. L., and BROUSE, D., *Casein Glues in Casein and its Industrial Applications,* A.C.S. Monograph No. 30, Reinhold, New York, (1939).

KNIGHT, R. A. G., *Adhesives for Wood,* Chapman & Hall, London (1952).

CELLULOSE DERIVATIVES

Cellulose may be chemically modified to produce thermoplastic materials which are partially synthetic derivatives suitable for adhesive formulations. Cellulose nitrate, cellulose acetate, and ethyl cellulose are the most extensively used derivatives, and each material has unique basic properties which favour particular applications as an adhesive. The main forms, properties and applications of the important cellulosic compounds are given here. Available as solvent solutions or solids, the setting action usually depends on solvent release by evaporation or solidification by cooling of a solid melt.

Three water soluble cellulose derivatives are employed as adhesive ingredients or binder materials: hydroxy ethyl cellulose, methyl cellulose and sodium carboxymethyl cellulose. Hydroxy ethyl cellulose is a binder and water retaining agent for ceramic dry-moulding powders, luminescent pigments, and is used to promote green strength in tungsten carbide.

Methyl cellulose is a binder material for the paper and cardboard industry; it is also used as a pigment binder in water-based paints.

Carboxymethyl cellulose is used as a binder in various ceramic glaze formulations and foundry core materials.

CELLULOSE ACETATE

Physical form
Solid (hot-melt type) or solvent solution in esters, ketones or hydrocarbons (chlorinated).

Properties
Compared with cellulose nitrate, it is less inflammable and water resistant but has better colour stability and heat resistance. Maximum service temperatures may approach 50°C. Resistance to biodeterioration is excellent but to solvents and cold exposure is only moderate.

Applications
Cellulose acetate plastics and photographic film; balsa wood (model making); porous materials including, leather, paper, textiles.

Remarks
Subject to cold flow and is unsuitable for use in structural (load bearing) applications. Adhesion to rubber, metals and glass is poor, but fair with ceramic materials.

CELLULOSE ACETATE-BUTYRATE

Physical form
Solid (hot-melt type) or solvent solution in esters, ketones, hydrocarbons (chlorinated), nitro paraffins. Available in powder form.

Properties
Water-clear material with better water resistance and heat stability than cellulose acetate. Excellent resistance to biodeterioration but will not withstand prolonged immersion in organic solvents.

Applications
Heat-sealing of porous and semi-porous materials based on paper, textiles, wood, leather.

Remarks
Similar to cellulose acetate but adhesion to rubber, metals and glass is slightly better. Compatible with a wider range of plasticisers than cellulose acetate.

CELLULOSE CAPRATE

Physical form
Hot-melt solid resin (in stick form)

Properties
Exceptionally stable to light and (unlike Canada balsam) does not discolour with time. Good resistance to water and aqueous solvents but is soluble in many organic solvents. Service temperature range is $-70°C$ to $70°C$ with some softening occurring at 100°C. Tensile strengths approach 700 N/cm^2 with glass and metals, and joints withstand considerable mechanical shock.

Applications
Optical cement for lenses, mirrors, cover slips. (Refractive index $N_0^{20°C}$ is 1·473–1·474).

Remarks
Usually applied to glass surfaces which have been preheated to 140°C. Adhesive sets on cooling but an annealing operation is recommended to remove residual joint strains. Has replaced Canada balsam for lens applications (advantage of thermoplasticity for repair purposes requiring disassembly and re-assembly).

CELLULOSE NITRATE (NITROCELLULOSE OR PYROXYLIN)

Physical form
Solvent solution in esters, ketones, or ketone-alcohol mixtures.

Properties
Inflammable material which is available in opaque or transparent forms; subject to discoloration in sunlight. Fair resistance to water, oils, organic solvents, weak acids and alkalies. Withstands moderate heat exposure and has excellent resistance to biodeterioration. Shear strengths approach 1 000 N/cm^2 with metal, wood and glass joints.

Applications
General purpose household cement for glass, paper, leather, textiles and some plastics (cellulose acetate); shoe manufacture for sole to upper bonding; leather power-transmission belts; cardboard packaging materials; heat sealing of glassine paper; chemical binder for nitro-glycerine in cordite and dynamite; honing stones (carborundum) to zinc alloys.

Remarks
Thermoplastic nature of material precludes use as structural adhesive above 60–70°C for most formulations. Adhesion to rubber and many synthetic footwear materials is poor. Is a binder material for aniline based inks employed with packaging materials.

ETHYL CELLULOSE

Physical form
Solid (hot-melt type) or solvent solution in esters, ketones, alcohols or aromatic hydrocarbons.

Properties
Light-stable materials which are resistant to water, alkalies but not most organic solvents. Heat resistance is generally fair, unless adhesive is stabilised, but low-temperature durability is excellent. Not subject to biodeterioration

Applications
Applications requiring retention of toughness at low temperatures; paper; rubber, leather; textiles; metals.

Remarks

Has low flammability. Adhesion to metals is moderate and only fair with wood. Unsuitable for glass and ceramics.

HYDROXY ETHYL CELLULOSE

Physical form

Solid (hot-melt type) or water solution or emulsion.

Properties

Excellent resistance to organic solvents but does not withstand water immersion. Good retention of adhesion on exposure to hot or cold conditions. Unlike ethyl cellulose, is prone to biodeterioration.

Applications

Porous materials such as paper, leather, fabrics; Decalcomania fabrication; binder material for improving wet strength of paper tissues; thickeners and stabilisers for pigmented lotions and cosmetics.

Remarks

Unsuitable as adhesive for wood, metals, glass, but gives fair adhesion to rubber. Often an ingredient of polyvinyl acetate emulsions for thickening purposes.

METHYL CELLULOSE AND SODIUM CARBOXY METHYL CELLULOSE

Physical form

Water solutions or emulsions

Properties

These materials have permanence properties similar to hydroxy ethyl cellulose but are less heat resistant. Methyl cellulose has excellent wet tack and gives dried films which are strong and resilient. Sodium carboxy methyl cellulose lacks the excellent biodeterioration resistance of methyl cellulose.

Applications

Methyl cellulose is used as: a wallpaper paste material; leather paste adhesive to prevent shrinkage on drying; corneal adhesive for ophthalmic contact lenses.

Remarks

Methyl cellulose gives poor adhesion to metals, glass and ceramics and is suitable only for porous materials and powders. Sodium carboxy methyl cellulose has similar properties as a binder but is not used as an adhesive by itself.

REFERENCE

YARSLEY, V. E. *et al*, *Cellulosic Plastics*, Iliffe Books Ltd, London, (1964).

CERAMIC OR REFRACTORY INORGANIC ADHESIVES

The uses of glasses as adhesives, to replace brazing, in aerospace applications and particularly for the bonding of refractory metals, has been referred to in an earlier Sira publication[1]. Extensive work was done at the University of Illinois[2] and by Aeronca Inc[3]. One of the best glasses found to be suitable for use with stainless steels had the following composition by weight: SiO_2 38%, Na_2O 5%, B_2O_3 57%. Tensile shear strengths at 540°C of about 1 400 N/cm^2 were claimed for this type of glass. The work was abandoned in the early 'sixties as glasses have the following disadvantages compared with brazing materials:

1. Glasses are brittle at room temperature and, therefore, have poor shock resistance. Furthermore, the coefficient of expansion must closely match that of the adherends, making the bonding of many dissimilar materials by glasses impossible. Attempts were made, in the USA, to increase the ductility and strength of the bond by adding powdered metals to the frit[4]. It was claimed that lap shear strengths of 2 800 N/cm^2 at room temperature were obtainable in this way.

2. Glasses are viscous liquids rather than true solids and are subject to creep, particularly at higher temperatures in the region 500–600°C.

3. The surface preparation of metal adherends is elaborate. Aeronca found that when using glassy adhesives, the most effective surface preparation was vapour degreasing followed by oxidising the metal to form a scale, acid etching to remove the scale and then flame-spraying with 0·076–0·125 mm of Cr-Ni alloy. This spray provided surface roughness and porosity and the necessary oxide film.

It is essential to have an adherent oxide film on the metal[5]. This oxide partially dissolves into the glass giving a strong bond. If, due to over-oxidisation of the metal in preparation, too much oxide remains after the seal is made, the oxide layer may fracture. If the seal is kept at a high temperature for too long, all the oxide will be dissolved in the glass and a weak bond will result. A typical surface treatment for ferrous alloys used in the electronic valve industry is to heat in wet hydrogen at 1 100°C, to remove surface carbon, and then to oxidise with a flame. Clearly, the exact degree of oxidation needs to be established by trial for any given metal.

For these reasons applications of glass adhesives are virtually limited to the field of radio valve and television tube manufacture. Here, a considerable technology has been built up[6] and much effort is justified in meeting the required conditions as the seals need to be electrically insulating and temperature-resistant, so as to permit baking to de-gas the electronic tubes.

Research now appears to be turning to the use of glass ceramics for bonding[7–10] (e.g. at Corning, USA and English· Electric, England). These are glasses which, with a suitable heat treatment, devitrify into a strong material with an interlocking crystalline structure, and have the following advantages over ordinary glasses:

1. The mechanical strength at room temperature is greater; this is said to permit greater expansion mismatches than do glasses.
2. The service temperature of the glass ceramic bond is nearly as high as the maturing or treatment temperature, whereas with glasses there is a gap of about 150°C between these temperatures.
3. The creep properties are better than those of normal glasses.

The adhesive is applied as a suspension of finely powdered glass ceramic in a vehicle such as a 1% solution of nitrocellulose in amyl acetate.

Corning, (New York) manufacture three types of glass ceramic cement, the most generally useful probably being 'Pyroceram 95' cement of which the properties and suggested bonding applications are:

firing temperature range, 400–450°C;
maximum service temperature, 425°C;
designated expansion range, 85 to 100×10^{-7} cm/cm/°C;
metals, chromium-iron, stainless steel, platinum, 50% nickel alloys, Dumet;
glasses, most lime and lead glasses;
ceramics, Forsterite, Steatite (aluminium silicates).

'Pyroceram 95' cement was tested at Sira by coating two strips of stainless steel with a thin layer of the cement and firing, under pressure, at 450°C. While adhesion of the cement to the metal was good the fusion of the cement layers into each other was poor, resulting in a weak bond. Pre-heating the bond to 700°C resulted in little improvement.

In the vast majority of cases, for the production of refractory bonds, industry uses adhesives of the alkali silicate, aluminium phosphate and allied types which set at relatively low temperatures (up to 600°C). Silicate cements are dealt with elsewhere in this book; other cements in this category are mostly of American origin and include the following.

'Ceramabond 503'

Single part, acidic, alumina base ceramic adhesive with service temperature up to 1 400°C. Curing temperature 120°C. Suggested application: bonding ceramics, graphite, quartz and glass.

'Ultra-temp 516' ceramic adhesive

Single part zirconia based adhesive with service temperature up to 2 400°C. Curing temperature 600°C. Suggested applications: bonding ceramics, glass, quartz, graphite and metals.

The above two adhesives are supplied by
Aremco Products Inc.,
P.O. Box 145,
Briarcliff Manor, NY 10510

'Astroceram' high temperature cements (A and B)

Supplied by
American Thermocatalytic Corporation,
216E Second Street,
Mineola, LI.,
New York 11501

A type cements (basic)

For use with metallic systems, but may also be used with non-metallic systems. Good unfired strength so as to be handleable. Maximum service temperature 2 400°C.

Type A for thick coatings.
Type A-LP for low porosity.

B type cements (acidic)

Preferred for use with ceramic systems and not to be used with metallic systems. Must be fired to achieve a bond (temperature not specified). Maximum service temperature 2 900°C.

Type B for thick coatings.
Type B-LP for low porosity.

REFERENCES

1. HURD, J., *Adhesives Guide*, Sira Research Report M39, (1959).
2. LEFORT, H. G. and BENNETT, D. G., 'High Temperature Resistant Ceramic Adhesives', *J. Am. Ceram. Soc.*, **41**, No. 11, 476, (1958).
3. BENZEL, J. F., 'Ceramic–Metal Adhesive Combinations', *Bull. Am. Ceram. Soc.*, **42**, No. 12, 748, (1963).
4. 'Research on Inorganic High Temperature Adhesives for Metal Composites', USAF Contracts AF33(616)–53338 and AF33(616)–7196.
5. WHITNEY, J. B., 'High Temperature Ceramic Adhesives', *Symposium on adhesives for structural applications,* Interscience, Sept. 27–28, 113–118, (1961).
6. PARTRIDGE, J. H., *Glass to Metal Seals*, (Soc. of Glass Technology), (1949).
7. MCMILLAN, P. W., *Glass Ceramics*, Academic, New York, (1964).
8. KINGERY, W. D., *Introduction to Ceramics*, John Wiley, New York, (1960).
9. PATRICK, J. B., 'Glass Engineering', *J. Br. Soc. Sci. Glass-blowers*, **5**, No. 1, (1968).
10. *Special Ceramics*, (Popper, Ed) Academic, New York, (1960).

CYANOACRYLATES

Type

Synthetic polymers (alkyl 2-cyanoacrylates)

Physical form

One component, low viscosity liquids.

Shelf life
Up to 12 mth in sealed containers stored under refrigeration between 1 and 5°C.

Working life, 20°C
Indefinite, before bonding pressure is applied. Materials polymerise when pressed into a thin film between two adherends.

Closed assembly time, 20°C
Within a few minutes.

Setting action
Polymerisation at room temperature, without addition of a catalyst, which is promoted by presence of moisture or weak bases (hydroxyl materials such as alcohols) on substrate surface.

Processing conditions
Substrates are cleaned with alkaline solvents such as ketones or toluene (acidic solvents such as trichloroethylene must not be used) and the adhesive dispensed sparingly from the container (usually a polythene bottle). Parts are brought together quickly under contact pressure, and held in position for a few minutes. Thin glue-lines give the best results. Setting time is governed by nature of adherend and surface condition; minor factors are adhesive viscosity, humidity and temperature. Alkaline substrates (glass) set in 10–30 s whereas acid ones (wood) require at least 3 min to set. Accelerators (surface primers) are available to promote curing.

Permanence
Generally good resistance to organic solvents such as alcohols, ketones, aliphatic and aromatic hydrocarbons. Notable exceptions are nitromethane N-dimethylformamide and hydrazine. Dilute alkali solutions weaken bonds but dilute acids have less effect. Resistance to humidity and weathering is generally satisfactory but constant immersion in water or steam will gradually weaken bonds. Continuous exposure to temperatures above 80°C is not recommended but short time resistance to temperatures approaching 180°C is possible. Bonds frozen in dry ice (-8°C) for several hours show good retention of strength.

Applications
Light structures requiring fast assembly such as, instrument components and transducers, electronic and optical units. Smooth non-porous materials where there is intimate surface contact since gap-filling properties are poor. Inclusion of rubber sandwich sheet between irregular surfaces is often effective. Wood, glass and ceramics, ferrites, rubber, leather and many plastics.

Specifications
Satisfies T.S. 468 and MIL-A-46050A

Remarks
Relatively expensive adhesives with advantage that they eliminate the need for expensive heating and pressure equipment. Instant adhesion to skin and tissues represents a major hazard. Bonds formed with glass, rubber, and wood are generally stronger than the material bonded. For some assemblies (e.g. optical lenses, small electrical or mechanical units, etc.) the design of the unit relies on the use of these rapid setting adhesives. Joint strengths usually improve with ageing up to 2 wk; typical results for various cure times are:
Steel-steel, 1 000 (15 min), 1 900 (2 d), 2 200 (7 d), N/cm^2 Shear (ASTM D.1002–64); Aluminium-aluminium 950 (15 min), 1 700 (2 d), 2 100 7 d), N/cm^2 Tensile (ASTM D. 2095–62T)

REFERENCES

COOVER, H. W., Jr., 'Cyanoacrylate Adhesives', *Handbook of Adhesives*, (Skeist, Ed), Reinhold, (1962).
MACPHERSON, C. J., 'High-strength adhesive', *Indust. Fastening and Assembly Conf.* Report, paper D1(6), 17, (1964).
ANON., 'Spacemen to bond self to vehicle with adhesive capsules', *Adhes. Age,* 10 (1), 35, (1967).
PAGE, R. C., 'Tissue adhesives-eliminates sutures and staples in many types of surgery', *Adhes. Age,* 9 (12), 27, (1966).
HOGGE, J. W., 'Tensile properties of alpha cyanoacrylate adhesives', 470312, *Inf. Bull. Alumin. Dev. Ass.,* 6, (1965).
MATSUMOTO, T., *et al.,* 'Method of arterial anastomosis using cyanoacrylate tissue adhesives', *Archs Surg. Lond.,* 94 (3), 388, (1967).

EPOXY ADHESIVES

Type
Thermosetting synthetic products derived from the reaction of a polyepoxide resin and a basic or acidic curing agent (hardener).

Physical form
Available as one or two component systems. One component products include solvent-free liquid resins, solutions in solvent, liquid resin pastes, fusible powders, sticks, pellets and paste, supported and unsupported films, preformed shapes cut to fit a particular joint. Two component adhesives usually comprise the resin and curing agent which are mixed just before use. Components may be liquids, putties, or liquid and hardener powder. May also contain plasticisers, reactive diluents, fillers, pigments and resinous modifiers.

Shelf life, 20°C
3 mth to 1 yr according to the system. Refrigeration extends the storage life.

Working life, 20°C
Depends on formulation. Some one-part systems

have indefinite working life. Pot-life of two part systems can vary from a few minutes to several hours. Temperatures extend the work life appreciably.

Closed assembly time, 20°C
Determined by the working life; can range from 5 min to 4 h for two part systems.

Setting action
Polymerisation.

Processing conditions
These are determined by the curing agent employed. The large variety of formulations available require close attention to the manufacturer's instructions with respect to mixing and methods of use. In general, two part systems are mixed, applied within the pot-life (minutes–hours) and are cured at room temperature (up to 24 h possibly) or at elevated temperature to reduce cure-time (typical cures range from 3 h at 60°C to 20 min at 100°C).

With an aliphatic amine (e.g. diethylenetriamine), at room temperature, the resin is cured sufficiently in 4–12 h to permit handling of the assembly. Full strength develops over several days. A compromise between cure-rate and pot-life has to be reached where too rapid a cure at room temperature results in the formulation of an unspreadable mixture in the mixing pot. Restriction of heat build-up (exothermic reaction) by lowering temperature, limiting the size of batch, or using shallow mixing container, extends the pot-life. Contact bonding pressures usually suffice, but small pressures from 1·5–2 N/cm² ensure more uniform joints having maximum strength. One-part systems incorporate a hardening agent which requires heat to activate curing (30 min at 100°C is typical).

Curing agents
The properties of the cured resin depend on the degree of cure and the type of curing agent used. The use of catalysts or reactive hardener to effect curing results in an evolution of heat as a by-product of the reaction (exotherm). For thin glue-lines, exothermic heat is largely conducted away into the adherends. It follows that for room temperature curing, the fastest hardeners are preferred where no heat is to be supplied to the resin from without.

Some of the curing agents commonly used with epoxy resins are given here.

Aliphatic amines

Triethylene-tetramine	(TETA)
Tetraethylene-pentamine	(TEPA)
Diethylene-triamine	(DETA)
Dimethylaminomethyl phenol	(DMP30)

Cold-setting systems which may be warm—or hot-set whereby curing times are reduced and end-product properties are important. A proportion of 10 parts catalyst to 100 parts resin is common but the ratio may be varied.

DMP30 will cure epoxies without external heat application but optimum performance is given by a post cure.

Aromatic amines

m-phenylene diamine	(MPDA)
Diaminodiphenyl methane	(DDM)
Methylene dianiline	(MDA)

Hot-setting systems which may be liquid or solid resins (can be modified by solvent addition), or one or two component solids or pastes. All these systems require heat and cannot be cured at room temperature. Pot-lifes of several hours at room temperature are usual.

The selected curing agent can have a marked effect on the shear strengths. Adhesive strength is improved by the employment of aromatic amine instead of aliphatic amine hardeners.

Polyamides

Versamids (derived from amines such as DETA and di-carboxylic acids).

These are slow curing at room temperature and are heated to reduce viscosity for easy blending with the epoxy resin. DMP30 is substituted for part of the polyamide where faster curing is required.

Aliphatic amines and polyamides are effective flexibilising agents and cured epoxies have excellent mechanical, physical and chemical properties. Cure shrinkage is small. Heat resistance is inferior to other systems and usually limited to 80°C.

Aromatic amine cured systems have better heat stability and these systems perform well up to 120–175°C. For some systems, there is a rapid loss of adhesion at 150°C.

Acid anhydrides
In addition to the above curing agents, aromatic and cyclic acid-anhydrides such as, phthalic anhydride (PA), pyromellitic anhydride, (PMDA) and chlorendic anhydride (HET), have been used as hardeners for special purposes. Although, thermal stability (up to 150°C continuously, or intermittently at 260°C) is better than for amine cured systems, processing difficulties militate against their general usage as curing agents for adhesive systems. Anhydrides are usually solids or viscous liquids which require heat to disperse them effectively within the epoxy resin. Thus, the pot-life of such systems is considerably lowered. High curing temperatures are required to achieve high temperature properties.

Boron trifluoride: monoethylamine (BF3:MEA) is a curing agent for one-component adhesive systems and is present as a latent catalyst which is activated when heated. Not recommended for adhesive formulations although widely used in electrical applications. Shear strengths have poor reproducibility. Water sensitivity is a problem too. Pot-life for 1–2 % catalyst is about 1 y. Cures in 4–6 h at 110°C.

Permanence
Depends on the particular system (modifying or curing agent used); it is advisable to check with the manufacturer or carry out an evaluation. In general, bonds maintain excellent strength and durability under many environments. Bond strength is generally little changed over several years' exposure to weathering or contact with oils, greases, hydrocarbon fuels, alkalies, aromatic solvents, acids, alcohols and hot or cold weather. Resistance to ketones and esters is usually poor. Some formulations are vulnerable to greases and oils, and loss of strength on prolonged immersion in water (especially hot) can occur, e.g. poor resistance to hot water and alkalies is a feature of polyamide cured epoxies. Polyamine and anhydride systems have poor cold resistance whereas polyamide types are satisfactory even at cryogenic temperatures (to $-200°C$). Anhydride and polyamine systems display the best heat resistance whereas polyamide types are limited to about 70°C.

Applications
Versatile structural adhesives for many materials based on metals, glass and ceramics, wood, concrete and thermosetting plastics (polyesters, phenolics). These industrial applications are typical:

Aircraft	Bonding of aluminium to itself or other metals; foam panels and honeycomb sandwich materials; fibre-glass polyester fuel tanks; cyclised rubber to metals; substitute for rivets.
Automotive	Attachment of reinforcements to light metals for greater rigidity where welding would cause distortion and painting problems; body solder substitute; fibre-glass body assembly; car battery cases.
Electronic-electrical	Implosion protection coatings for vacuum tubes; lamination of electric motor armatures and transformers; encapsulants for printed circuits; printed circuit laminations, e.g. copper to phenolic for printed circuit boards; transducer and instrument assembly; optical applications.
Building	Ceramic tiles, walls, floors, concrete repair, rubber and plastic to ceramic tiling; repair of roads and bridges.
Woodworking	Lamination of beams. The use

of resorcinol resins predominates here but epoxies are used to reinforce laminas which are not in contact.

Miscellaneous	Repair kits for factory and domestic use; marine structures; abrasive paper manufacture; artistic materials for bonding glass, metals, etc. Silicone and butyl rubbers, vinyl plastics (flexible), polyethylene, polypropylene and fluorocarbons give poor bonds with epoxy resins. The last three materials will give adequate bonds with special surface pretreatments.

Remarks
Epoxy resins have several advantages over other polymers as adhesive agents and these are briefly listed here:
(*a*) high surface activity and good wetting properties for a wide variety of materials,
(*b*) high cohesive strength for cured polymers which often exceeds adherend strength,
(*c*) lack of volatile reaction minimises shrinkage and permits bonding of large areas without pressure build-up during cure (other adhesives often need to be vented to allow volatile by-products to escape),
(*d*) low shrinkage compared with polyesters, acrylics and vinyl types; cured glue-lines are less strained,
(*e*) low creep with better retention of strength under sustained loading than thermoplastic adhesives,
(*f*) may be modified by (i) selection of base resin and hardener, (ii) addition of another polymer, (iii) addition of filler.
Specific epoxy systems are discussed on the following pages.
Tensile shear strengths for metal joints range from 700–5 000 N/cm^2 according to the adhesive composition; heat-cured bonds are considerably stronger than cold-setting ones. Impact strengths are fair, while peel and cleavage strengths are poor unless the epoxy resin is compounded with flexible polymers such as polysulphide, polyamides, or polyurethanes that enable the cured polymer to withstand vibrations, shock and peel stresses. However, the thermoplastic nature of modifiers decreases the heat resistance.

REFERENCES

MCINTYRE, J. M., 'Epoxy adhesives', *Indust. fastening and assembly Conf.*, Report paper C1(a) (1964).

DELMONTE, J., *et al*, 'Influence of application variables on properties of epoxy adhesives', *Appl. polymer symp.* **3**, 397, (1966).

IRVING, R. R., 'Adhesive bonding—its day is coming', *Iron Age*, **201** (15), 83, (1968).

SHERIDAN, M. B., 'Epoxy resins in building applications', *Appl. Plast.*, **9** (9), 28, (1966).

GOLDING, J. H. and JOHNSON, R. P., 'Use of epoxies for bonding concrete', *Plastics Inst. Plastics in bldg. structures—Proc.*, Paper 11, (67 and 103), (1965).

GOLDING, J. H., 'Recent developments in epoxy resins', *Sh. Metal Inds.*, (549), (1964).

LEWIS, A. F. and RAMSEY, W. B., 'Mechanical behaviour of polymers and adhesive joint strengths with amine cured epoxy resins', *Adhes. Age*, **9** (2), 20, (1966).

Light Prod. Engng, **4** (7), 27, (1966).

LEE, H. and NEVILLE, K., *Epoxy Resins*, McGraw-Hill, London, (1967).

EPOXY-NYLON

Type
Thermosetting blend of epoxy resin with compatible nylon.

Physical form
One-part, 100% solids, tape or film supported on glass cloth, or paste.

Shelf life
Usually supplied under refrigeration to prevent it curing and flowing. 6–8 wk at 0°C to 6 h at 20°C.

Working life, 20°C
Used within 1 h since flow commences at room temperature.

Closed assembly time, 20°C
Usually applied at once and processed immediately within 30 min.

Setting process
Polymerisation.

Processing conditions
Cured for 3 d at 20°C to 1 d at 150°C under bonding pressure from 16–32 N/cm². Cure temperatures for some formulations can be increased to 200°C with corresponding reduction in cure time. No volatiles are released during cure so that large areas can be bonded without venting.

Permanence
Service temperature range is −75° to +95°C with rapid loss in strength above the maximum temperature. Resistance of stressed joints to moisture is poor.

Applications
Structural assemblies requiring very high peel strengths. Bonding aluminium skins to honeycomb sandwich cores in the aircraft industry. Low temperature adhesive for metals.

Remarks
These adhesives have been specially developed for aircraft applications demanding high peel strengths and for general industrial bonding and offer no particular advantages over other structural adhesives. Limited shelf life and moisture instability have not favoured their production in the U.K. Impact, peel and shear strengths are usually higher than normally attainable in epoxy based adhesives. Very high shear

Table 4.1.

°C	Peel (ASTM D1876-61T) N/cm[1]	Shear (ASTM D1002-64T) N/cm²
−90	1·7	4 500
24	200	4 130
82	88	2 500
121	44	1 790
149	8·8	970

and peel strengths up to 120°C are displayed by epoxy-nylon but performance falls off rapidly above this temperature. Typical results for aluminium assemblies with a film adhesive are given in Table 4.1.

Shear strengths for steel bonds are usually lower than for aluminium; 2 000 N/cm² at −40°C is typical.

REFERENCES

KUNO, J. K., 'Comparison of Adhesive classes for structural bonding at ultra-high and cryogenic temperature extremes', *Natn. SAMPE Symp.*, Los Angeles, California, May (1964).

HEINE, E., 'Strengthening adhesives with nylon', *Adhes. Age*, **6**, 20, (1963).

GORTON, B. S., 'Interaction of nylon polymers with epoxy resins in adhesive blends', *J. appl. Polym. Sci.*, **8** (3), 1287, (1964).

RIEL, F. J., 'Adhesive Bonding, a new concept in structural adhesives', *S.P.E. Tech. Paper*, **7**, 27, (1961).

HERTZ, J., 'Epoxy-nylon adhesives for low temperature applications'. *Proc. cryogenic Engng Conf.*, Ann Arbor, Michigan, August (1961).

EPOXY-POLYAMIDE

Type
Synthetic thermosetting product from the reaction of epoxy resin with a polyamide resin (branched chain polymer type containing reactive aliphatic amino groups).

Physical form
Two-part products consisting of liquid epoxy

resins and polyamide resins which vary in consistency from viscous liquids to semi-solids.

Shelf life, 20°C
6 mth or longer (if refrigerated).

Working life, 20°C
Considerably longer than amine cured epoxy systems. Depends on ratio epoxy: polyamide in formulation, temperature, mass of the mix and any fillers present. Gel times (200 g masses) might range from 2–4 h.

Closed assembly time, 20°C
Components mixed shortly before use and applied well within gel time of formulation.

Setting process
Polymerisation (reaction is exothermic).

Processing conditions
Cured at room or elevated temperatures. Rate of cure will depend on adhesive blend and is usually established on test samples by determining hardness or solvent resistance (percentage of uncured material extracted by boiling ketone solvent). 36–48 h at 20°C or 3 h at 65°C to 20 min at 150°C is representative of high temperature curing. Only contact bonding pressure is required. Accelerators (triphenyl phosphite/mod. epoxy or phenol) can be used to increase curing rate.

Coverage
Ranges from 0·07–0·12 kg/m^2

Permanence
Good resistance to aromatic and aliphatic solvents, fuels, lubricants and oils, water, brine, and weak alkalies and oxidising acids. Thermal shock resistance is good; temperature cycling between −70°C and +200°C is usually without effect on mechanical properties. Limited to about 120°C for continuous service but display high strength properties at sub-zero temperatures down to −100°C and below. Are more sensitive to temperature changes than unmodified epoxies.

Applications
Structural adhesives for bonding metal to similar and dissimilar metals, glass and ceramics, leather, wood; metal to plastic, rubbers; repair of masonry and building materials; metal seam and weld sealing. Not suitable for polythene, fluoropolymers or polyvinyl alcohol plastics unless these are specially treated. Aircraft, automotive and general industrial applications. Low-temperature adhesives.

Remarks
Polyamides are particularly suitable as curing agents for epoxy resins and in this respect have certain advantages over volatile amines; display a reduced

sensitivity to minor variations in epoxy: hardener ratio; show improved flexibility and adhesion; have better resistance to impact and mechanical shock than unmodified epoxy resins. Pot-lifes are longer and the exotherm is reduced. Strength properties can be altered by varying the epoxy: polyamide ratio to make compositions more flexible, with greater elongation and higher peel strengths at the expense of lower shear strength and reduced creep resistance. Strength data for epoxy-polyamides which are typical are given here.
Aluminium 1 170 (4°C), 1 550 (25°C), 38 (177°C), 17 (260°C) N/cm^2 Shear (ASTM D.1002–64). Steel 1 270 (4°C), 1 260 (25°C), 41 (177°C), 3 (260°C), N/cm^2 Shear (ASTM D.1002–64).

REFERENCE

'Use of room temperature curing adhesive film on 727 airplane', *Structural Adhesive Bonding*, (Bodnar, Ed.) Interscience, (1966).

EPOXY-POLYSULPHIDE

Type
Product of reaction between epoxy resin and liquid polysulphide polymer (usually catalysed by an added tertiary amine).

Physical form
Two-part product based on components in liquid form.

Shelf life, 20°C
6 mth to 1 y for unmixed constituents.

Working life, 20°C
15 min to 1 h for mixed adhesive.

Closed assembly time, 20°C
Used within working life period.

Setting process
Polymerisation.

Processing conditions
Cured for 24 h at 20°C to 20 min at 100°C. Bonding pressures are within the range 7–16 N/cm^2. Disagreeable odour during processing makes ventilation mandatory.

Coverage
Usually ranges from 0·07–0·12 kg/m^2

Permanence
Good resistance to water, salt spray, hydrocarbon fuels, alcohols and ketones. Weathering properties are excellent. They are suitable for low temperature service down to −100°C and below (some blends have been employed for service at liquid nitrogen

temperatures, $-198.5°C$); maximum high service temperature of about 50°C or less. Moisture resistance is very good but may be less satisfactory for stressed bonds. Some formulations are liable to corrode copper adherends.

Applications
Structural assemblies where some degree of resilience in the bond is required. Bonding (old to old or new) concrete for floors, airport runways, bridges and other concrete structures where adhesion often exceeds that of concrete. Metals, glass and ceramics, wood, rubber and some plastics. Low temperature adhesives (outdoor assemblies, castings, coatings, encapsulants, subject to freeze–thaw cycles).

Remarks
Modification of epoxy resins with polysulphide (up to 50% w/w) effectively increases shear, peel and bend strengths, flexibility and elongation characteristics in various environments involving temperature changes or immersion media. Impact resistance is improved and the materials are less brittle (particularly at low temperatures). Moisture vapour transmission can be reduced by up to 90% by modifying. Electrical properties are slightly lower than for unmodified epoxies as is the heat distortion temperature for blends exceeding a 1:1 ratio of polysulphide:epoxy. According to the formulation employed, strength data within the following ranges are typical for joints tested in Shear (ASTM D1002–53T).
Aluminium-aluminium 1 700 (20°C) N/cm^2. Aluminium-platinum 650 ($-200°C$), 600 ($-80°C$), 600 (20°C) N/cm^2. Stainless steel-platinum 600 ($-200°C$) 800 ($-80°C$), 1 000 (20°C) N/cm^2.

Peel strengths (ASTM D1876–61T) fall within the range 14–32 N/cm^1.

REFERENCE

SORG, E. H. and MCBURNEY, C. A., 'Polysulphide epoxy compositions; liquid polysulphide polymer—epoxy resin formulations for coatings and adhesives', *Mod. Plast.*, **34**, 187, (1956).

EPOXY-POLYURETHANE

Type
Epoxy resin blended with urethane (carbamate) polymer.

Physical form
One-component paste with latent catalyst (dicyandiamide) and aluminium filler.

Shelf life, 20°C
Exceeds 1 yr.

Working life, 20°C
Indefinite.

Closed assembly time, 20°C
Indefinite.

Setting action
Polymerisation.

Processing conditions
Cured for 1 h at 182°C. (Maximum adhesive properties are not obtained by curing below 180°C). Bonding pressures up to 14–20 N/cm^2 are adequate.

Permanence
The durability of these materials for various service conditions has not been evaluated to any great extent. Retention of flexibility at low temperatures is responsible for the high peel strength performance noted below.

Applications
Structural adhesive for metals such as stainless steel, aluminium, copper, titanium where a combination of high peel and tensile strength is required (particularly for low temperature applications).

Remarks
Urethane modified epoxies are experimental materials which are being developed to overcome the peel strength deficiency of epoxy adhesives, and as alternatives to epoxy-nylon compositions which have limited high temperature and humidity resistance. Data taken from recent trade literature indicate the potential of the new system, 40 ($-73°C$), 150 (25°C), 70 (82°C), 22 (121°C) N/cm^1 Peel (ASTM 1781–62). 2 600 ($-73°C$), 3 870 (25°C), 3 000 (82°C), 860 (121°C) N/cm^2 Shear (ASTM 1002–64)
For other metals, high peel strengths are obtained provided high temperature curing is observed,

Metal (thickness in mm)		Peel strength N/cm^1
Aluminium	(0.61)	220
Copper	(0.67)	186
Mild steel	(0.48)	88
Titanium	(1.5)	79

Impact strengths (Izod) are about $1.1J/2.54 cm$ notch and the heat distortion temperature is 82°C. Ease of formulation and the handling characteristics of unmodified epoxy resins are features which recommend this system.

REFERENCES

'Characteristics of an experimental urethane modified standard epoxy adhesive (experimental epoxy resin CX-3599)'. *Technical Report, Dow Chemical. Europe SA* (1968).

CLARKE, J. A., and HAWKINS, J. M., 'Characteristics of a urethane modified standard epoxy adhesive', *Tech. Report Dow Chemical Europe SA* (1968).

FISH GLUE

Fish glues are a by-product of desalted fish skins (usually Cod) and have similar properties to the animal skin and hide glues which have largely replaced them as woodworking adhesives.

Fish glue is available in cold setting liquid form which does not set to a gel at room temperature. Solvents such as ethanol, acetone or dimethylformamide may be added to assist the penetration of the glue into substrates which may already be coated or finished (e.g. certain paper, leather and fabrics). The glue withstands repeated freezing and thawing without adverse effects on adhesion performance. Instant tack with good adhesion develops readily on re-moistening dry glue films with cold water. Water resistance is conferred on dried glue films by exposure to formaldehyde vapours which insolubilise the fish gelatin component.

The main uses concern the preparation of gummed tapes with animal/fish glue compositions, and the bonding of stationery materials. Latex, animal glues, dextrines and polyvinyl acetate adhesives are sometimes modified with fish glues to improve wet tack properties. High purity fish glues are important photoengraving reagents.

REFERENCE

WALSH, H. C., 'Fish Glue', *Handbook of Adhesives*, (Skeist Ed.) Reinhold, (1962).

FURANES

These are dark-coloured synthetic thermosetting resins of which furfuraldehyde (furfural) and furfuryl alcohol are the types commonly employed in adhesive formulations. The resins are available as hot- or cold-setting liquids which are cured by acid catalysts, such as sulphuric acids, to infusible solids. Low volatile loss during cure means that bonding pressures need not be high.

Resistance to boiling water, organic solvents, oils, weak acids and alkalies, is good but strong oxidising acids attack furane resins. High temperature resistance depends on the type and quantity of catalyst used; for continuous exposure service temperatures up to 150°C are acceptable.

Furane resins are employed as bonding agents or as modifiers for other adhesive materials. Typical adhesive applications include the following:

Surfacing and bonding agents for flooring compositions and acid-resistant tiles; chemically resistant cements for tank lining; phenolic laminate bonding (with shear strengths up to 4 000 N/cm^2); binder resins for explosives and ablative materials used in rockets and missiles (1 250°C service temperatures); alone and with other resins for foundry core boxes; binder resins for carbon and graphite products.

Furane adhesives are suitable for gap-filling applications because strength is maintained with thick glue-lines; for this reason the resins are extensively used as modifiers for urea-formaldehyde adhesives to improve gap-filling and craze-resistance qualities. Compatibility with a variety of other resins has led to the use of the resins in mixtures with silicates and carbonaceous materials for chemically-resistant grouting compositions.

REFERENCES

DELMONTE, J., 'The Furane Resins', *Modern Plastics Encyclopaedia*, McGraw-Hill, New York, (1968).
BOQUIST, E. R., *et al*, 'Alumina–Condensed Furfuryl Alcohol Resins', *WADC Tech. Rep.*, **15**, 61–72, (1963).
BROWN, L. H., and STIGGER, E. K., 'Furfural alcohol as a resin modifier', *Mod. Plast.*, **39** (4), 135, (1961).
DELMONTE, J., *The Technology of Adhesives*, Reinhold, New York, (1947).
DUNLOP, A. P., and PETERS, F. N., *The Furans*, Reinhold, New York, (1953).

HOT-MELT ADHESIVES

These materials are bonding agents which are applied as melts and which achieve a solid state and resultant strength by cooling.

Type

Hot-melts are thermoplastic (100% solids) materials which melt at temperatures ranging from 65–180°C; the following polymers, resins and waxes have been used in hot-melt formulations:

Alkyds (modified polyesters)
Asphalt and coal-tar bitumens
Coumarone—indene resins
Phenolic resins (heat stable)
Rosin (colophony) and (derivatives)
Terpene resins
Waxes (mineral, vegetable and petroleum)

Synthetic polymers are being increasingly employed as hot-melt materials and current technology is concerned with a wide range of thermoplastic resins, such as:

Butyl methacrylate
Ethyl cellulose
Ethylene—vinyl acetate copolymer
Polyethylene

Polyvinyl acetate and derivatives (e.g. with acrylic acid.)
Polystyrene and copolymers
Polypropylene
Polyisobutylene
Polyamides
Polyesters
Phenoxy resins (plasticised)
Polyisoprene (trans)

Hot-melt adhesives usually contain three basic materials:
1. A high molecular weight polymer to provide viscosity to the melt and cohesive strength to the solid.
2. A synthetic elastomer to increase tack, elasticity and strength; waxes may be added to reduce cost and lower viscosity of the mixture for easier application.
3. A resin (synthetic or natural) to add tack and fluidity, and promote wetting action.

The formulation is dependent on the adherends involved and also the application conditions of time, temperature and pressure required.

Physical form
Available in a variety of forms. Some of these are: Tapes, ribbons, films, pellets, cylinders, cubes and blocks. Many materials are applied as solvent solutions or emulsions to be dried and heat-activated later.

Shelf life, 20°C
Indefinite.

Working life, 20°C
Indefinite and depends on the formulation; may set almost instantly on cooling or can have its 'open-time' (period of tack) extended almost indefinitely to impart permanent pressure sensitivity.

Closed assembly time, 20°C
Within seconds.

Setting action
Solidification on cooling.

Processing conditions
Dependent on the composition of the adhesive. Hot-melts are predominantly adhesives for mass production industries such as packaging, bookbinding, laminating and footwear manufacture. Consequently, special equipment has been developed to apply them simply and rapidly. Two principal types of applicator are; 'melt reservoirs' which involve melting the solid adhesive in a pot and delivering by metering pump to a hot applicator nozzle; 'progressive feed' applicators which rely on feeding the solid adhesive (in cord-like form) continuously through a hot nozzle with an orifice profile of the required band width. Melt temperatures up to 180°C are commonly employed; bonding pressures vary, with the application, from contact to 70 N/cm^2.

Permanence
Durability of a hot-melt will obviously depend on the composition of the adhesive. Some formulations produce joints which have high tensile strength and are flexible and highly shock-resistant. Resistance to water and solvents is often very good. Thermal stability is determined by the melt-temperature so that service temperatures are usually restricted to 50–70% of the fusion temperature. Decomposition of the material occurs on prolonged heating above the melt temperature. Many hot-melts are suitable for low-temperature environments; polyamides and ethylene-vinyl acetate copolymers retain resilience down to −100°C and below.

Applications
Fast assembly of the materials used in the bookbinding, paper and packaging, footwear and plastics industries, viz: paper, board, leather, rubber, textiles, films and coatings, polyester, polypropylene, polyethylene, etc. Bonding of ceramics, cork and metal; paper to metal foil laminating; low temperature adhesives.

Remarks
Hot-melts are sought for the economic reasons associated with fast assembly, simplicity or automation. Where such factors are less important, the use of a solvent cement or thermosetting adhesive will usually provide bonds of greater strength and durability. Advantages of hot-melts are the following:

Non-inflammable, non-toxic, and do not freeze and embrittle. Develop fast bond strength, without need of drying (solvent removal), with many impervious materials. Good storage stability.

The disadvantages include:

Need for special equipment for application; temperature and viscosity factors limit strength; degradation above the melt temperature; limited control over the coating-weight; some adherends require preheating before application of melt.

REFERENCES

MAIOROVA, E. A. and OVCHINNIKOVA, T. V., 'Two-stage method of bonding polyamides to metal', *Soviet Plast.*, **11** 67, (1966).
MUNN, R. H. E., 'Polyamide resins for hot melt adhesives', *Assembly and Fastener methods*, **5** (2), 42, (1967).
FLOYD, D. E., *Polyamide Resins* (2nd Edn), Reinhold, (1966).
HAWKES, C. V., 'Using one-shot hot-melt adhesives for unsewn bookbinding: effects of spine roughing and notching.' *Pira Printing J.*, **1** (1), 22, (1968).
P.I.R.A., *The use of hot melt adhesives for the coating of board.*, PIRA Bibliography No. 248, (1968).

ESSER, H. O., 'Hot melts for paper packaging applications', *Plast., Paint Rubb.,* **12** (1), 78, (1967).

MARTIN, R. A. and TRAVER, C. R., 'Using pressure sensitive hot-melts composed of terpene and resins', *Adhes. Age,* **11** (1), 33, (1968).

WILLIAMSON, D. V. S., 'Hot melt adhesives in Europe', *Adhes. Age,* **8** (8), 24, (1965).

INORGANIC ADHESIVES AND CEMENTS

Adhesives based on compounds such as sodium silicate, magnesium oxychloride, lead oxide (litharge), sulphur, and various metallic phosphates, are typical of the inorganic adhesives group. These materials form strong resistant bonds for special applications, and are still widely used. The advent of synthetic organic polymer adhesives, during the last two decades, has led to a decline in use of many of the older inorganic adhesives 'laboratory' recipes.

The characteristics of the more important commercial materials are summarised below.

SODIUM SILICATE

Commonly known as water-glass, this colourless low cost inorganic material is usually supplied as a viscous water solution (the composition is normally expressed in terms of the ratio $SiO_2 : Na_2O$ and usually varies from 2 to 3·5 with viscosities suitable for most commercial bonding formulations). These adhesives display little tack and positioning pressure must be applied to hold substrates together until the bond is sufficiently dry.

The dry adhesive is brittle and water sensitive and until atmospheric carbon dioxide forms an insoluble material, the glue-line may be dissolved out by water. Water resistance can be improved by applying suitable aluminium salts to substrates such as paper, prior to bonding. To a certain extent, the addition of sugar, glycerine, sorbitol and other materials promotes the retention of moisture in the film and increases its flexibility; at the same time, tackiness, setting time and film toughness are increased. Sodium silicate is very resistant to high temperatures (some formulations withstand + 1 100°C) and, being inorganic, is naturally resistant to mould growth and bacterial attack.

The main use of soluble silicate adhesives is for bonding paper and in the manufacture of corrugated boxboard, boxes and cartons, where rapid absorption of water into the paper permits the use of high-speed machine operations. Kaolin clay (8–10%) is frequently added to raise the viscosity and prevent excessive penetration of the adhesive into the cardboard. Other applications for silicates include the following:

> Wood bonding (shear strengths up to 2 000 N/cm^2 are possible) and manufacture of low grade plywood;

Metal sheets to various substrates (copper to walls, plywood to steel, aluminium foil to paper in packaging industry; aluminium to asbestos for oven doors, metals to glass);

Glass to glass, porcelain, leather, textiles, stoneware, etc;

Glass fibre assemblies; optical glass applications; manufacture of shatter-proof glass; binding solutions (potassium silicate) for phosphor coatings and pigments on glass substrates;

Insulation materials based on wood, metals, ceramics, asbestos, fibre-glass, mica, kieselguhr, etc. where fire-resistant properties of silicates are useful;

Refractory cements for tanks, boilers, ovens, furnaces; acid-proof cements;

Fabrication of foundry moulds; briquettes and abrasive polishing wheel cements;

In addition to this, the soluble silicates may be reacted with silicofluorides or silica, to produce acid-resistant cements. These products have low shrinkage properties and a thermal expansion approaching that of steel. Compressive strengths approach 5 000 N/cm^2 but the strength is lost at 400°C. The cements have poor resistance to strong alkaline solutions.

PHOSPHATE CEMENTS

These cements are based on the reaction products of phosphoric acid with other materials such as sodium silicate, metal oxides and hydroxides, and the salts of basic elements. Zinc phosphate, formed by reacting zinc with phosphoric acid, is the most important metal phosphate cement and is widely used as a dental cement. This material is also modified with silicates to produce the so-called 'permanent materials' used for fillings. Compressive strengths up to 20 000 N/cm^2 are typical of these cements, which are formulated to have good resistance to water (saliva). Copper-phosphates are similarly employed but have a shorter use-life and are used primarily as dental fillers for their antiseptic qualities.

Metallic phosphates of aluminium, magnesium, chromium and zirconium are produced by reacting phosphoric acid with the metal oxide or halide. When heated to 300°C, the metallic phosphates develop excellent thermal stability and high temperature strengths and are virtually insoluble in hot water. Aluminium phosphates containing silica as an additive are used as strain-gauge cements for high temperature conditions. Other applications concern the bonding of refractory materials.

BASIC SALTS (SOREL CEMENTS)

Magnesium oxychloride is an inorganic adhesive which is notable for its heat and chemical resistance.

It is usually supplied as a two-part product (magnesium oxide and magnesium chloride) which is mixed at the time of use. The addition of copper powder overcomes a tendency to dissolve in water and promotes weathering resistance. These materials set rapidly within 2–8 h to give a resilient material which bonds to many refractory materials, ceramics and glass. They conduct static electricity from flooring and similar materials. The set materials are three times as strong as Portland cement (compressive strengths up to 7 000 N/cm^2); tensile adhesion to shot-blasted glass exceeds 70 N/cm^2. Sorel cements resist oils and greases and are not prone to biodeterioration.

Zinc oxychloride is similar in properties to the Sorel cements and is used as a dental cement. Zinc oxide-eugenol compositions are also employed as temporary dental cavity filling materials.

LITHARGE CEMENTS

Litharge-glycerine mixtures (1 glycol: 2–3 lead oxide, PbO) set in 24 h and provide adhesives which are chemically resistant to weak acids (not sulphuric acid) and hydrocarbons. Typical uses include the repair of sinks, pipe valves, glass, stoneware and ammonia gas conduit pipes. They have been used as ceramic seals and in the potting of electronic equipment.

SULPHUR CEMENTS

Liquid sulphur (m.pt. 110°C) may be regarded as an inorganic adhesive of the hot-melt type. Service temperatures cannot exceed 93°C because of a marked change in the expansion coefficient at 96°C as a result of a phase change. Physical properties are improved by the addition of carbon black and polysulphides (Thiokols). Tensile strengths of about 400 N/cm^2 falling to 300 N/cm^2 (after 2 yr in water at 70°C) have been reported.

The principle use of sulphur cements is for acid-tank construction, where high resistance to oxidising acids (e.g. nitric and hydrofluoric acid mixtures at 70°C) is required. Resistance to oleic acid, oxidising agents and strong basic materials (lime) is poor. Adhesion to metals (particularly copper) is usually good.

HYDRAULIC CEMENTS

These materials set by hydration and include, Portland cement (calcium silicate), calcium aluminates, lime-silica cements, barium aluminate and barium silicate cements, lime and gypsum (Plaster of Paris). The cements are important bonding materials for buildings, roads, bridges, and other outdoor construction work rather than adhesives as the term is conventionally understood. The reader is referred to a review (Taylor) of the chemistry and other properties of these cements for further information.

INORGANIC POLYMERS

Recent attempts to synthesise polymeric inorganic materials, comparable to the organic polymers but with improved thermal stability, have not been outstandingly successful. Hydrolytic instability and the tendency for polymers to re-arrange to smaller units on heating, are problems which hamper the preparation of tractable systems. Most of the totally inorganic polymers do not possess exceptional thermal stability when compared with organic polyaromatic materials.

Apart from glass, the silicone resins are perhaps the most successful of the inorganic polymers although these do not provide the heat stability associated with recent polyaromatic materials.

Phosphonitrilic polymers are an example of polymers consisting of inorganic chains containing organic groups or substituents. These are based on the so-called 'inorganic rubbers' and display improved hydrolytic stability together with continuous heat resistance to above 260°C. Research quantities are commercially available although their potential as high temperature adhesives has not been determined. Carboranes are compounds derived from decaborane and substituted acetylenes. Polyesters, silicones, and polyethers have been synthesised with carborane side groups. The silicone polymers are stable to 470°C. The high cost of decaborane has not prevented industry from marketing these materials which show good thermal properties and are easily processed. However, adhesive applications remain to be explored. Generally, the introduction of organic substituents into inorganic polymers improves processability but imposes limits on the thermal stability. Efforts to overcome these disadvantages may be worthwhile but the future for high temperature adhesives, at present, appears to be with organic polymers of the polyaromatic type.

REFERENCES

Technical Data Sheets on Dexil, 200, 201, 202, 'Carborane polymers', Olin Chemicals, 745, 5th Ave. New York, N.Y. 10022.
SKINNER, E. W., *Science of Dental Materials*, (3rd Edn), W. B. Saunders Co. Philadelphia, (1946).
HUNTER, D. N., *Inorganic Polymers*, Blackwell Sci., (1963).
TAYLOR, H. F. W., *The Chemistry of Cements*, R. Inst. Chem. Lecture Series No. 2, (1966).
WILLS, J. H., 'Inorganic Adhesives and Cements', *Adhesion and Adhesives*, (Houwink and Salomon Eds), Elsevier (1965).

VAIL, J. S., *Soluble Silicates,* ACS Monograph No. 116, Reinhold, New York, (1952).

IONOMER RESINS

Ionomer is the generic term for thermoplastic polymers containing ionic crosslinks in the inter-molecular structure. Introduced commercially in 1965, these polymers are predominantly used for moulding and extrusion applications but specific resins are available as film adhesives and metal-foil laminating resins for food packaging materials. Ionomer foams are also available.

Ionomers are characterised by a unique combination of properties which indicate that their potential as adhesives deserves further consideration. Flexibility, resilience, high elongation and impact strength with good retention of toughness at low temperatures down to $-200°C$, typify the resins. Optical transparency is excellent ($\mu = 1.51$). Mechanical and optical properties deteriorate on exposure to weather or ultra-violet radiation although stabilisers improve resistance to degredation.

Dielectric properties are good over a broad frequency range and these materials are used as insulating and jacketing materials for wires and pipes. Ionomers are insoluble in all the common organic solvents but softening and swelling can occur. Resistance to alkalies is excellent but weak acids slowly attack these materials.

Moisture permeability and water resistance is good. Low temperature flexural properties are retained down to $-140°C$. The upper use temperature is 70°C and heat distortion temperatures in the range 37–44°C at 45.4 N/cm² limit the use of ionomers in structural applications. Oxidative resistance is good but exposure to air at high temperatures approaching 316°C leads to yellowing of the polymer unless anti-oxidants are present in the formulation.

Ionomer foams or films may be used as hot-melt adhesives for metals. Shear strengths (ASTM D1002–64T) steel (stainless) to steel for joints of at least 800 N/cm² are realised where the film form is used as a hot-melt adhesive.

REFERENCE

'Ionomer Resins', *Modern Plastics Encyclopaedia,* 45 (14A), 208, McGraw-Hill, (1968).

ISOCYANATES

Isocyanate based adhesives are highly reactive materials which adhere strongly to a variety of materials such as metals, rubber, plastics, glass, leather and textiles. Di- and poly-isocyanates are the most important adhesive materials and owe their effectiveness as bonding agents to a strong chemical affinity for many functional groups (particularly those containing active hydrogen such as amino, imino, carboxyl, amide, sulphonic and hydroxyl groups). Isocyanates may be employed as adhesive agents in the three following ways:

1. Alone, or admixed with an elastomer solution, as adhesives for bonding rubber to metal, or fabric,
2. As modifying agents for rubber-based adhesives for general bonding purposes
3. As reactants with polyesters or polyether to produce polyurethane adhesives for special applications.

ISOCYANATE ADHESIVES

These adhesives depend on the use of the isocyanate (in a solvent) as a primer which is applied and dried for one of the adherends to be bonded.

Diphenylmethane diisocyanate in chlorobenzene has been used to produce bonds between metals and elastomers, which are heat, solvent, impact, and fatigue resistant, by hot press (bonded at 48 N/cm²) curing the rubber against the isocyanate primed metal. Shear strengths within the range 550–900 N/cm² are possible for joints based on natural, nitrile or polychloroprene rubbers and metals such as steel, aluminium or copper and its alloys.

A 2% solution of the same isocyanate in an aromatic hydrocarbon promotes adhesion between synthetic fibre textiles and rubber dip-coatings when used as a primer.

ISOCYANATE—MODIFIED ADHESIVES

A di- or poly-isocyanate is mixed with a rubber-based adhesive (up to 20% w/w and generally in a dry solvent vehicle) and applied to the substrates which are then contact bonded, after drying to a tacky state, and cured.

Diphenyl diisocyanate and natural or synthetic rubbers in aromatic solvents act as effective adhesive-primers for bonding rubbers to fabrics based on cotton, polyamide (nylon) or rayons. Curing procedures follow those employed for processing unmodified elastomer adhesives; low bonding pressures usually suffice to produce good interfacial contact between bonding surfaces. Diisocyanates may be used to increase the adhesion between polyester fibres (e.g. terylene) to natural rubber by incorporating the isocyanate into the stock material during rubber processing. Peel strengths approaching 80 N/cm (initial) to 39 N/cm (average) are obtainable. Other typical peel strengths

(N/cm[1] ASTM D413–99) for fabric-rubber combinations, which result from the use of isocyanate-rubber adhesive compositions (*see* Table 4.2).

Higher peel strengths are often obtained by increasing the quantity of adhesive-primer applied.

Isocyanates improve the adhesion of polychloroprene, natural and butyl rubbers to metals and alloys when employed in weight ratios up to 2:1 of isocyanate:rubber in polychloroprene adhesives. Tensile strengths up to 580 N/cm^2 are achieved for natural rubber–steel (mild) bonds with diisocyanate–chlorinated rubber adhesives.

Di- or poly-isocyanates may be stabilised to give temporarily inactive solutions or suspensions (possibly aqueous) by 'blocking' with agents such as phenol. The isocyanate is regenerated by heat decomposition of the blocked isocyanate during the subsequent processing of the bonded assembly and is released to effect bonding.

The rubber-based adhesives develop bonds rapidly which have good strength and heat resistance. The highly reactive nature of isocyanates dictates the use of dry solvents which are free of reactive impurities. The isocyanates are affected by moisture and have limited working lives (often of the order of a few hours) and this has restricted their use in industrial applications.

ISOCYANATE—POLYESTER METHANE ADHESIVES

Poly functional isocyanates are reacted with polyhydroxy compounds (such as unsaturated polyesters or phenols) to form polyurethanes containing free isocyanate groups which are available for reaction with the adherend. The reaction may be partial and only reach completion during an adhesive curing process (i.e. involving the use of a catalyst and/or heat). These materials, which may be thermoplastic or thermosetting, display excellent adhesion to most surfaces. (*see* Polyesters, Unsaturated).

Caution in the handling of these materials is mandatory by virtue of the reactivity and toxicity of some types of isocyanate. Triphenylmethane triisocyanate (in methylene chloride solution) is non-volatile and easy to use but other materials require adequate ventilation and prompt cleansing of contaminated body areas.

REFERENCES

SANDLER, S. R., 'Polyisocyanurate adhesives', *J. appl. Polym. Sci.*, **11** (6), 811, (1967).
ZALIKIN, A. A., *et al*, 'Employment of polycyclic polyisocyanates as components in cold-hardening bonding compositions', *Soviet Plast.*, **6**, 47, (1967).
Vulcabond TX for Rubber-to-Metal Bonding, Imperial Chemical Industries, Ltd., (1957).
LASURE, R. M., 'Isocyanates in bonding rubber to synthetic fabrics', *Adhes. Age*, **5** (3), 26, (1962).

MELAMINE FORMALDEHYDE

Type
Synthetic thermosetting resin.

Physical form
Film, liquid or powder form. Powder is mixed with water or water-alcohol mixture and sometimes supplied with a filler powder.

Shelf life, 20°C
Film, 3 mth; powder, up to 2 yr.

Working life, 20°C
Film, 3 mth; powder mixture, up to 24 h.

Closed assembly time, 20°C
Film, 3 mth; powder mixture, up to 24 h.

Setting process
Condensation polymerisation with elimination of water.

Processing conditions
Suitable for use with radio-frequency equipment. Application dependent, and within the range, 10 h at 50°C to 2 min at +90°C. Bonding pressures from 70–100 N/cm^2.

Coverage
0·15–2 kg/m^2 general purpose bonding. 0·075–0·125 kg/m^2 plywood fabrication.

Permanence
Melamine resins have superior resistance to water and better temperature stability than the related

Table 4.2.

Rubber adhesive (isocyanate modified)		Adherends								
		Cotton to			Rayon to			Nylon to		
		NR	PCR	SBR	NR	PC	SBR	NR	PC	SBR
Natural	(NR)	45	62	48	33	69	49	27	41	35
Polychloroprene	(PCR)	45	42	51	62	53	65	35	32	39
Styrene butadiene	(SBR)				53	58	62			

urea resins. Hot pressed melamine resins are especially resistant to boiling water, and are generally equivalent to hot pressed phenolics and resorcinol resins in water resistance and durability; low temperature, acid catalysed, melamines are less durable than the phenolic resins. Resistance to biodeterioration is good. Melamine resins are comparable with many other materials and are used as fortifiers for urea-formaldehyde resins to promote their resistance to boiling water. Melamine-urea adhesives are intermediate between melamine and urea adhesives with durability proportional to melamine content.

Applications
Non-structural woodworking assemblies for exterior use. Scarf joint manufacture for casein bonded wood laminates. Fabrication plywood which is resistant to boiling water; melamine is employed as a urea-formaldehyde fortifier.

Specification
BS 1204:1965 Adhesives for constructional work in wood.
BS 1203:1963 Adhesives for plywood.

Remarks
Bond strengths exceed the strength of the wood for very thin glue-lines. Film types display poor gap filling properties; powders benefit from the use of extender additives. Adhesives are light coloured, non-staining and suitable for light woods and veneering. High temperature curing resins have long assembly periods and retain their low viscosity better than the low temperature curing types which develop tack quickly. For structural bonding, it is advisable to employ less expensive adhesives with more easily controlled curing processes. Bonded assemblies are usually conditioned for 1 d before handling.

REFERENCES

CHUGG, W. A., *The Structural glued laminated timber industry in North America*, E/1B/5, Timber Research and Development Association, Tylers Green, (1962).
BLAIS, J. F., *Amino Resins*, Reinhold, (1959).
KNIGHT, R. A. G., 'The efficiency of adhesives for wood', *Forest Prod. Res. Bull.*, **38**, (1963).

NATURAL RUBBER

Type
Thermoplastic elastomer of plant origin.

Physical form
Translucent, light brown, high viscosity solvent (12–20% solids) solutions, usually inflammable. Low viscosity, milky white, non-flammable latices (35% solids) in aqueous vehicle. Available as two-part vulcanising adhesives.

Shelf life, 20°C
6 mth to 1 yr in sealed containers.

Working life, 20°C
Open time 5–30 min or longer. Vulcanising types, within 8 h.

Closed assembly time, 20°C
Latices, indefinite with long open time. Good retention of high initial tack.

Setting process
Solvent or water evaporation or vulcanisation by heat or catalyst.

Processing conditions
Porous materials, wet bonded and solvent allowed to migrate into adherend. Non-porous materials, contact bonding of tacky substrates. Curable dispersions (by vulcanisation) require bonding pressures from contact to 70 N/cm^2 at 20–90°C. Full cure in about two weeks.

Coverage
0·15–0·25 kg/m^2.

Permanence
Good resistance to water and biodeterioration (when protected). Poor resistance to oils, solvents and oxidising agents. Unvulcanised rubbers deteriorate above 66°C whereas vulcanised ones withstand 93°C. Vulcanised latex has better resistance to water and solvents although swollen by hydrocarbons. Becomes brittle at −34°C and the latex forms are damaged by freezing. Sunlight exposure properties are not very good for unvulcanised types although ageing properties are moderate.

Applications
Low strength applications. Paper and porous materials. Footwear trade (leather and rubber) and upholstery bonding of fabrics, rubbers and leathers. Textiles, carpets, felts. Packaging materials. Lamination of metal foil to paper or wood and sponge rubber to itself or metals and wood. Pressure sensitive adhesive application.

Remarks
Natural rubber adhesives display excellent flexibility, high initial tack and good tack retention. Coated surfaces may be contact bonded weeks after adhesive application. Latices and water dispersions are usually alkaline and formulated with a higher solids content than solvent types.

Adhesive films from latex are stronger and age better than those from solvent cements. Latex types are corrosive towards copper and alloys. Natural rubbers are frequently compounded with resins and other modifying agents in adhesive compositions.

REFERENCES

BOOTH, C. C. and DUNCAN, T. F., 'Elastomeric adhesives—characteristics and applications', *Adhes. Age*, **34**, (1961).
BLACKLEY, D. C., *High Polymer Latices*, (Vol. 1 Fundamental principles, Vol. 2 Testing and applications), Maclaren, London, (1966).
BATEMAN, L., (Ed.), *The Chemistry and Physics of rubber-like Substances*, Maclaren, London, (1963).
BRYDSON, J. A., (Ed.), *Developments with Natural Rubber*, Maclaren, London, (1967).
STERN, H. J., *Rubber: Natural and Synthetic*, (2nd ed.), Maclaren, London, (1967).
BLOW, C. M., *Natural Rubber Latex and its Applications*, National Rubber Development Board, London, (1956).

NITRILE RUBBERS

Type
Synthetic thermoplastic elastomers.

Physical form
Viscous liquids based on acrylonitrile-butadiene copolymers in solvents (usually ketones and esters, chlorinated hydrocarbons, nitromethane) or latex foams, with additives. Inflammable and non-inflammable types are available. Light or dark coloured. Solids content ranges from 15–40%.

Shelf life, 20°C
3 mth to 1 yr in sealed containers.

Working life, 20°C
Equals the shelf life where no solvent loss takes place.

Closed assembly time, 20°C
10–20 min for wet bonding. Indefinite where dried adhesive films are stored for later solvent or heat reactivation bonding.

Setting process
Solvent release from adhesive coated adherends, heat or solvent reactivation of dried adhesive films before joining.

Processing conditions
Porous materials, wet bonded and solvent allowed to absorb into the adherend with time. Non-porous materials, removal of bulk solvent by evaporation for 20 min to several hours (drying accelerated by warm air or infra-red, etc.) followed by contact bonding of tacky substrates with minimum of pressure. Vulcanisation by heat curing or catalyst addition to improve strength and heat resistance of the bond.

Coverage
Dependent on viscosity of adhesive; for 30% w/w solids content 0·15–0·25 kg/m² is typical.

Permanence
Good resistance to oils, water, many organic solvents (especially aliphatic hydrocarbons) and most acids (except strongly oxidising ones). Resistant to plasticisers. Limited heat resistance up to 120–150°C with minimum service temperature of −50°C. Latex types are similar to solvent types but water and heat resistance is inferior.

Applications
Plastics and plasticised vinyl materials and many rubbers compounded for low temperature flexibility; rubber to vinyl plastics; neoprene and nitrile rubbers to metals (often primed with chlorinated rubber); fabrics; wood; paper, etc., to metals and glass. Leather and rubber (footwear trade). Latex types have similar application. Especially useful for painted metal, thin plastic films. Nitrile adhesives are the most versatile of general purpose rubber types but are less popular than neoprene based ones.

Remarks
Bond strengths and creep resistance vary considerably with compounding agents used. Shear strengths 100–1 400 N/cm². Unsuitable for structural applications under continuous loading exceeding 70 N/cm². Products range from tacky low-shear, high peel strength to high shear, high peel strength. Tack and tack ranges are usually inferior to natural rubber adhesives; many dry to a tack free state. Heat reactivation at 85°C is recommended where maximum bond strength is desired. Nitrile rubber is blended with phenolic resins to form important structural adhesives for metals (*see* Phenolic-nitrile).

REFERENCE

BROWN, H. P. and ANDERSON, J. F., 'Nitrile Rubber Adhesives', *Handbook of Adhesives* (Skeist Ed.), Reinhold, (1965).

NYLON ADHESIVES

Nylons are synthetic thermoplastic polyamides of relatively high molecular weight which have been employed as the basis for the various adhesive systems briefly described below.

SOLUTION ADHESIVES

Unsubstituted nylon homopolymers are insoluble in most organic solvents but soluble in phenols and certain organic acids. However, such solutions are inconvenient to use as adhesive systems.

Substituted nylons are soluble in some alcohols and alcohol/water mixtures and have been used to provide solution adhesives with good heat resistance. Water- and solvent-resistance is poor. Copolymerised Nylons 6, 610 and 6:6, and N-methyl methoxy polyamide are suitable polymers for solution adhesives. Soluble nylons may be compounded with many thermosetting resins to improve the resilience and peel strength properties. (*see* EPOXY NYLON).

HOT-MELTS

The nylons may be used as hot-melt adhesives for bonding metals, wood and other rigid materials. The nylons have high fixed melting points which must be exceeded to ensure adhesion, e.g. Nylon 6:6 (260°C), Nylon 610 (220°C), Nylon 6 (215°C), Nylon 11 (186°C). The soluble nylons are sometimes included in hot-melt adhesive formulations.

PHENOLIC-NYLON

The adhesion of nylon to metals is greatly improved by employing a phenol formaldehyde resin as a primer for the adherends. The phenolic resin is applied thinly to suitably pretreated metal surfaces and allowed to dry for at least 30 min; the nylon film is then placed between the coated surfaces and the bond cured for 2–3 min at 220°C (e.g. assuming here Nylon 610 is concerned), under a bonding pressure of 7–70 N/cm². It is important to reach the melting point of the nylon for adhesion to take place. Alternatively, a one-component solution adhesive based on N-methyl methoxy polyamide, or another soluble nylon, and the phenolic resin can be prepared and used.

Other thermosetting resins may be substituted for phenol as primers to advantage, e.g. resorcinol formaldehyde, melamine resin (butylated) and epoxy resins (+aliphatic amine hardener). Creep resistance and fatigue strength of nylon-resin adhesives are generally good as is high temperature performance. For phenolic resins with Nylon 6, shear strengths with aluminium exceed 3 500 N/cm² at 20°C and 2 000 N/cm² at 150°C. Similarly, for melamine resin, shear strengths of about 3 000 N/cm² at 20°C and 1 300 N/cm² at 150°C are realisable.

REFERENCES

RAYNER, C. A. A., 'Polyamide resins and Nylons', *Adhesion and Adhesives*, (Houwink and Solomon, Eds), Elsevier, (1965).
GORTON, B. S., 'Interaction of nylon polymers with epoxy resins in adhesive blends', *J. appl. Polym. Sci.*, **8** (3), 1287, (1964).

PHENOLIC-EPOXY

Type
Blend of thermosetting phenolic and epoxy resins.

Physical form
Viscous liquids (which may contain solvents) or glass-cloth or fabric supported films or tapes. Often modified with filler or thermal stabilisers.

Shelf life
Liquid, films and tapes 3 wk (20°C) to 3 mth (0°C).

Working life, 20°C
Formulation dependent (few hours—indefinite).

Closed assembly time, 20°C
Within 30 min.

Setting process
Formation of a thermoset on heat cure.

Processing conditions
Solvent blends are usually force-dried at 80–90°C for 20 min before assembly of adherends. Generally cured for 30 min at 95°C with contact pressure followed by 30 min to 2 h at about 165°C and 7–40 N/cm² pressure. Post cured for optimum strength at elevated temperatures.

Permanence
Good resistance to weathering, ageing, water, weak acids, aromatic fuels, glycols and hydrocarbon solvents. Service temperature range is usually −60°C to 260°C, but special formulations are suitable for cryogenic temperatures down to −260°C.

Table 4.3.

	−75°C	25°C	260°C	400°C	Remarks
Stainless steel	1 700	1 620	900	250 N/cm²	Subject to degradation with stainless steel above 260°C.
Aluminium	2 400	2 200	1 000	480 N/cm²	Good stability with aluminium up to 300°C where metal loses yield strength.

Applications
High temperature structural bonding of metals (including copper and alloys, titanium, galvanised iron and magnesium); glass and ceramics, phenolic composites. Honeycomb sandwich materials in aircraft fabrication (liquid adhesives favoured for high peel strengths). Liquid types often used as primer for tapes.

Remarks
Relatively expensive materials. Display excellent shear and tensile strengths over a wide temperature range. Inferior to many other thermosetting structural adhesives for strength and handling properties but are used most advantageously above 160°C. Better strength performance with film than liquid systems. Peel and impact strengths are usually poor. Typical shear strengths (ASTM D1002–53T) for metals are shown in Table 4.3.

REFERENCE

BLACK, J. M., and BLOMQUIST, R. F., 'Metal-bonding adhesives for high temperature service', *Mod. Plast.*, **33** (10), 253, (1956).

PHENOL FORMALDEHYDE (ACID CATALYSED)

Type
Synthetic thermosetting resin.

Physical form
High viscosity resin and liquid acid-catalyst.

Shelf life, 20°C
Up to 6 mth.

Working life, 20°C
From 1–6 h or 15–90 min at 30°C.

Closed assembly time, 20°C
Within 1 h or 15 min at 30°C.

Setting process
Condensation polymerisation with elimination of water.

Processing conditions
For general purposes, 3–6 h at 20°C. For timber (Hardwood), 15 h at 15°C and 120 N/cm^2; (Softwood), 15 h at 15°C and 70 N/cm^2. Curing time effectively reduced by an increase in temperature.

Coverage
0·05–0·15 kg/m^2 general purpose bonding. 0·15–0·3 kg/m^2 timber construction work.

Permanence
Resistant to weather, boiling water and biodeterioration. Resistance to elevated temperatures is good but inferior to that of heat curing phenolic and resorcinol resins. Excess acidity, due to poor control over acid-catalyst content, often leads to wood damage on exposure to warm humid air. Durability of joints at high and low temperatures for extended periods is usually good.

Applications
Woodwork assemblies where service temperatures do not exceed 40°C. Furniture construction and, to a minor extent, plywood fabrication; metal to wood joining for exterior use.

Specification
BS 1204:1965 Adhesives for constructional work in wood. BS 1203:1964 Adhesives for plywood.

Remarks
Joints require conditioning period from 1–7 d according to end-use. Metals require priming with a vinyl-phenolic or rubber resin adhesive before bonding. Good gap filling properties. Not recommended as structural adhesive unless glue-line acidity exceeds pH 2·5; acidic nature of adhesive makes use of glass or plastic mixing vessels mandatory. Moisture content of the wood should fall within range 6–15%. Mixed adhesive is exothermic and temperature sensitive; strict control of room temperature is an important factor where reliable assembly times are required for laminating process.

REFERENCES

Durability of water-resistant woodworking glues, Bulletin 1530, US Forest Products, Laboratory, Madison 5, Wisconsin, (1956).
CHUGG, W. A., and GRAY, V. R., '*The durability of acid-setting phenolic resin adhesives*', C/1B/5. Timber Research and Development Association, Tylers Green, (1960).

PHENOLIC FORMALDEHYDE (HOT SETTING)

Type
Synthetic thermosetting resin.

Physical form
Spray-dried powder to be mixed with water; alcohol, acetone or water solvent solutions; glue film. May be compounded with fillers and extenders.

Shelf life, 20°C
Powders and resins, 1–2 yr; film type up to 1 yr.

Working life, 20°C
Powders and resins, 1–24 h; film type, indefinite.

Closed assembly time, 20°C
24 h or more according to wood employed.

Setting process
Condensation polymerisation with elimination of water.

Processing conditions
Resins, up to 15 min at 100–150°C and 70–170 N/cm² bonding pressure. Film, up to 15 min at 120–150°C and 70–140 N/cm².

Coverage
0·05–0·075 kg/m².

Permanence
Resistant to weather, boiling water, and biodeterioration. Temperature stability is better than for acid-catalysed phenolic resins.

Applications
Fabrication of exterior-grade weather and boil-proof plywood. Glass to metal adhesive for electric light bulbs.

Specification
BS 1203 : 1963 Adhesives for plywood.

Remarks
Gap-filling properties are poor and inferior to acid-catalysed phenolic resins. Joints require conditioning period of up to 2 d. Though durable and resistant to many solvents, bonds are brittle and prone to fracture under vibration and sudden impact. Alkaline nature of adhesive precludes the use of brass or copper mixing vessels. Phenol-formaldehyde is employed as an additive to other materials to form adhesives for glass and metals; as a modifying agent for thermoplastic elastomer adhesives or as a component of thermoplastic resin-elastomer adhesives for metal bonding.

REFERENCE

WHITEHOUSE, A. A. K., PRITCHETT, E. G. K. and BARNETT, G., *Phenolic Resins*, (3rd ed), Iliffe, (1967).

PHENOLIC-NEOPRENE

Type
Thermosetting phenolic resin blended with poly-chloroprene (neoprene) rubber.

Physical form
Liquid (solvent solution in toluene, ketones, or solvent mixture); Film with glass—or nylon—cloth interliner.

Shelf life, 20°C
Liquid, 6 mth to 1 yr in sealed container.
Film, up to 6 mth in cold storage.

Working life, 20°C
Liquid, 2–60 min open time.

Closed assembly time, 20°C
Less than 1 h.

Setting process
Formation of a thermoset on heat cure.

Processing conditions
Film, cured at 150–260°C for 15–30 min at 35–180 N/cm² bonding pressure. Liquid, dried at 80°C and cured for 15–30 min at 90°C and contact to 70 N/cm² pressure. Bonds may be removed from hot press while hot. Liquid adhesive may be used as a metal primer for film types.

Permanence
Excellent resistance to biodeterioration, water, glycol, petroleum and common chemicals. Resistant to continuous temperatures up to 93°C with about 50% decrease in strength at this temperature. Better retention of strength at low temperatures (approaching −50°C) than phenolic-nitrile adhesives. Resistance to creep and cold flow is excellent and joints sustain high loads for long periods.

Applications
Structural adhesive for metals such as aluminium, magnesium and stainless steel (not always satisfactory for zinc, copper and chromium); metal honeycombs and facings; plastic laminates; wood to metal bonds are often primed with phenolic-neoprene before bonding with resorcinol-phenolic adhesives; glass and ceramics; rubber.

Remarks
Compounding with neoprene rubber increases flexibility and peel strength of phenolic resins and extends high temperature resistance. Film form preferred where solvent removal is a problem. Higher curing temperatures give strongest bonds. Typical joint strengths with metals according to the formulation are:
1 400–3 500 (Film), 130–480 (Liquid) N/cm² Shear (ASTM D1002–53T); 1 700–3 500 (Film), 130–600 (Liquid), N/cm² Tensile (ASTM D897–49); 8–50 (Film), 17–100 (Liquid) N/cm¹ Peel (ASTM D903–49).
Impact strength and fatigue resistance is good. Less important as structural adhesive than phenolic-nitrile adhesives. Isocyanates are often added to improve adhesion to some surfaces.

REFERENCES

KERAMEDIJIAN, J., 'Heat-reactive phenolic resins in neoprene contact cements', *Adhes. Age*, **5** (6), 34, (1962).

BLOMQUIST, R. F. and OLSON, W. L., 'Rubber-base adhesives for wood', *Forest Prod. J.*, **10**, 494, (1960).

PHENOLIC-NITRILE

Type
Thermosetting phenolic resin blended with acrylonitrile butadiene rubber.

Physical form
Liquid, (solvent solution in toluene, ketones, or solvent mixtures). Film with glass—or nylon-cloth interliner.

Shelf life, 20°C
Liquid, 6 mth to 1 yr in sealed container. Film, up to 6 mth in cold storage.

Working life, 20°C
Liquid, 2–15 min, Film, 2–90 min.

Closed assembly time, 20°C
Within 10 min (liquid) to 60 min (film).

Setting process
Formation of a thermoset on heat cure.

Processing conditions
Film, cured at 150–260°C for 15–30 min with bonding pressures from 12–180 N/cm^2. Liquid, dried at 80°C and cured for 15–30 min at 90°C and contact to 70 N/cm^2 pressure.

Permanence
Excellent resistance to biodeterioration, water, oils and plasticisers, salt spray and many solvents. Resistant to continuous temperatures up to 150°C; short exposures may be tolerated at 250°C. Poor to good performance at low temperatures with embrittlement at −57°C which reduces impact and shear strength. Best salt spray resistance of structural adhesives.

Applications
Structural adhesive for metals, rubbers, plastics, wood, glass and ceramics. High temperature structural applications such as automobile brake linings and clutch facings. Attachment of abrasive grinding stones to metal holders.

Remarks
Nitrile rubber extends flexibility, peel strength and high temperature resistance of phenolic resins. Higher curing temperatures produce the strongest bonds. Typical joint strengths with metals are:

1 400–3 800 (Film), 200–1 400 (Liquid) N/cm^2 Shear (ASTM D1002–53T); 2 000–5 500 (Film), 350–1 400 (Liquid) N/cm^2 Tensile (ASTM D897–49); 100–130 (Film), 35–80 (Liquid) N/cm^1 Peel (ASTM D903–49).
Shear strengths can approach 1 400 N/cm^2 at 150°C and 700 N/cm^2 at 250°C. More favoured as structural adhesives than phenolic-neoprene types.

REFERENCES

BOBER, E. S., 'Bonding metals with adhesives', *Adhes. Age*, **2** (12), 30, (1959).

RIDER, D. K., 'Adhesives in printed circuit applications', *Symposium on Adhesives for Structural Applications*, Picatinny Arsenal, Interscience (1961).

PHENOLIC-POLYAMIDE

Type
Blend of phenolic thermosetting and polyamide (thermoplastic nylon) resins.

Physical form
Two component adhesive. Phenolic resin with polyamide film.

Shelf life, 20°C
Indefinite.

Working life, 20°C
15–30 min open time for phenolic resin.

Closed assembly time, 20°C
Less than 40 min.

Setting process
Formation of a thermoset on heat cure.

Processing conditions
Pretreated surfaces are primed with phenolic resin and dried for 15–30 min; the polyamide film is sandwiched between adherends and cured for 2–3 min at 220°C and 7–70 N/cm^2 bonding pressure.

Permanence
Durability is similar to vinyl acetal-phenolic resin adhesives. Resistance to weathering, oils, petroleum and cold exposure is good. Less resistant to water than vinyl formal-phenolic types. Fair to moderate resistance to heat with a service temperature range from −60°C to 150°C.

Applications
High temperature structural adhesive for metals and sandwich materials. Chief use, at present, in aircraft industry. Has had a limited use as a high temperature structural adhesive for metals and

sandwich materials; main applications in the aircraft industry in countries other than the U.K.

Remarks
These materials are apparently not available as commercial products in this country. They are not favoured because of processing difficulties and poor reproducibility of performance.

REFERENCE

HURD, J., *Adhesive Guide,* Sira Research Report M39, (1959).

PHENOLIC-VINYL BUTYRAL

Type
Blend of phenolic (thermosetting) and polyvinyl butyral (thermoplastic) resins which is predominantly thermosetting.

Physical form
Liquid (solvent solutions based on alcohol-toluene or ketones). Film, with fabric cloth interliner. Coated paper tapes.

Shelf life, 20°C
Not less than 12 mth.

Working life, 20°C
Not less than 2 d.

Closed assembly time, 20°C
Up to 1 h at 50–70°C usually allowed for the evaporation of volatile solvent before assembly.

Setting process
Formation of a thermoset on heat curing.

Processing conditions
Film, cured at 150°C for 15–30 min at 10–20 N/cm^2. Similar curing conditions apply to other forms.

Permanence
Generally similar properties to the phenolic-vinyl formal types, with excellent resistance to biodeterioration, water, oils and aromatic fuels and weathering. Shear strengths drop as the thermoplastic temperature of the vinyl is approached. Better resistance to creep and fatigue than vinyl formal-phenolics.

Applications
Metal or reinforced plastic facings to paper (resin impregnated) honeycomb structures; cork and rubber compositions; cyclised and unvulcanised rubbers; steel to vulcanised rubber; electrical applications; primer for metals that are to be bonded to wood with phenolics.

Remarks
Lack the shear strength and toughness of the phenolic-vinyl formal types. Bond strengths given below are typical for metal structures: 1400–3 500 N/cm^2 Shear (ASTM D1002–53T); 700–2 800 N/cm^2 Tensile (ASTM D897–49); 13–20 N/cm^1 Peel (ASTM D903–49).

REFERENCES

WHITNEY, W. and HERMAN, S. C., 'Polyvinyl acetal/phenolic structural adhesives', *Adhes. Age,* **3** (1), 22, (1960).
WHITNEY, W. and HERMAN, S. C., 'Polyvinyl acetal/phenolic structural adhesives', *Adhes. Age,* **9,** 32, (1966).
Properties and Uses of Butvar (polyvinyl butyral) and Formvar (polyvinyl formal), Tech. Publ. Shawinigan Resins Corporation, (1967).

PHENOLIC-VINYL FORMAL

Type
Blend of phenolic (thermosetting) and polyvinyl formal (thermoplastic) resins which is predominantly thermosetting.

Physical form
Liquid (solvent solution). Fabric tape coated on both sides with adhesive. Two-part products based on liquid phenolic resin and powdered vinyl-formal.

Shelf life, 20°C
Film types: 12 mth: Liquid/Powder types: not less than 12 mth (when stored in a cool dry place).

Working life, 20°C
Liquid types (usually with 3–5% accelerator component) not less than 3 d.

Closed assembly time, 20°C
Up to 1 h at 50–70°C usually allowed for evaporation of volatile solvent before assembly.

Setting process
Formation of a thermoset on heat curing.

Processing conditions
Film, cured at 177°C for 5 min, or 150°C for 30 min, with 35–350 N/cm^2 bonding pressure. Powder-resin form; powder is sprinkled on air dried resin coated material and cured as for film. Accelerators are available for materials requiring lower curing temperatures to retain stability. Removal of volatiles by force drying is important for liquid types.

Coverage
Approximately 37 5 g of liquid +75 g of powder per m^2.

Permanence

Dependent on formulation. These adhesives retain adequate strengths when exposed to weather, mould growth, salt spray, humidity and chemical agents such as water, oils and aromatic fuels. Generally good resistance to creep although temperatures up to 90°C produce creep and softening of some formulations. Fatigue resistance is excellent with failure generally occurring in the adherends rather than in the adhesive. Useful service temperature range is −60°C to 100°C.

Applications

Structural adhesive for metal to metal in aircraft assembly; metal honeycomb panels and wood to metal sandwich construction; cyclised rubber and, in some cases, vulcanised and unvulcanised rubbers; primer for metal to wood bonding with resorcinol or phenolic adhesives; copper foil to plastic laminates for printed circuits.

Remarks

One of the most suitable thermosets for metal-honeycomb and wood-metal structures. Generally equivalent to phenolic-nitrile adhesives for strength but have slightly better self-filleting properties for honeycomb assembly. Superior to epoxy types where strength is desirable in sandwich constructions. Typical bond strengths are: 2 000–3 500 N/cm^2 Shear (ASTM D1002–53T); 700–2 800 N/cm^2 Tensile (ASTM D897–49); 60 N/cm^1 Peel (ASTM D903–49) depending on the metal joined.

REFERENCES

Properties and Uses of Butvar (polyvinyl butyral) and Formvar (polyvinyl formal), Tech. Publ. Shawinigan Resins Corporation, (1967).

WHITNEY, W. and HERMAN, S. C., 'Polyvinyl acetal/phenolic structural adhesives', *Adhes. Age*, **3** (1), 22, (1960).

Redux (phenolic-vinyl formal), Technical Data Sheets, Ciba (A.R.L.) Ltd. Duxford, Cambs., (1967–1968).

PHENOXY

Type

Polyhydroxy ether, synthetic thermoplastic.

Physical form

One-component in powder, pellet or film form. May be dissolved (e.g. cellulose acetate or toluene: acetone mixtures) or supplied in special shapes.

Shelf life, 20°C

Exceeds 1 yr.

Working life, 20°C

Indefinite at room temperature but limited stability at melt temperatures.

Closed assembly time, 20°C

Within minutes.

Setting action

Hot-melt, sets on cooling.

Processing conditions

Liquid forms require removal of solvent by drying before bonding. Time and temperature are important factors in obtaining optimum strengths, but bonding pressure is not critical. (Contact to 17 N/cm^2). Bond for 30 min at 192°C or 2–3 min at 260°C, or 10 s at 300–350°C.

Permanence

Withstands weathering and resists biodeterioration. Excellent resistance to inorganic acids, alkalies, alcohols, salt spray, cold water and aliphatic hydrocarbons. Swells in aromatic solvents and ketones. Good thermal stability with a service temperature range from −62°C to 82°C. Resistance to cold flow and creep is uniquely high for a thermoplastic, even at temperatures as high as 80°C.

Applications

Structural adhesive for rapid assembly of metals and rigid materials; continuous lamination of metal to metal (cladding) or wood, also flexible substrates, paper, cloth, metal foil and plastic lamination applications; pipe jointing (steel or aluminium) with film type; assembly of automobile components; component of hot-melt adhesives for conventional applications; bonding of polymer materials such as polyester film, polyurethane foam, acrylics, phenolic composites but not polyvinyls, polyamides, polyacetals, or polyolefines. Phenoxies are predominantly employed for surface coatings or primers and, for this purpose, may be combined with coumarone-indene or ketone-formaldehyde resins.

Remarks

Phenoxy adhesives give rigid, tough glue-lines with high adhesive strength. Shear strengths are similar to epoxy resin systems and for metals generally exceed 1 700 N/cm^2 (and may approach 2 750 N/cm^2); for many other rigid substrates failure occurs cohesively within the glue-line. Typical joint shear strengths (ASTM D 1002–53T) for aluminium joints are, 2 170 (−55°), 2 420 (25°), 1 930 (82°), 965 (104°C) N/cm^2. Creep is less than 0·0013 mm for an 8 d dead load of 1 100 N/cm^2 and impact strengths (ASTM 0950–54) exceed 34 N/2·54 cm notch.

Film thickness is usually not critical and can be as little as 0·012 mm. Attention to adherend surface preparation is important for realisation of optimum strength; chromate etching of metals or vapour blasting is usually satisfactory. Liquid adhesives do not usually provide optimum bond strength (complete solvent removal may be a problem). Hot-melt adhesive systems can present difficulties; thermal degradation may occur before resin is completely melted unless plasticisers are present in the composition.

REFERENCES

LEE, H., STOFFEY, D., and NEVILLE, K., 'Phenoxy Resins', *New Linear Polymers,* McGraw-Hill, (1967).
Bakelite Phenoxy Resins PKHH and PAHJ for Solution Coatings and Adhesives, Bulletin J-2421-B, Union Carbide Corpn.
BUGEL, T. E., *et al.,* 'Phenoxy Resin: a New Thermoplastic Adhesive', *ASTM Symp.,* Atlantic City, N.J. June (1963).
NORWALK, S., *et al.,* 'How to Process the New Phenoxies', *Plast. Technol.,* **9** No. 2, 34, (1963).

POLYAMIDES

Polyamide resins are synthetic thermoplastic resins which are of two types: 'Linear' polymers which are usually neutral and chemically non-reactive solids, or 'Branched chain' polymers, containing unreacted amino groups, which are highly reactive liquids or semi-solids. The latter type are used in thermosetting structural adhesive compositions (see EPOXY-POLYAMIDE); the former are thermoplastic heat-activated adhesives and these are described here.

Type
Thermoplastic reaction products of dibasic acids with diamines.

Physical form
Solids with melting point within the temperature range 100–190°C in cake, pellets, cubes, film, granules, stick form, etc. Available as solvent solutions (usually alcohol-hydrocarbon mixtures).

Shelf life, 20°C
Exceeds 1 yr.

Working life, 20°C
Indefinite for melted solid provided material is maintained close to melting point during application.

Closed assembly time, 20°C
Melted adhesive is usually applied and bonded within seconds, no limit to assembly time where resin fluidity is maintained.

Setting process
Solidification of molten resin on cooling.

Processing conditions
Solvent types, release of solvent by evaporation and subsequent heat-activation of dry adhesive film. Solid types, melted and applied to substrates (which may be heated where high heat conductivity is likely to lower temperature of melt sufficiently to reduce flow and wetting properties). Rapid bonding properties means that these materials are usually applied by hot-melt dispensing equipment for fast assembly work. Many hot-melt polyamides are applied at temperatures approaching 200°C. They have a fairly narrow bonding range unless modified with additives.

Permanence
Polyamide resins and adhesives based on them are tough, high cohesive strength materials displaying strong adhesion. Good tensile and elongation properties are retained after exposure to high temperatures for some time. Loss of strength with temperature increase is dependent on the formulation (*see* Remarks) but low temperature performance (below −50°C) is good. Resistance to attack by water, vegetable oils, mineral oils and grease is moderately good but poor towards alcohols, hydrocarbons, ketones, brine (5% salt), oxidising acids, and alkalies. Weather and salt-spray resistance is good.

Applications
Fast assembly processes where high production speeds are required, e.g. footwear industry for shoe assembly; automotive industry for metal to metal components such as radiators (in place of solder); electrical industry for coil insulation; printing industry for bonding aluminium to lead printing plates and silk-screen ink formulations; paper and packaging industry for production of laminated paper, metal-foil, plastic and fabric products, sealing of packages and cartons moisture resistant coatings; structural bonding of wood, cork, metals, plastics (polyolyfine and polyester foils, cellophane, cellulose acetate) leathers, ceramics and textiles.

Remarks
Polyamides are particularly useful for thermoplastic adhesive compositions because they have sharp melting points and low-melt viscosities with quick grab properties when heat or solvent activated. Compatibility with rosin, phenolic resins, waxes, plasticisers, shellacs, coumarone-indene resins is basis for many hot-melt formulations. Tensile strengths (ASTM D 1 248–60T) for aluminium joints [2 600 (−198°), 2 200 (−80°), 770 (20°C) N/cm^2] give an indication of the potential performance over a temperate range.

REFERENCES

FLOYD, D. E., 'Polyamide Adhesives', *Handbook of Adhesives,* (Skeist Ed.), Reinhold, (1962).
FLOYD, D. E., *Polyamide Resins,* Reinhold, (1958).
MAIOROVA, E. A. and OVCHINNIKOVA, T. V., 'Two-stage

method of bonding polyamides to metal', *Soviet Plast.* **11**, 67, (1966).

MUNN, R. H. E., 'Polyamide resins for hot melt adhesives', *Assembly and Fastener Methods*, **5** (2), 42, (1967).

POLYAROMATICS

During the past decade, much progress has been made in improving the thermal and oxidative stabilities of organic resins at high temperatures. Heat-resistant resins and polymers have been developed as adhesives to meet the needs of the aircraft industry (supersonic aircraft) and space vehicles (missiles, satellites, rockets) where resistance to temperatures approaching 320°C is required throughout the life of bonded assemblies based on metals and reinforced plastics composites.

The oxidative stability of organic polymers is improved by the incorporation into the molecule of aromatic units such as benzene rings and heterocyclic rings such as substituted imide, imidazole and thiazole. Resins incorporating the first three heterocyclic groups specified are presently available for use as adhesives in high temperature structural applications. The resins are usually referred to as polyimides (PI), polybenzimidazoles (PBI) and polybenzothiazoles (PBT). They are supplied in the form of prepolymers, containing open heterocyclic rings, which are soluble and fusible. At elevated temperatures, the prepolymers undergo condensation reactions leading to ring closure and the formation of insoluble and infusible cured resins.

POLYIMIDES (PI)

Type
Synthetic thermosetting resin; reaction product of a diamine and a dianhydride.

Physical form
Solvent solution of the polyimide prepolymer or film (usually containing a filler such as aluminium powder) with glass-cloth interliner.

Shelf life
Less than a month at ambient temperatures. Storage under refrigeration (0°C) extends shelf life to 3 mth. Polyimides are moisture sensitive.

Working life, 20°C
Indefinite, unstable at room temperature.

Setting action
Polycondensation at high temperatures.

Processing conditions
Liquid types, removal of solvent, by heat or under reduced pressure, and pre-curing resin to desired degree (B-staging) usually at 100–150°C. Volatile content may vary from 8–18% w/w after B-staging. Final cure (C-staging) is carried out by stages over the range 150–300°C or higher. Film types, may require B-staging process. Typical cure schedule involves heating to 250°C over 90 min period and maintenance at 250°C for 90 min. Post curing at higher temperatures (up to 300°C and beyond) is recommended where maximum high temperature mechanical properties are required. Bonding pressures are within the range 26–65 N/cm^2. It is important to follow recommended cure cycles closely.

Permanence
Good hydrolytic stability and salt spray resistance. Excellent resistance to organic solvents, fuels and oils; resistant to strong acids but attacked slowly by weak alkalies. After exposure to 2% ozone for 4 000 h, strength retention is about 50%. Service temperature range is −196°C to 260°C, for long-time exposure but materials will withstand short exposures up to 350°C (200 h) and 377°C (10 min). Polyimides are exceptionally good high-temperature electrical insulating materials and have exceptional resistance towards atomic radiation (electrons and neutrons).

Applications
Structural adhesive for high and low temperature applications; metal to metal bonding of stainless steel, titanium, aluminium, etc; aircraft industry; preparation of glass-cloth-reinforced composites for electrical insulation; bonding of ceramics.

Remarks
Polyimide adhesives require higher cure temperatures than phenolic-epoxy adhesives; curing at 260°C is usually adequate where service temperatures do not exceed 260°C. Volatiles are released during the cure of polyimide adhesives and the best adhesion is obtained where these are free to escape (e.g. honeycomb or perforated core structures). For long term ageing at temperatures in the range 204–316°C, polyimides are superior to polybenzimidazole and phenolic-epoxy adhesives. Using film types cured at 260°C, typical values for stainless steel bonds are: 2 900 (−196°), 1 900 (24°), 1 240 (204°), 690 (427°), 345 (540°C), N/cm^2 Shear (ASTM D 1002–64T). With titanium substrates, retention of shear strength (1 400 N/cm^2) is good after 2 000 h ageing at 260°C; after the same ageing period at 316°C shear strengths of about 900 N/cm^2 are obtained. Peel strengths for metal to metal bonds fall within the range 43–53 N/cm^1.

REFERENCES

DARMORY, F. P., 'Extreme high temperature polyimide adhesive for bonding titanium and stainless steel', *Adhes. Age,* **17** (3), 22–24, (1974).

PASCUZZI, B., 'Bonding supersonic aircraft with polyimide resin systems—a study in thermal ageing', *Adhes. Age*, **14** (2), 26–33, (1971).

POLYBENZIMIDAZOLES (PBI)

Type
Synthetic thermosetting heterocyclic linear polymer.

Physical form
Film composed of glass-cloth impregnated with resin containing filler (commonly aluminium) and anti-oxidants (e.g. arsenic thioarsenate).

Shelf life, 20°C
90 d.

Working life, 20°C
Indefinite—unstable at room temperature.

Setting action
Polycondensation at high temperatures.

Processing conditions
Assembly is placed in a pre-heated press at 370°C and pressed at 3 N/cm^2 for 30 s. Pressure is then increased to 60–140 N/cm^2 and glue-line temperature maintained at 370°C for 3 h. Temperature is then reduced to 260°C or less and assembly removed from press; alternatively, the adhesive can be cured by an autoclave technique. For improved mechanical properties, post-curing in an inert atmosphere (nitrogen, helium or vacuum oven) is recommended. This involves 24 h at each of the temperatures, 316°, 345°, 370°, 400°C followed by 8 h at 427°C. The cycle is followed by an additional cure of 3 h at 370°C in air for ultimate properties.

Permanance
Good resistance to salt spray, 100% humidity, aromatic fuels, hydrocarbons, and hydraulic oils. About 30% loss of strength occurs in exposure to boiling water for 2 h. Electrical properties are fairly constant throughout up to 200°C. Thermal stability at high temperatures for short periods is good (540°C for 10 min or 260°C for 1 000 h). Useful service temperature range as adhesives is −250°C to 300°C.

Applications
High temperature structural adhesives and laminating resins for the aerospace industry; honeycomb bonding in supersonic aircraft; bonding of stainless steel, titanium, aluminium and beryllium alloys; high strength cryogenic adhesive.

Remarks
Industrial applications for these adhesives have not been investigated in great detail although these polymers display a thermal and oxidative stability which is superior to that of conventional adhesive systems. High cost and the requirement for very high temperature cures and post cures have discouraged widespread use. For applications requiring long-time ageing to temperatures in the 200–320°C range, the polybenzimidazoles are inferior to the polyimides but more stable than modified epoxy adhesives. The following shear strengths are typical for stainless steel bonds (cured for 1 h at 221°C then 1 h at 332°C) tested at 260°C. 1 900 (1), 2 100 (200), 1 300 (500), 0 (1000), N/cm^2 (h), Ageing period at 260°C. Shear (ASTM D 1002–64T).

REFERENCES

Narmco Imidite 1850, Tech. Bulletin Whittaker Corpn. Narmco Materials Div. California.
REINHART, T. J., 'Mechanical Properties of Imidite Adhesives' *Structural Adhesives Bonding*, Interscience, (1966).
LITVAK, S., 'Research on Polybenzimidazole Structural Adhesives for Bonding Stainless Steel, Beryllium and Titanium Alloys,' *Structural Adhesives Bonding* (Bodnar, Ed) Interscience (1966).
HILL, J. R., 'Process development of polybenzimidazole adhesives,' *Adhes. Age*, **9,** 32 (1966).

POLYBENZOTHIAZOLES (PBT)

These polymers are similar to the polybenzimidazoles but do not appear to have been extensively evaluated as adhesive systems. Solubility and fusibility properties are inferior to those of the polyimides and polybenzimidazoles and applications have been restricted to the fabrication of glass-cloth composites. The composites have outstanding resistance to tensile stress at 288°C (21 000 N/cm^2 after 1 000 h at this temperature) and display high flexural strengths up to 430°C (21 000 N/cm^2) after 6 h. Oxidative stability is superior to similar composites with PBI resins and an ability to withstand exposure to high service temperatures for short periods (540°C for 10 min or 343°C for 200 h) indicates some potential as an elevated temperature adhesive system provided the processing difficulties can be overcome.

REFERENCE

CRAVEN, J. M. and FISCHER, T. M., 'Film-forming Polythiazoles', *J. Polym. Sci.,* (B), **3,** (1), 35 (1965).

POLYPHENYLENES (PP)

Polyphenylenes appear to offer the maximum thermal stability within the family of organic polymers. Poly-p-phenylene is infusible up to 530°C and withstands prolonged heating at 240°C without change. Insolubility and general intractibility are experienced with polyphenylene systems and have imposed difficulties on the development of these materials as adhesive systems. Recently, modified polyphenylene prepolymer resins have become available as development materials which may be cross linked and reinforced with fibres, such as carbon and asbestos, to yield useful composites. The processing of these materials involves removal of component solvents and volatiles at about 50–60°C or under vacuum, followed by heat curing (under pressure) at 130–140°C and post-curing up to 325°C. The cured resins have shown good hydrolytic and solvent resistance.

REFERENCES

Polyphenylene Resins, Data Sheet, Monsanto Chemicals Ltd. London, (1968).
LONG, F., MILLWARD, B. B. and ROBERTS, R. J., 'Thermal-oxidative stability of polyphenylene resins in asbestos reinforced laminates', *10th National SAMPE Symposium*, San Diego, (1966).

POLYCHLOROPRENE (NEOPRENE) RUBBERS

Type
Synthetic thermoplastic elastomers.

Physical form
Viscous liquids composed of polychloroprene rubber in solvents (usually inflammable mixed aromatic-aliphatic hydrocarbons), with additives. Solids content 20–60%. White, brown or tan coloured; may be translucent or opaque. Compounded with fillers such as calcium silicate, silica, clay, carbon black and anti-oxidants.

Shelf life, 20°C
3 mth to 1 yr in sealed containers.

Working life, 20°C
Provided no solvent loss corresponds to shelf life.

Closed assembly time, 20°C
10–20 min for wet bonding. Indefinite where dried adhesive films are stored for later heat or solvent reactivation bonding.

Setting process
Solvent evaporation followed by assembly of tacky adhesive coats on adherends, or, heat or solvent reactivation of dried adhesive films before assembly.

Processing conditions
Porous materials, wet bonded and solvent allowed to migrate through adherend with time. Non-porous materials, bulk of solvent removed by evaporation (over 20 min to several hours, drying is accelerated by hot air, infra-red, etc.) and contact bonding of tacky substrates with minimum of pressure. Improved strength results where bond is vulcanised by heat curing or by addition of catalyst (e.g. isocyanate) with curing for 20 min at 80°C and 35 N/cm² for bonding pressure.

Coverage
Depends on viscosity of formulation; for 30% w/w solids content about 0·15–0·25 kg/m² is a typical range.

Permanence
Good resistance to water, salt spray, biodeterioration, aliphatic hydrocarbons, acetone and ethyl alcohol, lubricants, weak acids and alkalies. Unsuitable for contact with aromatic and chlorinated hydrocarbons, certain ketones, and strong oxidising agents. For continuous exposure, limited to service temperature range from −50°C to 95°C. Deterioration occurs at high temperatures where acidic by-products are released. Acid-acceptors such as ZnO and MgO are often added to reduce degradation. Durability properties depend, to some extent, on the additives and modifying agents commonly present in neoprene adhesives. Mechanically, bonds absorb vibration and display good shear and peel strength. Within loading limitations (20–70 N/cm²) the bonds withstand continuous stress better than other thermoplastic rubbers.

Applications
General purpose adhesives for wide range of materials: decorative plastic laminates; natural and synthetic rubbers; leather and rubber to metals; thin sheets of aluminium and steel; linoleum; fabrics and synthetic textiles (polyester and polyamide); woodworking, plywood and hardboard panelling to walls; automobile industry, sponge rubber weather strips to metal doors and upholstery (PVC, leather, fabrics to metal and wood); aircraft industry, vulcanised rubber to metal primed with chlorinated rubber. Building industry, wallboards and decorative phenolic laminates, footwear, temporary and permanent adhesion of soling materials based on rubber, resin-rubber and leather. Gap-filling properties are good. Joints may require several weeks conditioning period to realise optimum properties.

Remarks
Unsuitable for structural applications demanding shear strengths exceeding 200 N/cm² because they are liable to cold flow under relatively low loads. For structural applications, neoprene rubbers are blended with synthetic resins to promote mechanical strength and heat stability. (*see* PHENOLIC-NEOPRENE). Rate of strength development is rapid and, unlike

other rubber adhesives, neoprenes can sustain low loads at high temperatures soon after bonding. Tack retention is generally inferior to natural rubber. Solvent, or heat reactivation is less successful than with other rubber adhesives. They can support loads of 20–70 N/cm^2 for extended periods soon after bonding.

REFERENCES

ANON., 'Most versatile of industrial adhesives', *Aust. Plast. Rubb. J.*, **22** (267), 25, (1967).
CRAIN, G. W., 'Contact adhesives bond parts quickly and economically', *Mater. Des. Engng*, **64** (2), 76, (1966).
MARTIN, R., 'Obtaining higher bond strengths with neoprene adhesives', *Adhes. Age*, **9** (12), 31, (1966).
DOMBAY, S., 'The behaviour of polychloroprene-based adhesives in relation to fabric tendering', *Rubb. Plast. Age*, **45** (10), 1200, (1964).
ANON., 'Neoprene's growing role in industrial adhesives', *Adhes. Age*, **10** (12), 36, (1967).

POLYESTERS

Polyesters comprise a large class of synthetic resins (including thermosetting and thermoplastic types) which have a wide variety of applications. Polyesters are produced by reactions between dihydric alcohols and dicarboxylic acids and the resultant products are classified as unsaturated or saturated, depending on the absence or presence of chemically reactive double bonds in the linear polymer. Saturated polyesters (e.g. ethylene glycol terephthalate) are mainly employed for fibre and film production and are not used as adhesive systems. Unsaturated polyesters are employed principally as laminating resins and in combination with fibrous reinforcements as composite moulding formulations; their use as adhesives has been limited, possibly because of their high cost and the high volume shrinkage which occurs on curing. Unsaturated polyesters may be modified with oils and fatty acids to produce Alkyd (or Glyptal) resins, or used as modifying agents for other types of resin in adhesive compositions (e.g. polyurethanes are the products of isocyanates used in conjunction with certain polyesters).

ALLYLS

These form a special type of unsaturated polyester. Allyl resins are mainly employed for moulding applications involving electrical and electronics components. They appear to have had limited use as adhesive materials although the monomers are commonly used as crosslinking agents for polyesters.

Polymerisation of allylic resins is extremely slow at room temperature and catalysed systems often have shelf stability up to a year at ambient temperatures. The curing rate increases rapidly with temperature and the resins set quickly at about 140°C.

The thermosetting optical cement CR–39 (Columbia Resin 39) is one example of an allyl resin adhesive (*see* ALLYL DIGLYCOL CARBONATE). The cured material exhibits improved abrasion and chemical resistance over other transparent adhesive resins, and displays the good heat resistance and dimensional stability associated with thermosetting systems. These properties are retained on prolonged exposure to severe environmental conditions.

ALKYDS (OR GLYPTALS)

The term 'alkyd' is sometimes regarded as being synonymous with the term 'polyester' in published literature. These are basically unsaturated polyesters modified with oils or fatty acids; they are extensively used as surface coating materials which is their main application. Although adhesion to many materials, including metals, is strong, the alkyds have had only limited use as adhesives. They have been used for adhesive bonding of metal laminations (electric motor and transformer cores) and mica sheets. Good insulating properties favours the use of alkyds in electrical applications and flexible types effectively reduce vibration in electric motor laminations. Alkyd resins are employed as primers for metal surfaces prior to bonding them to wood and other surfaces with suitable adhesives. They are also used as modifiers in other adhesive systems based on nitrile rubber or phenolic resins.

This type of resin is often free of volatile components and bonds require minimum pressures to effect bonding. High volume shrinkage on curing restricts the application of these resins as adhesives.

Alkyd resins display moderately good resistance to water, strong or weak acids and alkalies, and organic solvents. Light stability is good and maximum service temperatures approach 95°C.

POLYESTERS (UNSATURATED)

Type
Synthetic thermosetting resin.

Physical form
Resin (unsaturated polyester) in solvent (unsaturated, e.g. styrene), and hardener (organic peroxide). Resins may be extended with filler powder.

Shelf life, 20°C
3 mth to indefinite.

Working life, 20°C
5 min to 24 h.

Closed assembly time, 20°C
5 min to 1 h.

Setting process
Co-polymerisation of resin and solvent.

Processing conditions
Vary with formulation, 5 min to 1 h at 20–110°C and contact to 12 N/cm² bonding pressure. With catalyst added, the adhesives usually set in a relatively short period and do not require drying. Presence of air inhibits curing of some polyester formulations.

Coverage
0·05 kg/m².

Permanence
Cured adhesives have good durability and resistance to weather, water, biodeterioration and moderate temperatures. Poor resistance to fire, high temperatures exceeding 100°C and certain chemicals unless modified. Resistance to solvents is generally good.

Applications
Manufacture of glass-fibre composites is a major application. Bonding of ceramics, some metals and rubbers, and to a limited extent, wood. Grouting materials for joints in concrete.

Remarks
Resins set between impervious surfaces at low bonding pressures. Marked volume shrinkage on curing restricts the use of these materials for certain adherends (e.g. wood). Cured films can be flexible or rigid and good impact strengths can be obtained. Electrical properties vary with the formulation. Polyester adhesives are most advantageously used where high and low temperature extremes are not encountered. Tensile and shear strengths fall within range 200–2 200 N/cm² according to formulation and adherend.

REFERENCES

MARTENS, C. R., *Alkyd Resins*, Reinhold, (1962).
PATTON, T. C., *Alkyd Resin Technology*, Interscience, (1962).
ROSATO, D. V. and WEIDMAN, R. A., 'Polyester's potential as an adhesive', *Plast. Wld*, **26** (2), 44, (1968).
PARKYN, B., LAMB, F. and CLIFTON, B. V., *Polyesters*, Vol. 2. Unsaturated polyesters and polyester plasticisers, Iliffe (1967).
GOODMAN, I. and RHYS, J. A., *Polyesters*, Vol. 2. Saturated polyesters, Iliffe, London, (1965).
Polyester Handbook, Scott Bader & Co., Wollaston, Wellingborough, (1967).
VINOGRADOVA, S. V., *Polyesters*, Pergamon, Oxford, (1965).

POLYSTYRENE

Polystyrene is a transparent, colourless thermoplastic resin which is available in solvent solution or aqueous emulsion form. Solvent solutions will set at room temperature, and contact pressure is usually sufficient for bonding. It has not found wide application in the adhesives field and is mainly employed as a solvent solution for joining polystyrene plastics. Other uses concern the bonding of porous materials such as plaster, paper, wood, cork and leather. With some woods, shear strengths up to 1 300 N/cm² may be obtained. Adhesion to plastics other than polystyrene, and metals, is poor to moderate.

Polystyrene is used as a modifier for other adhesives such as unsaturated polyesters, hot melt materials and in optical cements ($N_D^{20°C} = 1·59$).

Resistance to high temperatures is limited and the heat distortion temperature is about 77°C. Electrical insulating properties are very good. It has good resistance to water, nuclear radiation, and biodeterioration but generally poor resistance to chemicals. High flammability and a tendency to brittleness and crazing are additional undesirable properties. The brittleness is reduced where the adhesive is based on the copolymer of styrene with butadiene. These copolymers are more useful adhesive materials and are available as aqueous latices or dispersions containing up to 50% copolymer (*see* STYRENE-BUTADIENE).

REFERENCES

KOHN, L. S. and ZAVIST, A. F., 'Adhesives in the Electrical Industry', *Handbook of Adhesives*, (Skeist, Ed.), Reinhold, (1962).
BOUNDY, R. H. and BOYER, R. F., *Styrene: Its Polymers, Copolymers and Derivatives*, Hafner, (1965).

POLYSULPHIDE (THIOKOL)

Type
Synthetic thermoplastic elastomer.

Physical form
High viscosity liquid, two-part products consisting of the compounded liquid elastomer and a setting agent (sometimes dispersed in solvent or plasticiser, to reduce viscosity) with filler to improve heat resistance. Lead dioxide is commonly used as a catalyst.

Shelf life, 20°C
Up to 6 mth.

Working life, 20°C
15 min to several hours, after mixing.

Closed assembly time, 20°C
Within pot-life period.

Setting process
Polymerisation to elastomeric state with catalysts.

Processing conditions
Cures at room temperature and attains maximum strength in 3–7 d. No drying is required where solvents are absent from formulation. Contact bonding pressure is sufficient for these materials which are primarily sealants and usually applied with an extrusion gun.

Coverage
About 0·1 kg/m².

Permanence
Resistance to water, organic solvents, greases, oils, salt, water, ageing and weathering is better than that of the thermoplastic rubber adhesives. Low temperature properties are excellent with retention of flexibility at −62°C. High temperature resistance is poor and the material usually softens at 70–94°C with little strength retention above 120°C (addition of epoxy resin to formulation invariably improves heat resistance).

Applications
Polysulphide compounds appear to have been used more as sealants and caulking agents for outdoor application than as adhesives proper. Function as adhesives for certain applications where low strength and tendency to cold flow under load is acceptable. Typical uses include: the bonding/sealing of perspex glass windows, to aluminium or steel frames in aircraft or automobiles, and instrument cases; pressure cabin sealant; fuel tanks and fuel cell cavities; caulking wooden decks (marine); hose repair compounds; building trade sealants for metals, wood, glass, stone, etc.; rubber to metal; plastic to glass in periscopes; potting agents in electronic equipment.

Remarks
Polysulphide polymers do not display appreciable adhesion when cured on solid substrates unless modified with additives such as epoxy or phenolic resins. Primers based on polychloroprene, vinyl compounds, and chlorinated rubber are also employed to improve adhesion. Peel strengths increase with film thickness but tensile and shear strengths are a maximum for minimum film thickness. Typical strengths for these materials are within the range, 34–120 N/cm[1] Peel (flexible adherents), 40–200 N/cm² Shear (rigid).

REFERENCES

PANEK, J. R., 'Polysulphide Sealants and Adhesives', *Handbook of Adhesives*, (Skeist, Ed.), Reinhold, (1962).
GEORGE, D. A., STONE, P., DUNLAP, L. A. and ROTH, F., 'Polysulphide sealants', *Adhes. Age,* **6** (3), 35, (1965).

POLYURETHANES

Type
Thermoplastic or thermosetting. Reaction products of polyfunctional isocyanates and poly-hydroxy alcohols or certain polyesters.

Physical form
One component thermoplastic systems in solvents (ketones, hydrocarbons) often containing catalysts in small amounts to introduce a degree of thermosetting properties. Two-part thermosetting products in liquid form, with or without solvents. The second part is a catalyst.

Shelf life, 20°C
3–6 mth for solvent types where solvent loss is restricted.

Working life, 20°C
Solvent types have open time of 30 min or more. Two-part products set by catalytic action within 8 h. Tertiary amines are suitable catalysts for room temperature curing.

Closed assembly time, 20°C
Within 30 min.

Setting process
One-component type, solvent release and room temperature cure (catalysts may be added to reducing curing time). Two-component types, polymerisation by catalyst addition.

Processing conditions
Solvent types, contact bonding of tacky adherends following solvent release or heat solvent reactivation of dried adhesive coating. Two-part products are mixed and fully cure in 6 d at 20°C. May be heat cured in 3 h at 90°C to 1 h at 180°C. Bonding pressures range from contact to 35 N/cm².

Permanence
Good resistance to water and most chemicals, solvents, oils, greases, salt spray, ozone, radiation, and weathering is obtained with some formulations. Maximum service temperature is almost 177°C for thermosetting types (thermoplastic compositions are softened by heat reactivation above 70°C). Excellent shear strength down to −200°C with retention of elasticity and shock resistance. Electrical properties are good and the adhesives exhibit low combustibility. Resistance to biodeterioration is excellent.

Applications
Non-structural applications subject to moderate loads. Bonding of metals, rubbers, most plastics (except polyolefines and fluorocarbons), wood, cork, paper, textiles, glass and ceramics. Particularly

suitable for flexible and rigid polyvinyl chloride plastics (where compatability with plasticisers is good). Assembly of metal structures for service under cryogenic conditions. Specific uses include, bonding of urethane rubber to itself and other substrates; leather to itself and plastics (footwear materials); flock attachment to rubber and textiles; lamination of vinyl plastics to synthetic fabrics.

Remarks

Modifying agents for rubber based adhesives to promote adhesion. Restricted to applications where they have definite advantages for specific purposes mentioned above, although forming strong flexible bonds with a wide variety of materials. Relatively expensive compared with other rubber adhesives and have the disadvantage that isocyanates are toxic, unpleasant to handle (skin and vapour hazard). Exhibit high tack, often giving instant strength on contact bonding. Toughness and abrasion resistance are outstanding properties of urethane materials. Joint strengths which are typical of two-part products are: Steel-steel 3 000 ($-200°C$), 800 (25°C) N/cm^2 Shear (ASTM D 1002–53T); 5 000 ($-200°C$), 800 (25°C) N/cm^2 Tensile (ASTM D 897–49); Aluminium-aluminium 6 000 ($-200°C$), 1 600 (25°C) N/cm^2 Tensile (ASTM D 897–49). Peel strengths (ASTM D 903–49) for plastics substrates are generally about 20 N/cm^1 but with polyvinyl chloride can be as high as 70 N/cm^1.

REFERENCES

SAUNDERS, J. H. and FRISCH, K. C., *Polyurethanes: chemistry and technology*, Interscience, (1964).
DELMONTE, J., 'Bonding thermoplastics with improved urethanes', *Adhes. Age*, **11** (4), 27, (1968).
PHILLIPS, L. N. and PARKER, D. R. V., *Polyurethanes: chemistry, technology and properties*, Iliffe, (1964).
DOMBROW, B., *Polyurethanes*, Reinhold, (1957).

POLYVINYL ACETALS

These thermoplastic synthetic resins are the products formed by the reaction between polyvinyl alcohol and aldehydes. The most important materials are those from formaldehyde (polyvinyl formal) and butyraldehyde (polyvinyl butyral) although they have had limited use as adhesives when unmodified.

Polyvinyl butyral is the more flexible and soluble polymer and yields higher peel strengths. It is available as a clear colourless solvent solution or as an aqueous dispersion. Adhesion to glass is excellent and the manufacture of laminated safety glass is a major application. In this respect, transparency, stability to sunlight and sufficient resilience to retain glass fragments together on impact are factors which favour its employment. It is also used for the bonding of mica and as a modifier for hot-melt adhesives in the packaging industry. Strippable protective coatings for non-porous surfaces such as glass and metals are provided by films cast from dispersions.

Polyvinyl formal is less frequently employed as an adhesive than the butyral polymer but has been used in hot-melt formulations and as a bonding agent for wood, metals and electrical wire coil enamels and binders.

The polyvinyl acetals have moderate resistance to high temperatures, water and solvents, but excellent resistance to mould growth. When used as hot melt adhesives, typical shear strengths (N/cm^2) for wood and metal are the following: Aluminium-aluminium 1980 (Polyvinyl formal), 2 230 (Polyvinyl butyral); Aluminium-wood (beech) 490 (Polyvinyl formal), 412 (Polyvinyl butyral).

In conjunction with phenolic resins, they form important structural adhesives for metals (*see* PHENOLIC-VINYL BUTYRAL and PHENOLIC-VINYL FORMAL); the addition of the vinyl constituent improves the flexibility and adhesion of the phenolic resin. Polyvinyl formal appears to be more widely used for this purpose than does the butyral.

REFERENCES

Properties and Uses of Butvar (polyvinyl butyral) and Formvar (polyvinyl formal), Tech. Publication Shawinigan Resins Corporation.
'Hot-melt adhesives', *Packag. Rev.*, **80**, 55, (1960).
MOSER, F., 'Polymeric adhesives for glass', *Plast. Technol.*, **2** (12), 799, (1956).
LAVIN, E., and SNELGROVE, J. A., 'Polyvinyl Acetal Adhesives', *Handbook of Adhesives*, (Skeist, Ed.), Reinhold, (1962).

POLYVINYL ACETATE

Type
Synthetic thermoplastic resin.

Physical form
Low solids content and high viscosity solvent solutions or emulsions which may be plasticised.

Shelf life, 20°C
Indefinite.

Working life, 20°C
Indefinite where solvent loss is negligible.

Closed assembly time, 20°C
10–15 min.

Setting process
Loss of water (solvent) by evaporation or absorption into adherend.

Processing conditions
For wood, 10 min to 3 h at 20°C and contact to 100 N/cm^2 bonding pressure. Solvent removal is difficult for some applications and it may be necessary to force-dry the open glue-line, and heat weld or employ solvent activation with polymerising solvents. Generally, freezing has an adverse effect on processing.

Coverage
0·15–0·2 kg/m^2

Permanence
Low resistance to weather and moisture restricts use to interior applications. Solvent solutions have slightly better resistance to water than emulsion types. Resistance to most solvents is poor, but the adhesives withstand contact with grease, oils and petroleum fluids, and are not subject to biodeterioration. Cured films are light stable but tend to soften at temperatures approaching 45°C.

Applications
Furniture and joinery industries but not suitable for timber structures. Porous and non-porous adherends such as paper, cloth, leather, glass and plastics. Concrete additive to promote compressive strength.

Specifications
BS 3544:1962 Methods of test for polyvinyl acetate adhesives for wood.

Remarks
Low cost adhesives with high initial tack properties, but quick setting to give almost invisible glue-lines. 1–7 d conditioning period is recommended before handling bonded assemblies. Maximum bond strengths up to 1 400 N/cm^2 are achieved by baking the adhesive films followed by solvent reactivation and assembly. The tendency to creep under sustained load precludes the use of these materials for many structures. Gap filling properties are good; some improvement in creep, moisture and heat resistance properties is effected by modifying polyvinyl acetate emulsions with urea or phenol formaldehydes.

REFERENCES

Polyvinyl Acetate Emulsions, Monograph No. 1. Vinyl Products Ltd. Carshalton, (1960).
BUTTS, G. T., 'Polyvinyl acetate copolymers for adhesives', *Adhes. Age*, **2** (10), 20, (1959).
Properties and Uses of Gelva Polyvinyl Acetate, Tech. Publ. Shawinigan Resins Corporation, (1967).

POLYVINYL ALKYL ETHERS

These are non-crystalline colourless vinyl polymers which range in form from high tack viscous gums to tough rubbery solids, and have useful adhesive properties. Polyvinyl ethyl ether (PVE) polyvinyl methyl ether (PVM), polyvinyl-*n*. butyl ether (PVB) and polyvinyl isobutyl ether (PVI) are the commonest examples of vinyl alkyl ethers presently in use. The elastomeric types are soluble in a wide range of organic solvents (e.g. ethers, hydrocarbons, esters, ketones and lower alcohols); PVM is unique in that it is soluble in water below 32°C but precipitates out at higher temperatures. They have a limited heat resistance but are not embrittled at very low temperatures. Light stability is good and superior to rubber-based adhesive types.

The rubbery types are extensively used for applications where high tack is important, such as pressure sensitive adhesives and in footwear sole bonding. Vinyl alkyl ethers are components of hot-melt adhesives formulated with the cheaper resins (e.g. wood rosin, and are used in protective surface coating formulations based on nitrocellulose and polyvinylchloride sols.

PVE based adhesives are suitable for cloth, paper, non-fibrous substrates such as cellophane, polyester film, flexible vinyl films ceramics, wood and plaster. PVE is also used to improve the peel strength and lower the bonding temperature of phenoxy based adhesives. Storage stability of PVE solutions and films is excellent, with only slight changes in colour, tack and cohesive strength after 2 years.

PVM is employed as a re-moistenable adhesive for paper labels which show non-curling properties and adhere well to glossy paper surfaces. Another application concerns its formulation with polyvinyl butyral to decrease the water sensitivity of interlayers in laminated safety glass.

REFERENCES

SCHILDKNECHT, C. A., 'Vinyl Ethers' *Monomers*, Interscience, (1951).
DUFFET, N. B. and WAKEFIELD, H. F., 'Vinyl alkyl ether pressure sensitive adhesives', *Adhes. Age*, **3** (8), 28, (1960).
Bakelite Vinyl Ethyl Ether Resins, Bulletin of Union Carbide Plastics Co.

POLYVINYL ALCOHOL

Polyvinyl alcohol is a water soluble thermoplastic synthetic resin which has had a limited application as an adhesive. The chief uses are in bonding porous materials such as leather cloth, and paper in food packaging and as a re-moistenable adhesive. It is available as a water solution with good wet tack properties; setting action is by loss of water to give a flexible transparent bond having good resistance to oils, solvents and mould growth but poor resistance to water. Cured films are impermeable to most gases. The maximum service temperature is about 66°C.

Polyvinyl alcohol is also used as a modifier for other aqueous adhesive systems to improve film-forming properties or promote adhesion. Low-cost paper laminating adhesives are based on polyvinyl alcohol formulations with dextrines and starches. Other applications include its use for envelopes and stamps, as a binder for ceramics, sand-moulds for metal slip coating, non-woven fabrics and additive for concrete in the building industry.

REFERENCE

Grades, Uses, Modification, Properties of 'Elvanol' polyvinyl alcohol, Technical Publication. E.I. DuPont de Nemours Co. Inc.

POLYVINYL CHLORIDE

Although polyvinyl chloride is employed to some extent in adhesive formulations, it has had limited application as a basic adhesive material. It is insoluble in many organic solvents but solvent carrying polyvinyl chloride adhesives may be used to bond rigid polyvinyl chloride to itself and acrylic materials. Copolymers with polyvinyl acetate have been used to bond vinyl plastics to metals, glass, and as laminating adhesives for vinyl-cloth products. Other copolymers with polyisobutylene and vinyl alkyl ethers are included in some hot-melt adhesive compositions.

Vinyl chloride-synthetic rubber formulations (e.g. polychloroprene or nitrile rubbers) are also employed to bond metals and for laminating protective vinyl polymers to steel. More recently, vinyl plastisols have been incorporated in oil absorbent adhesives intended for direct application to oily steel surfaces.

REFERENCES

Cementing rigid P.V.C., Data Sheet I.S.714, Imperial Chemical Industries, Ltd.
BROOKMAN, R. S., 'Combining Vinyl and Cloth', *Adhes. Age,* **2** (11), 30, (1959).
'New vinyl-steel laminates', *Mod. Plast.,* **31** (9), 107, (1954).

RECLAIM RUBBER

Type
Thermoplastic elastomer based on reclaimed devulcanised rubber which may be compounded with latex or styrene butadiene.

Physical form
Hydrocarbon solvent or water dispersions with additives (fillers, anti-oxidants and tackifiers). Solids content, 20–80%. Black, grey and red coloured.

Shelf life, 20°C
Up to 1 yr in sealed containers.

Working life, 20°C
Open time 2–30 min or longer.

Closed assembly time, 20°C
Indefinite, long open time.

Setting process
Solvent evaporation from adhesive.

Processing conditions
Porous materials, wet bonded and solvent allowed to absorb into adherend. Non-porous materials, partial drying 5–30 min at 90°C followed by contact bonding of tacky substrates. Curable dispersions (by vulcanising agents) available. Bonding pressures from contact to 70 N/cm^2 at 20–90°C.

Coverage
Dependent on viscosity of formulation.

Permanence
Comparable to natural rubber for resistance to heat, water and solvents. Moderate ageing properties with deterioration in sunlight or on heat exposure above 70°C. Heat resistance varies widely with formulation. Becomes brittle at 70°C.

Applications
Minimum strength applications for wood, paper, panel boards, fabrics, painted metal, leather and some natural rubbers. Surgical tape manufacture. Compounded with latices for impregnation of tyre-cord fabrics, and with asphalt for fabric to metal and glass fibre to metal. Building and automobile assembly work such as weatherstrip attachment and sound proofing lines. Sealers and caulkers. Plastic foam bonding.

Remarks
Unsuitable for structural purposes. Strengths in shear (30–140 N/cm^2) and peel (10–30 N/cm^1) may be high for short loading periods. Excellent tack properties although tack range and cohesive strength differ widely with formulation. Less resilient than natural rubber but adhesion to metals is better. Low cost, easy application and rapid strength attainment after long open time favours reclaim rubber instead of natural rubber for many uses. Often blended with natural rubber. More amenable to spray application than other rubber adhesives.

REFERENCE

NOURRY, A., (Ed), *Reclaimed Rubber: its Development, Applications and Future,* Maclaren, London, (1962).

RESORCINOL FORMALDEHYDE AND PHENOL-RESORCINOL FORMALDEHYDE

Type
Synthetic thermosetting resins. Phenol-resorcinol formaldehyde is a copolymer of resorcinol formaldehyde and phenol-formaldehyde.

Physical form
Red-brown resin and hardening agent (liquid or powder). Resin may be diluted with alcohol or water or extended with filler.

Shelf life, 20°C
Resin and liquid hardener, 1 yr. Powder, indefinite.

Working life, 20°C
3–4 h to 1 h at 38°C.

Closed assembly time, 20°C
30 min to 2 h.

Setting process
Condensation polymerisation with elimination of water.

Processing conditions
For general purposes, 4–8 min at 100°C or 8–10 h at 20°C and 35–100 N/cm^2 bonding pressure. Timber (Hardwood) 8–15 h at 20–40°C and 100 N/cm^2; (Softwood) 8–15 h at 20–40°C and 70 N/cm^2. Moisture content of wood should be below 15%. Plastics, cold cured or heat cured below softening point of plastic. Contact to 35 N/cm^2 bonding pressure. Press time decreases with increasing temperature; some assemblies bonded at 45°C require 1 h press time, those glued at 90°C require 3 min press time. Resins set with difficulty under winter temperature conditions.

Coverage
0·05–0·1 kg/m^2 general purpose bonding and plastics, 0·2–0·25 kg/m^2 timber construction work, 0·05–0·125 kg/m^2 plywood fabrication.

Permanence
Durability of bond is generally comparable to that of heat setting phenolic adhesives and limited by the durability of the adherend material. Resistance to severe weathering, humidity, boiling water, temperature and biodeterioration is excellent. Maximum chemical resistance after 6 d. Maximum durability for hardwood assemblies is achieved by curing at moderately high temperatures.

Applications
Plywood and timber structures for exterior use, ship and marine constructions and wooden structural assemblies such as roofs, bridges and frameworks, wood to metal laminates where the metal is primed with metal bonding adhesives (epoxies). Bonding of acrylic, polyamide, moulded urea and phenolic plastics.

Specification
BS 1203 : 1963 Adhesives for plywood.
BS 1204 : 1965 Adhesives for constructional work in wood.

Remarks
Bond strengths for wood usually exceed the strength of this material; neutral character of resorcinol resins avoids damage to cellulosic fibres. Long storage life, and non-reactivity until catalyst addition is made, are valuable properties. Hot-cured assemblies should not be handled or machined for at least 24 h. Low temperature cured bonds require a 7 d conditioning period. Gap-filling properties are improved by filler additives; phenol-resorcinols are suitable for glue-lines up to 0·4 mm thick. Staining power of resins often precludes bonding light coloured woods and veneers. Although phenol-resorcinol resins cost less than resorcinol resins, both types are more expensive than the urea resins used for similar applications. Long storage life, and non-reactivity until catalyst addition is made, are valuable properties.

REFERENCES

Resorcinol formaldehyde adhesives, Bulletin No. 134. Aero Research Technical Notes. CIBA (A.R.L.) Ltd. Duxford, (1954).
VAN GILS, G. E., 'Reaction of resorcinol and formaldehyde in latex adhesives for tire cords', *Ind. Engng Chem. Prod. Res. Dev.*, **7** (2), 151, (1968).

RUBBER DERIVATIVES

Adhesive formulations based on chemically treated natural or synthetic rubbers are available in solvent form and include the following:

CHLORINATED RUBBER

Mainly employed for bonding natural and synthetic rubbers to metals and rubber to rubber bonding. Serves as a primer for metal surfaces before bonding vulcanised or unvulcanised rubber with nitrile or neoprene based adhesives. Neoprene or nitrile rubbers are applied directly to the metal coated with a tacky chlorinated rubber primer and the assembly is heat cured. Natural unvulcanised rubbers are similarly bonded to metal except that a neoprene intermediate layer is required. Chlorinated rubber is used as an additive for nitrile or neoprene rubber adhesives to improve specific adhesion to metals.

Resistance to water, salt spray, oils, fuels, ageing and biodeterioration is good. Resistance to solvents is limited and towards heat and cold exposure moderate. The maximum service temperature is about 140°C.

CYCLISED RUBBER

A resinous, relatively hard material with better cohesive strength, durability and chemical resistance than untreated rubber. Cyclisation is produced by treatment of natural rubber with strong acids (e.g. sulphuric) or certain salts. Used primarily to bond rubber to metal or itself. For bonding unvulcanised rubber to metal, the adhesive coated surfaces are joined when dry and the assembly vulcanised under pressure. Cyclised rubber is used as a primer for metal surfaces prior to bonding them to rubber with rubber based adhesives.

Resistance to water, salt spray, ageing, and biodeterioration is good, but poor towards oils and fuels. Heat resistance is inferior to that of chlorinated rubber and there is moderate deterioration on exposure to cold.

RUBBER HYDROCHLORIDE

A resinous material derived from the treatment of rubber with hydrochloric acid and used mainly for packaging film fabrication (e.g. Pliofilm). Limited adhesives applications but has been employed for rubber to metal bonding where vulcanisation processing of the rubber is required.

Resistance to moisture and some chemicals is good; maximum service temperature is about 110°C.

REFERENCE

GRANT, L., 'Chlorinated rubber formulations for contact adhesives', *Adhes. Age,* **11** (3), 32, (1968).

SILICONES

Silicones are semi-inorganic polymers (polyorganosiloxanes) which may be fluid, elastomeric or resinous according to the type of organic groups on the silicon atoms and the extent of cross-linkage between polymer chains.

The use of silicones as adhesives has not been extensive but certain silicone resin and elastomer polymers have been developed with sufficient adhesive properties for certain applications. Although suitable for continuous service up to 250°C, silicone bond adhesives have low initial adhesive strength. Silicones may also be used to improve the heat stability of other adhesive materials such as epoxy and phenolic resins.

SILICONE RESINS

Physical form
These are available as water-white to straw-coloured aromatic solvent solutions with a solids content of about 60% or higher according to end use.

Shelf life, 20°C
Usually exceeds 6 mth.

Processing conditions
For pressure-sensitive adhesive tape applications, solvent is removed by drying coated adherends for 15–30 min at 70–100°C. A tacky adhesive film results which may be contact bonded to another surface without further treatment. Some backing materials (other than glass cloth) require the use of a primer to improve adhesion, or a repellent to prevent migration of adhesive into the backing material. A 2% solution of tetrabutyl titanate in white spirit which is applied, hydrolysed and dried, provides the most satisfactory repellent. For certain formulations a post cure for 5 min at 250°C or 5 min at 125°C (where a peroxide or lead-octoate catalyst is added up to 2% w/w) ensures optimum balance of cohesive and adhesive strength. For insulation material bonding and sealing, typical processing involves air-drying followed by baking for 1 h at 100°C to remove solvent and then post-curing for 1 h at 250°C to produce a hard, dry film.

Permanence
Good resistance to moisture and weathering with low water absorption. Withstands effects of sunlight and ozone and is resistant to biodeterioration. Heat and oxidation resistance is excellent and service temperatures range from below −60°C to 250°C. Silicone resins withstand contact with many oils and chemicals, including dilute acids and bases, but are prone to dissolution in organic solvent media. Electrical properties are good with high dielectric strength and arc resistance, and low loss factor; silicones will not break down and allow carbon tracking or current leakage.

Applications
Pressure-sensitive tape manufacture; typical backing materials include glass cloth (plain or silicone-varnished), silicone rubbers, polyester film, polytetrafluoroethylene (or p.t.f.e. glass cloth) metal foils, synthetic fibre fabrics. Tapes are used as insulation wrappings for electrical windings and coils for service up to 180°C (Class H operation, BS 2757: 1956). Bonding mica and asbestos panel boards; adhesive-sealant for exposed magnesium oxide (heating element) insulation.

Remarks

Adhesive strength varies with the backing material for pressure sensitive tapes. Cohesive strength of cured resins is often increased by incorporating fillers such as clays, quartz, zinc oxide, calcium carbonate and acid-washed asbestos. Fillers will require evaluation before large-scale use. Typical peel strengths (ASTM D 1 000) for a silicone resin catalysed with 1% benzoyl peroxide and cured for 5 min at 150°C are, 10 (−70°C), 10 (−20°C), 5 (100°C), 4·5 (200°C), 3·5 (254°C) N/cm[1]. Peroxide catalysed silicone resins bond better to surfaces primed with amine organo silicon compounds.

REFERENCES

SOLODOVNIK, V. D., *et al., Investigation of the properties and possible applications of boron organosilicon polymers as components of heat-resistant adhesives*, J. Crosfield, Warrington, (1965).

IVANOVA, E. G. and DAVYDOR, A. B., 'VK-8 heat-resistant bonding agent of high strength and elasticity', *Soviet Plast.*, **10**, 59, (1966).

SILICONE RUBBERS

Physical form

Available as solvent-free white pastes in one or two part products which cure to form flexible silicone elastomers (silastomers). These materials are essentially adhesive/sealants. Fluorosilicone elastomers are also available as one-part products.

Shelf life, 20°C

6–12 mth.

Processing conditions

One part products (RTV type) cure at room temperature on exposure to atmospheric moisture. Thin films (0·6 mm) cure within 90 min while thicker films (13 mm) are fully cured within 7 d. May be applied with extruder gun. Two part products are by catalytic action. Pot-life is dependent on catalyst concentration and ambient temperature and typical values for curing times (20°C) are, 1% catalyst 10 h (Pot-life), 65 h (Setting-time), 1 wk (Cure-time); 5% catalyst 3 h (Pot-life), 22 h (Setting-time), 1 wk (Cure-time). Setting-times (after which partially cured silastomer can be handled) are increased by solvent or silicone fluid addition to the formulation. Curing is accelerated by temperature increase and retarded by temperature decrease.

Permanence

Properties are consistent with silicone rubbers in general. Highly resistant to moisture, hot water, oxidation, ozone, corona discharge and general weathering with good resistance to many chemicals including dilute acids and alkalies, e.g. resistance to sulphur dioxide is good but poor to hydrochloric acid gas. Cured silastomers swell in the presence of hydrocarbon solvents and fuels. Heat stable over the range −75°C to 250°C and retention of flexibility at low temperatures is good. Fluorosilicone rubbers retain the durability properties of silicone rubbers and in addition display excellent resistance to prolonged immersion in fuels, oils and solvents.

Applications

General purpose sealing, bonding and repair material for caulking, waterproofing and insulating electrical and mechanical assemblies (domestic and industrial); sealant for seams and welded joints in automotive industry; bonding between metals, glass, ceramics, wood and plastics requiring flexible, heat stable bond properties; repairing damaged silicone rubber surfaces in hot-air ducts, oven gaskets, encapsulants on electrical and electronic equipment, and wire or cable insulation. Natural and synthetic textile coating to produce heat-resistant, anti-stick surfaces (e.g. conveyor belting). Fluorosilicone rubbers may be used for similar applications where solvent and fuel exposure is involved. They are especially suitable for bonding fluorosilicone rubber to any clean surface.

Remarks

Shear strengths (N/cm^2) for metals are typified by the following data for a silicone elastomer. Steel-steel (mild) 170 (20°C), 104 (After 7 d at 250°C); Aluminium-aluminium 210 (20°C), 125 (After 7 d at 250°C). Peel strengths (ASTM D 1876–61T) of about 17–32 N/cm[1] are obtainable. Adhesion values for a fluorosilicone rubber to metal are usually within the range 20–40 N/cm[1]. Organosilicon primers are usually recommended to improve the adhesion of silicone and fluorosilicone elastomers.

EPOXY-SILICONE

Silicones may be blended with epoxy resins to combine the thermal stability of the silicones with the mechanical strength of the epoxies. One epoxy-modified silicone adhesive (Metlbond 311) retains about 80% of its shear strength (initially 1 170 N/cm^2 with stainless steel) at 400°C, with 30% retention after 1 000 h ageing period at 260°C. These materials may thus be useful for certain structural assemblies although the mechanical properties are inferior to modified phenolic or epoxy based adhesives.

REFERENCES

FREEMAN, G. G., *Silicones*, Iliffe, (1962).

MEALS, R. N. and LEWIS, F. M., *Silicones*, Reinhold, New York, (1959).

KINGSLEY, D., 'Silicones and Miscellaneous Adhesives', *Handbook of Adhesives*, (Skeist, Ed.), Reinhold, (1962).

SOY(A)BEAN AND VEGETABLE PROTEINS

Type
Nitrogenous protein soyabean is the commonest and most representative of plant protein adhesives derived from seeds and nuts, hemp, and zein.

Physical form
Yellow powder which is mixed with water, lime (hydrated) or sodium hydroxide.

Shelf life, 20°C
Dry powder, 1 yr.

Working life, 20°C
Several hours.

Closed assembly time, 20°C
Up to 30 min.

Setting process
Chemical reaction with elimination of water.

Processing conditions
Plywood (cold press), 4–12 h at 70–100 N/cm² bonding pressure. Plywood (hot press), 3–10 min at 100–140°C and 100–150 N/cm² and moisture content of wood should not exceed 20%.

Coverage
0·15–0·3 kg/m².

Permanence
Water resistance is limited, but like the casein glues they recover strength on drying. Is susceptible to biodeterioration under humid conditions unless inhibitors are present in the formulation. Poor resistance to heat. Durability of these adhesives restricts their use to interior applications.

Applications
Interior-grade plywood and softwood panelling. Component of a blood albumen-soyabean adhesive for interior plywoods based on softwood.

Remarks
Where maximum bond strength is unwanted, the glue may be extended with fillers such as clays and wood flows, to produce glues for low strength applications, e.g. paper and softboard laminating, cardboard box fabrication and as particle binders.

REFERENCES

PERRY, T. D., 'Soyabean Glues' *Adhesives for Wood*, Chapman and Hall, London, (1952).

CIRCLE, S. J., *Soybeans and Soybean Products*, Interscience, New York (1950).
LAMBUTH, A. L., 'Soybean Glues', *Handbook of Adhesives*, (Skeist, Ed) Reinhold, (1962).

STARCH

Physical form
Powder (blended flour) which is mixed with water or chemicals such as sodium hydroxide.

Shelf life, 20°C
Dry powder, up to 1 yr.

Working life, 20°C
Several days, to indefinite if protected from biodeterioration.

Closed assembly time, 20°C
10 min.

Setting action
Loss of water.

Processing conditions
For general purposes may be set at room temperature; develops tack rapidly. Plywood manufacture, 1–2 d at room temperature and 50–70 N/cm² bonding pressure.

Coverage
0·15–0·25 kg/m².

Permanence
Poor resistance to weathering, moisture and temperature changes. Under damp conditions, biodeterioration is a hazard and the adhesive is limited to use under dry interior conditions only.

Applications
Paper carton and bottle labelling; stationery; interior grade plywood fabrication.

Remarks
Addition of sodium hydroxide to starch adhesives increases the risk of staining timber substrates. Joint strengths are low compared with other vegetable adhesive types but are usually adequate for the applications mentioned above. Poor resistance to water and biodeterioration is improved by the addition of preservatives.

REFERENCE

KERR, R. W., (Ed.), *Chemistry and Industry of Starch*, Academic, New York, (1950).

THERMOPLASTIC RESINS
(MISCELLANEOUS)

In addition to the thermoplastic adhesives discussed already, many other synthetic and natural materials of this type find application for particular purposes. Some of the forms, main uses and properties are conveniently summarised here.

COUMARONE-INDENE

Physical form
Aromatic solvent solution (benzene toluene, xylene), or mastic.

Properties
Forms rigid, tough films which soften at high temperatures. Resistant to water and alkalies but not oils or greases. Bond strengths are low.

Applications
Bonding wood, fabrics, felt. Component of hot-melt adhesive formulations. Flooring applications.

Remarks
Predominantly an adhesive constituent rather than an adhesive.

SHELLAC

Physical form
Alcoholic solvent solution or as hot-melt mastic.

Properties
Good electrical insulating properties. Lacks flexibility unless compounded with other materials. Resistant to water, grease and oils. Softens on heating and is soluble in many solvents. Bond strengths are moderate.

Applications
Porous materials, metals, ceramics, cork and mica. Gasket cements. Primer for metals and mica intended for adhesive bonding. Insulating sealing waxes. Component of hot-melt adhesives. Electrical industry applications.

Remarks
Declined in use because of high cost. Basis of de Khotinsky cement. Available in various degrees of purity and colour.

ROSIN (COLOPHONY)

Physical form
Solvent solution or hot-melt mastic.

Properties
Poor resistance to solvents and oils; good resistance to water. Subject to oxidation and has poor ageing properties. Brittle and is usually modified with plasticisers. Bond strengths are moderate and develop fairly rapidly.

Applications
Paper (temporary adhesive and label varnish). Component of pressure sensitive adhesives based on styrene-butadiene co-polymers. Component of hot-melt adhesives.

Remarks
Synthetic materials have replaced many of the rosin adhesives where the base material is scarce. Esters and other rosin derivatives find application as adhesive components.

OLEO-RESINS (VEGETABLE OILS + ROSIN, PHENOLIC OR ALKYD RESINS)

Physical form
Viscous liquids with solvent or as mastics (with filler).

Properties
Excellent tack and rapid development of low bond strength. Soft plastic consistency results in low bond strength. Good ageing properties and retain flexibility for long periods (years). Fair resistance to water and oils but poor resistance to solvents and alkalies. Pliable at low temperatures ($-46°C$) but embrittle at 40–70°C over long exposure period.

Applications
Linoleum, felt, cork and similar materials to each other and metals; plastic tiles; acoustical wall coverings. Caulking and sealing.

Remarks
Usually require a solvent to reduce viscosity of oil/resin formulation. Good wetting properties for most surfaces.

BITUMEN (INCLUDING ASPHALT)

Physical form
Water emulsions, hot-melts or solvent solutions.

Properties
Dark coloured materials having poor strength. Depending on formulation, may be soft, sticky, or hard and brittle. Resistant to water, acids and alkalies but not oils and many solvents. Generally poor resistance to high (above 45°C) and low temperatures.

Applications

Concrete, glass, metals, felt, paper; flooring applications; tiles to walls; laminating paper and metal foil; corkboard insulation; waterproofing walls; road binding materials. Hot-melt component.

Remarks

Suitable for low stress applications where durability is important. Tendency to cold flow even at room temperature. Some emulsions resist flow up to 105°C. May be compounded with epoxy resins to improve strength properties.

REFERENCES

SEYMOUR, R. B., *Hot Organic Coatings*, Reinhold, New York, (1959).
KOPYSCINSKI, W., NORRIS, F. H., and HERMAN, S., *Adhes. Age*, **3**, May, (1960).
MANTELL, C. L., *The Water Soluble Gums*, Hafner, (1965).
HOIBERG, A. J., (Ed) *Bituminous Materials: Asphalts, Tars, and Pitches*, Vol. 3, Interscience, (1966).
ABRAHAM, H., *Asphalts and Allied Substances*, Vol. 5, Van Nostrand, (1962).
SAAL, R. N. J., 'Bituminous Binders and Coatings', *Adhesion and Adhesives*, Vol. 1, (2nd edn) Elsevier, (1965).
WHISTLER, R. L., *Industrial Gums*, Academic, New York, (1959).

UREA FORMALDEHYDE

Type

Synthetic thermosetting resin.

Physical form

Two part products consisting of the resin and hardening agent (liquid or powder). Also available as spray-dried powder, with incorporated hardener, which is activated by mixing with water.

Shelf life, 20°C

Resins, 3–6 mth; powders, at least 1 yr; catalysts, indefinite.

Working life, 20°C

From 0–48 h according to composition and setting temperature. Room temperature setting types range from 3–10 h.

Closed assembly time, 20°C

From 0–24 h depending on composition. From 0–30 min where resin and catalyst are applied separately to each adherend.

Setting process

Condensation polymerisation with elimination of water.

Processing conditions

Suitable for use with radio-frequency equipment.

For general purposes, 2–4 h at 20°C and 35–70 N/cm² bonding pressure. For plywood manufacture, 5–10 min at 120°C and up to 160 N/cm². Timber: (Hardwood) 15–24 h at 20°C and 140 N/cm²; (Softwood) 15–24 h at 20°C and 70 N/cm². Bonding pressures depend on type of wood, shape of parts and similar factors.

Coverage

0·1 kg/m² general purpose bonding. 0·075 kg/m² plywood fabrication. 0·125–0·2 kg/m² timber construction work.

Permanence

Urea formaldehyde adhesives lack the durability of combinations with melamine, phenolic, or resorcinol resins, and are unsuitable for service conditions which are extreme, e.g. high humidity, boiling water and temperatures above 60°C. Resistant to cold water and biodeterioration and may be compounded with additives to improve durability. Hardening agents used are important components and manufacturers advice should be followed. Resins cured with ammonium thiocyanate have superior hot water resistance to those hardened with diammonium phosphate. Use of starch extender reduces water resistance, increases resin viscosity to overcome adhesive absorption into porous materials (wood) and lowers cost of adhesive. Use of phenolic or resorcinol additives, promotes resistance to water, weather and temperature changes; durability improved but increases cost of adhesive. Use of melamine fortifier, increases resistance to water and temperature and improves durability. Melamine or resorcinol fortifiers do not enhance durability beyond that of phenolic resins.

Applications

Wood and related materials with moisture content of 7–15%. Manufacture of light woods, thin veneers and plywood furniture assembly and joinery. Timber construction subject to low stresses.

Remarks

Development of full bond strength is gradual over a few days but assemblies may be handled after 6–12 h. Bond strengths usually exceed the strength of the wood when glue-line thickness ranges from 0·05–0·16 mm. Gap-filling properties are poor where glue-lines exceed 0·37 mm. Thick glue films will craze and weaken bond strength unless special modifiers, such as furfural alcohol resins are used. Gap filling properties are important where the adhesive is extended with wood or walnut shell flow. Adhesives are non-staining and light coloured; unsuitable for exterior conditions or extreme temperatures.

Specification

BS 1204:1963 (Adhesives for constructional work in wood). BS 1203:1965 (Adhesives for plywood).

REFERENCES

BLOMQUIST, R. F., and OLSON, W. Z., 'Durability of urea resin glues at elevated temperatures', *Forest Prod. J.,* **8** (8), 266, (1957).

CHUGG, W. A. and GRAY, V. R., *The durability of room-temperature setting urea-formaldehyde adhesives in engineering structures.* C/1B/6, Timber Research and Development Association, Tylers Green, (1961).

ROY, D. D., *et al.,* 'Animal blood as extender for U.F. resin adhesives', *Pop. Plast.,* **12** (11), 11, (1968).

BUNE, A., 'Application of synthetic resin adhesives', *Assembly of Fastener Methods,* **6** (2), 30, (1968).

WATER AND SOLVENT BASED ADHESIVES

At some stage in their application, all adhesives must be fluid enough to wet the substrate prior to bond formation. Most adhesives are fluid when applied and these are represented by the water based, solvent based, prepolymer fluids (100% solids) and hot-melts. For water or solvent based adhesives, assembly of parts can be carried out without removing the liquid vehicle provided one of the substrates is porous; for non-porous substrates (e.g. metals) the removal of the vehicle is necessary before bonding the parts. Here, bond formation depends on the reactivation of the dried adhesive film under heat and pressure.

Water based adhesives are made from materials that can be dissolved or dispersed in water alone. Some of these materials are the basis for solvent based adhesives and the principal ones which are used for liquid adhesive formulations are given below.

Water based
Casein
Dextrin and Starch
Glue (animal)
Gums
Lignin
Polyvinyl alcohol
Sodium carboxymethyl cellulose
Sodium silicate

Organic solvent based
Cellulose acetate-butyrate
Cellulose nitrate
Cyclised rubber
Polyisobutylene

Water or organic solvent
Asphalt* and coal tar resins*
Amino resins* (melamine and urea formaldehyde)
Phenolic resins* (phenol and resorcinol formaldehyde)
Polyamides
Polyacrylate† (and polymethacrylate†)

Rubbers (natural and reclaimed,* synthetic elastomers based on polychloroprene,† acrylonitrile-butadiene†, styrene-butadiene† and butyl*)
Vinyl polymers and copolymers (polyvinyl acetate,† polyvinyl chloride,† polyvinyl ethers, polyvinylidene chloride†)

The reader is referred to the relevant pages for more detailed description of the properties and applications of specific adhesives. The advantages and disadvantages of liquid-based adhesives are conveniently summarised in Table 4.4.

REFERENCES

MANTELL, C. L., *The Water Soluble Gums,* Hafner (1965).

DAVIDSON, R. L. and SITTIG, M., (Eds), *Water Soluble Resins,* Reinhold, (1962).

WAXES

Waxes are thermoplastic materials of natural or synthetic origin which are usually applied as hot-melt adhesives above 54°C; setting action results from the cooling of the molten material. Numerous types are available which are derived from mineral, vegetable, petroleum sources or based on synthetic low-molecular weight polymers, and they may be compounded with elastomers to improve their flexibility.

Resistance to water and biodeterioration is usually good but solvent resistance is poor and the heat resistance very limited.

Adhesives of this type are useful for temporary bonds with non-porous materials such as metals and glass, and for packaging applications involving paper, cellophane, and foils and films based on metals and plastics. Sirawax softens at +30°C; is a typical and generally useful laboratory wax adhesive for temporary fixation of glass, ceramics, metals, etc., (e.g. as components of small assemblies and instruments).

Other uses of waxes concern their use as temporary encapsulants and formulation in hot-melt compositions (*see* HOT MELT ADHESIVES).

REFERENCE

HILDITCH, T. P., *Industrial Fats and Waxes,* Ballière, Tindall and Cox, London, (1943).

*Available as water dispersions (latices)—made by emulsion polymerisation.
†Available as water dispersions—made by emulsifying or dispersing solid polymers.

Table 4.4   ADVANTAGES AND DISADVANTAGES OF LIQUID-BASED ADHESIVES

Factor	Water-based	Solvent-based
Cost of production	Inexpensive, easily manufactured	Medium cost adhesives.
Inflammability	Non-inflammable	Often a fire and explosive hazard.
Toxicity	Non-toxic solvent	Usually toxic and constitute a health hazard. Venting equipment required.
Storage properties	Prone to metal contamination from container vessels	Shelf lifes up to 1 yr where solvent is retained.
Solids content	Wide range	Wide range.
Viscosity	Wide range	Moderately wide range.
Drying rates and open-times	Slow drying and long setting times	Quick drying; variable over a wide range.
Maintenance of applicator equipment	Easy clean-up of equipment	May present difficulties where adhesive is quick drying.
Development of bond strength	Slow. Generally poorer tack properties than solvent-types	Rapid. Tack properties are usually good.
Substrate compatibility	Unsuitable for hydrophobic surfaces (many plastics) where wetting is often inadequate. Causes shrinkage of some substrates (textiles, paper and cellulosics). Corrosive action towards some metals. Suitable for many substrates (including foamed plastics)	Suitable for many hydrophobic surfaces; good wetting properties. Often unsuitable for plastic foams where solvent attack occurs. Compatible with metallic substrates; basis for the contact adhesives which display generally good adhesion to a wide variety of materials.
Water resistance	Poor	Good.
Cold resistance	Subject to freezing with embrittlement of adhesive.	Generally good retention of resilience when frozen.
Electrical properties	Poor	Good.

CHAPTER 5

Adhesive Products Directory

5.1. INTRODUCTION

The data presented in this chapter have been selected from manufacturers' trade literature and supplemented with other published information which is believed to be accurate. In addition, some performance data derived from evaluation work at Sira have been included.

Throughout this handbook an attempt has been made to carry out tests and to quote literature or other sources of information correctly. It is appreciated that there is always the possibility of error and the author apologises to anyone harmed by any such errors which were not apparent at the time of publishing. Similarly an apology is made to those whose products have not been mentioned. It is, of course, impossible to deal with all products and selection of some must of necessity mean omission of many.

It should be noted that adhesives' manufacturers generally disclaim responsibility for the results obtained by the user, who should carry out his own evaluation to determine suitability of the product. Adhesives' properties vary between wide limits and their performance depends on the conditions under which they are prepared and used; it is preferable that the manufacturer should always be approached for detailed information about his product. The data given here should thus be regarded as an indication of the probable values obtainable under certain conditions rather than as absolute values obtainable under all conditions. Hence, it should be recognised from the outset that differences in results from those described here are often possible.

Only a small percentage of the total number of adhesives products available has been described and space limitations have restricted the amount of detail it has been possible to include. Some manufacturers supply such a large number of types that several pages would be required to describe them. As far as possible an attempt has been made to include a description of a representative sample of each type of adhesive. Whenever practicable, commercial products of a particular adhesive type have been grouped together in order to demonstrate the potential versatility of performance. Many manufacturers, although mainly concerned with a few basic adhesive types, will formulate specialised adhesives to meet their customers' requirements. The size of any entry in this section is not indicative of the manufacturers' importance in the adhesives' field since it is impossible to know whether the adhesives are produced as main or secondary products.

5.2. ADHESIVE PRODUCTS TABLE

Adhesive products have been listed in an alphabetical order which approximately corresponds to the sequence of adhesive materials described in the previous section. The information appearing in the table following is given under the following headings:

Trade source
A list of code numbers representing commercial suppliers of adhesive products will be found on p. 322 (Table 10.3).

Trade name or designation
Information on the main chemical ingredient of the adhesive where available. Some products are described in general terms where the manufacturer has not disclosed the composition of the product (e.g. synthetic rubber-resin, wax, etc).

Number of components
The number of ingredients which are supplied to the user must be mixed to constitute the adhesive

(e.g. components such as resins, hardeners and catalysts).

Physical form, consistency or viscosity
Viscosity values given refer to the mixed adhesive where multiple-component formulations are concerned.

Working life
The term refers to the pot-life at room temperature (20–25°C) of approximately 100–150 g batches for multiple-component adhesives unless otherwise stated. For one-component pre-mixed adhesives the working life is specified in terms of the time within which the joint must be assembled after adhesive application (e.g. open-time for solvent-based adhesives).

Method of application
The means by which an adhesive may be applied are briefly indicated. Mechanical applicators are feasible for many of the adhesives listed as hand-applied materials, but consultation with the manufacturer is recommended.

Processing
Curing cycles specify the shortest and longest time/temperature conditions for curing thermosetting adhesives; intermediate cycles are usually available from manufacturers. The column also indicates the drying time or heat-activation conditions for other types of adhesive (e.g. solvent-based resin-rubbers) where appropriate. Bonding pressures and times are stated in an adjacent column.

Main uses
Possible uses are restricted to the more important industrial, domestic and special purpose applications.

Adherends
Materials which may be bonded are listed in column form as a quick reference for adhesive selection purposes. Joints which are the subject of performance data described in the adjacent columns are indicated in italics below the adherends list.

Physical data
Wherever possible bulk physical properties of the adhesive and bond strength data are reported against test specifications, conditions and temperatures described in the adjacent columns. The key to the alphabetical code used to denote the properties will be found on p. 217 together with a standard table for conversions between the SI system and other units of measurement. The data are intended as a guide to general properties and performance of adhesive products. They should not be used to make comparisons between adhesives where the processing conditions and test methods employed are different. An asterisk * next to the Trade Name or Designation is used to indicate those products on which additional information is available from Sira.

Trade source	Trade name or designation	Basic type	No. of compo-nents	Colour	Physical form, consistency or viscosity Ns/m^2	Working life	Method of application	Processing		Service temp. range °C	Main uses
								Curing cycle	Bonding pressure N/cm^2		
144	Tensol cement No. 3	Acrylic	2	Trans-parent	Syrupy liquid. Viscosity depends on mixture proportions	Viscosity dependent 2 wk for syrup at 20°C in the dark	Brush Spatula	2 h at 55°C + 8 h at 80°C or 5 h exposure to u.-violet radiation	Contact to 4	−60 to 80	Structural work requiring maximum joint strength. Outdoor applica-tions. Bonding rigid acrylic sheets to other materials
144	Tensol cement No. 6	Acrylic	1	Trans-parent	Liquid 4.5×10^{-4} m^2/s	Solvent evaporates in 1 h at 20°C	Brush Spatula	20 h at 20°C to 6 h at 80°C	Contact to 4	−60 to 80	Acrylic sheet bonding requiring moderate strength with limited gap filling
144	Tensol cement No. 7	Acrylic	2	Trans-parent	Liquid 1.3×10^{-3} m^2/s	0.5−1 h	Brush Spatula	14 d at 20°C to 4 h at 20°C + 4 h at 80°C	Contact to 4	−60 to 80	High bond strength work with material opaque to u.-violet Especially suitable for lamination and outdoor application Edge jointing of acrylic materials using special techniques
144	Tensol cement No. 8	Acrylic	1	Trans-parent	Liquid 3.15×10^{-4} m^2/s	1 yr at 20°C in the dark	Brush, Liquid dispenser, Syringe	5 h exposure to u.-violet radiation	Contact	−60 to 80	For production line assembly of acrylic component

Adherends	Test spec.	Test conditions	Test temp. °C	Physical data	Key	Remarks
rylic materials lymethylmethacrylate— rubber lymethylmethacrylate— polymethylmethacrylate		Cured by heat Cured by heat Cured by heat Cured by u.-violet Cured by heat	−60 20 60 20 20	900 N/cm² 620 N/cm² 372 N/cm² 3 450 N/cm² 4 450 N/cm²	b b b b b	Liquid and powder components. Gap filling up to 6·3 mm. Joints may be handled after 1 h. High pressure mercury vapour lamps (HPMV) can be used to harden the cement when placed 30 cm from joint. Resistant to weathering and moisture
rylic materials lymethylmethacrylate— polymethylmethacrylate		Cured by solvent loss Cured for 6 h at 80°C	20 20	1 380 N/cm² 1 890 N/cm²	b b	Single liquid (Flash Pt., 10°C) curing by solvent evaporation or absorption into adherend. Non-gap filling (max. gap of 0·05 mm). Joints may be handled after 3 h. Joint failure possible after 2 yr exposure to weather
rylic materials ass ood (open grain) ood (open grain)—wood lymethylmethacrylate— polymethylmethacrylate		 Cold cured at 20°C Cured for 4 h at 80°C	 20 20 20	 Cohesive failure of wood 3 450 N/cm² 4 550 N/cm²	 b b b	Two liquid components curing by polymerisation (Flash Pt. of mixture, 11°C). Gap filling property depends on technique. Joints may be handled after 1 h. Smooth close-grained wood is poor absorber of the cement. Virtually unaffected by outdoor exposure; may yellow slightly. Glass should be primed with a 5% vinyl trichlorosilane in petroleum ether solution (the adhesive is mixed with dibutyl phthalate up to 25% v/v before using)
rylic materials lymethylmethacrylate— polymethylmethacrylate		Cured by u.-violet	20	2 750–4 120 N/cm²	b	Single liquid (Flash Pt., 11°C) cured without heat by u.-violet radiation; joint may be handled in 35 min after curing at 7·5 cm distance. Low viscosity enables narrow joints to be filled by capillary action. Less contraction on setting than Tensol 3. Gap filling limited to liquid-tight cavities. Unaffected by weathering and moisture

Trade source	Trade name or designation	Basic type	No. of compo-nents	Colour	Physical form, consistency or viscosity Ns/m^2	Working life	Method of application	Processing		Service temp. range °C	Main uses
								Curing cycle	Bonding pressure N/cm^2		
59	Agomet M	Acrylic + other resins	2		Liquid	For 2% catalyst, varies from 3 h at 5°C to 40 min at 30°C. For 5% catalyst, varies from 2 h at 5°C to 25 min at 30°C	Spatula	For 2% catalyst, 2 d at 15°C to 1 h at 80°C. For 5% catalyst, 2 d at 15°C to 25 min at 80°C	Contact to 70	−70 to 120	Metals, rigid plastics, ceramics and similar hard materials employed for large structures exposed to the weather. Building applications, roofing, aircraft components
59	Agomet U3	Acrylic + other resins	2		Liquid	For 1·5% catalyst 50 min at 18°C For 5% catalyst, 24 min at 18°C (10 g mass)	Spatula Brush	1–7 h at 20°C. At 10°C, cure times increase by 1·5 h over 20°C times. Heat curing up to 180°C feasible	Contact to 70	−70 to 120	Metals, rigid synthetic plastics and ceramics, wood (oak, ash, fir) and laminated paper

Adherends	Test spec.	Test conditions	Test temp. °C	Physical data	Key	Remarks
luminium/Copper/Magnesium alloy F44		Abrasion pretreatment	20	3 670 N/cm²	b	Not suitable for zinc for which Agomet U grade should be
l/Cu/Mg F44		Chromic/Sulphuric acid etch	20	4 650 N/cm²	b	employed. For copper/zinc alloys, copper, brass bronzes,
luminium/Silicon/Magnesium alloy F32		Abrasion	20	3 330 N/cm²	b	etc., the adhesive should be applied just before the end of
l/Si/Mg F32		Chromic/Sulphuric acid etch	20	3 670 N/cm²	b	the pot-life. Performance data on shear specimens obtained
eel (mild)—steel		Abrasion	20	3 830 N/cm²	b	with 10 mm single overlap
		Chromic/Sulphuric acid etch	20	3 720 N/cm²	b	joint on $100 \times 20 \times 2$ mm metal. Dynamic stressing
eel (stainless)—steel		Abrasion	20	3 920 N/cm²	b	determined on overlap speci-
		Chromic/Sulphuric acid etch	20	3 920 N/cm²	b	mens. Fairly good electrical insulator, similar to polyester
rass—brass		Abrasion	20	2 950 N/cm²	b	or acrylic resins.
l/Cu/Mg F44—Al/Cu/Mg F44		Chromic/Sulphuric acid etch	−70	3 720 N/cm²	b	Good resistance to ageing, weathering and water
			20	4 650 N/cm²	b	
			60	4 170 N/cm²	b	
			120	2 550 N/cm²	b	
l/Cu/Mg F44—Al/Cu/Mg F44		Abrasion pretreatment	−70	2 370 N/cm²	b	
			20	3 670 N/cm²	b	
			60	3 110 N/cm²	b	
			120	1 470 N/cm²	b	
luminium—aluminium		Chromic/Sulphuric acid etch	20	5 900 N/cm²	d	
eel (F55)—steel		Chromic/Sulphuric acid etch	20	5 000 N/cm²	d	
luminium—aluminium		Abraded, cured at 20°C	−20	29·3 N/cm	a	
			20	39·0 N/cm	a	
			120	34·2 N/cm	a	
		Abraded, cured at 140°C for 1 h	−20	34·2 N/cm	a	
			20	34·2 N/cm	a	
			120	39·0 N/cm		
eel—steel		10⁶ cycles for 5 mm overlap	20	1 860 N/cm² at 25 Hz	w	
Cu Mg alloy F44— Cu Mg F44		150 000 cycles for 5 mm overlap	20	2 930 N/cm² at 25 Hz	w	
eel				$2·6 \times 10^{10} - 1·2 \times 10^{8}$ (at $50 - 10^{6}$ Hz)	p	
uminium						
pper and alloys				$4·8 - 3·2$ (at $50 - 10^{6}$ Hz)	n	
nc					o	
rcelain				$0·309 - 0·050$ (at $50 - 10^{6}$ Hz).	b	
ass					b	
arble				Cohesive failure of		
od (4·5 mm thick)				these adherends when	b	
lyester resin (3mm)				bonded to them-	b	
rylic resin (2·7 mm)				selves, one another,	b	
lyvinyl chloride, rigid 2 mm)				or to steel		
elamine resin (1·4 mm)						
sein formaldehyde resin 5 mm)						
uminium/Magnesium/Copper alloy F44—(2 mm thick)		Cured at 20°C and post cured 1 h at 120°C after abrasion pretreatment	−60	1 470 N/cm²	b	
Cu/Mg F44			−20	3 420 N/cm²	b	
			0	3 420 N/cm²	b	
			20	2 450 N/cm²	b	
			80	880 N/cm²	b	

continued

Trade source	Trade name or designation	Basic type	No. of compo-nents	Colour	Physical form, consistency or viscosity Ns/m²	Working life	Method of application	Processing		Service temp. range °C	Main uses
								Curing cycle	Bonding pressure N/cm²		
Agomet U3 — continued											
75	C.M.W. cement	Acrylic	2		Thick paste	About 2 min after mixing	By hand	Cold curing within 10 min	Hand pressure		Surgical bonding femoral head prosthesis
171	Cement* PS-18	Methyl metha-crylate	2	None	Liquid	About 30 min after mixing at 25°C to 90 min at 10°C	Brush or dispenser tube	2 h at 21°C	Contact to 1·5		Acrylic resins and other materials. Fabric–acrylic impregnates
85	Jelly glue	Animal (hide)	1	Pale brown	Stiff jelly			Melted at 70–75°C Sets on cooling	Contact		Woodworking Carpet materials Paper Bookbinding
85	Ground glue	Animal (hide)	1	Pale brown to brown	Fine/coarse powder or cake		Applied as water solution, 0·8 kg/l. Sets on cooling and drying		Contact		Woodworking Paper Porous materials
88	Croid 409	Animal glue	1	Brown	High viscosity aqueous solution pH = 7–7·5	Unlimited pot life	Brush Roller	Applied at room temp. or preferably at 49–55°C	Contact to +10		Constructional woodwork
45	Calbar	Animal glue	1	Amber	Liquid	Indefinite. Assembly 10–15 min after ap-plication	Brush Spreader	Sets on storage at 20°C after pressing	34–120		Gluing veneers and decorative laminates

Adherends	Test spec.	Test conditions	Test temp. °C	Physical data	Key	Remarks
el—steel (0·3 mm thick)		Cured at 20°C and post cured 1 h at 120°C after abrasion pretreatment	120 −20 20 120 200	490 N/cm² 39·2 N/cm 34·2 N/cm 17·6 N/cm 4·8 N/cm	b a a a a	
el—steel		Butt joints of 1 cm² cross section	20	2 150 N/cm²	d	
/Cu/Mg alloy F44— Al/Cu/Mg F44		10⁶ cycles for 10 mm overlap	20	635 N/cm² at 25 Hz	b	
ne to metal in orthopaedic surgery						Fixation achieved by mechanical interlock of irregular bone surfaces. No adhesion to wet bone
rylic (sheet)—acrylic		Joints heated at 25°C for 1 d 25°C for 1 d 70°C for 1 d 70°C for 1 d	23 70 23 70	2 700 N/cm² 1 500 N/cm² 2 950 N/cm² 1 860 N/cm²	b b b b	Test data obtained after 1 yr outdoor exposure of joints. Recommended 4 h cure at 21°C before machining joints. Maximum strengths obtained by annealing at 70°C for 4 h. Low viscosity version of PS-18 is available with 90 min pot-life at 21°C
ass—vinyl plastics etal—polyester od—polystyrene (rigid) bber—phenolics						
per ood extiles						May be thinned with water (up to 30% w/w)
per ood extiles						Available in wide range of jelly strengths and viscosities
ood—wood	BS 745:2		20	420 N/cm²	d	May be diluted with water (25% w/w). Gel temperature, 20–22°C. Poor resistance to water
ood. Plastics (laminate sheets)						May be diluted with water up to 25% v/v for assembly work. Pressing time is 1–2 h under normal temperature and humidity conditions

Trade source	Trade name or designation	Basic type	No. of compo-nents	Colour	Physical form, consistency or viscosity Ns/m^2	Working life	Method of application	Processing		Service temp. range °C	Main uses
								Curing cycle	Bonding pressure N/cm^2		
2	Plastick	Animal glue + catalyst	1	Amber	Liquid	4 h at 15°C to 2 h at 27°C	Brush Machine	Hot setting 2½ min at 54°C to 1 min at 79°C	Contact to 100		Wood veneering
47	Solbit 319	Bitumen	1	Black	100% w/w solids solution	Indefinite	Dip coat	Applied as melt at 60–65°C	Contact	−20 to 32	Bonding of expand polystyrene and other low melting points expanded plastics to surfaces
47	Bostik 9015	Bitumen in petroleum solvent	1	Black	Solution 70% w/w solids	Indefinite	Spreader Trowel	Dried to tacky state at 20°C and bonded	Contact		For fixing thermo plastic asphalt tile and vinyl/asbestos tiles to floors and sub-floors
47	Bostik 4053	Bitumen-latex emulsion	1	Brown-black	Thin emulsion	Indefinite	Spreader	Dried to tacky state at 20°C and bonded	Contact		For fixing wood mosaic tiles to concrete
68	Certofix	Fish glue	1	Yellow	Thick liquid 50% w/w solids	Open time ½ h at 20°C	Dispensed from container	Dries at 20°C in 1 h	Contact		General purpose adhesive for bond absorbent materia
160	Samson liquid fish glue	Fish glue	1	Pale buff	High viscosity liquid	Open time ½ h at 20°C	Brush machine	Dries at 20°C in 1 h	Contact	60	General purpose glue for porous materials
77	Dunlop CT-S	Butyl rubber	1	Off white	Heavy paste	20 min at 20°C	Spatula Trowel	Contact bonded when wet	Contact		Ceramic tile adhesive. Inflammable, Flas Pt. 46°C, coverag 1·8 m²/l
45	Casco Casein glue 1562	Casein	1	White	Powder	7 h at 16–21°C to 3h at 26–32°C	Stiff brush Mech-anical spreader	Cold setting after 20 min standing period on mixing	14 to 35 for 4–16 h at 19°C		Laminated timber arches and beams plybox beams, an timber engineerin work
45	Casco 1060	Casein + 60% latex	2	White	Powder + emulsion	3 h at 20°C	Brush, Roll spreader	Cold setting after 20 min standing period on mixing	Contact to 14 for 8–12 h		Bonding of dis-similar materials give flexible, wate resistant bond

Adherends	Test spec.	Test conditions	Test temp. °C	Physical data	Key	Remarks
od (oak, pine, beech) od (beech)—wood			20	380 N/cm²	b	May be diluted with water. Not suitable for use with iron vessels or platens where oak veneering is undertaken (i.e. staining hazard)
ystyrene (expanded) to ckwork, concrete, asbestos— nent, metals. ous surfaces (primed with set 183)						Excellent resistance to humidity but not solvents. Bostik Primer 384 available for porous surfaces. Coverage 1 m²/kg. Softening Pt. 53 + 5°C
C—asbestos composites halt crete nent and other screeds						Bostik Thinner 6012 available. Unaffected by frost but may thicken at low temperatures. Coverage 2·5–3·5 m²/l. Flash Pt. exceeds 23°C
od crete						Non-flammable but must be protected from frost. Bostik Thinner 6013 available. Coverage 1–1·5 m²/l
er ther rics od—wood			20	965 N/cm²	b	Not resistant to water
od pboard er						Rapid setting glue; good flexibility. Moderate resistance to water. High tack
amic tile od ster (primed) crete ckwork						Flexible bonds. Resistant to humid service conditions. Dunlop S70, polyvinyl acetate primer recommended for plaster surfaces
ber with moisture content –18% w/w						Powder is mixed with water (16–21°C) in weight ratio 1 : 1·75 by wt. Low bonding pressures are adequate for softwoods. Full bond strength developed after seasoning period of 48 h. Coverage, 3 m²/kg
minium od nolic formaldehyde (rigid) ther ber						Powder: water: latex weight ratio variable to meet flexibility needs. Water: powder: latex of 2 : 1 : ¾ (low flexibility) or 1¾ : 1 : 3 (high flexibility). High humidity conditions may cause corrosion of aluminium. Avoid copper

Trade source	Trade name or designation	Basic type	No. of compo- nents	Colour	Physical form, consistency or viscosity Ns/m²	Working life	Method of application	Processing Curing cycle	Processing Bonding pressure N/cm²	Service temp. range °C	Main uses
38	Bexol 13 cellulose acetate cement	Cellulose acetate in solvents	1	Clear	0·26. Solids content 10% w/w	Open time indefinite (min at 20°C)	Brush	Wet bonded and allowed to set at room temp.	Contact to 4		Bonding cellulose acetate to itself and porous substrates
172	UHU Hard	Cellulose nitrate in mixed	1	Col- ourless	Liquid 8 Solids content 30% w/w	Un- specified	Spatula	Sets by solvent loss	Contact to 4		Domestic applications, model building
160	Samson C108	Nitro- cellulose in ester solvent	1	Trans- parent	Low viscosity liquid	Open time ½ h at 20°C	Brush				Labelling
160	Samson C110	Nitro- cellulose in ester solvent	1	Trans- parent	Medium viscosity liquid	Open time 10 to 15 min at 20°C	Brush	Heat set 1 h at 60°C after wet bonding	Contact to 15	60	Labelling. General bonding of inorg materials including metals
217	Polycell	Cellulose ether	1	Trans- parent	Liquid		Brush	Dries in air	Contact		Interior decorating applications
217	Heavy duty polycell	Cellulose ether + resin	1	White	Powder		Brush	Dries in air	Contact		As for Polycell
144	Cellofas B	Sodium carboxy— methyl cellulose in water	1	Trans- parent	Liquid		Brush	Dries in air	Contact		Decorating and general purpose
144	Methofas P	Hydroxy propyl methyl cellulose in water	1	Trans- parent	Liquid		Brush	Dries in air	Contact		Wallpaper and thickening of emulsion polyvinyl acetate adhesives
144	Methofas SA	Modified methyl- cellulose	1	Trans- parent	Liquid		Brush	Dries in air	Contact		Heavy duty adhesive. Decorate paper and plastic

Adherends	Test spec.	Test conditions	Test temp. °C	Physical data	Key	Remarks
ellulose acetate ellulosic materials per, board ather brics						Based on highly flammable petroleum solvents. Shelf life, 9 mth below 24°C
etals ass ellulosic plastics crylics ood—wood (beech)		After 1 h at 20°C After 24 h at 20°C After 7 d at 20°C	20 20 20	254 N/cm² 540 N/cm² 725 N/cm²	d d d	Joints have good resistance to water, mineral acids, alkalis, fuels, and flammable solvent base (Flash Pt. 24°C)
ellulosic materials per						Sets rapidly; fairly flexible when cured. High resistance to water vapour and oils. Inflammable, Flash Pt. below 23°C
per ather xtiles icone carbide etals						Sets rapidly to give tough rigid bond with good resistance to mineral oils. Has proved suitable for bonding abrasive honing stones to zinc holders.* Inflammable, Flash Pt below 23°C *Sira evaluation
per lystyrene (foam) aster uminium (foil)						Neutral adhesive which is non-staining. Inert towards metallic printing (gold or bronzes) on paper
per lystyrene (foam) 2 mm thick nyl (fabric backed sheet)						Water soluble powder. Contains fungicide. Not suitable for extra thin, high density lustre wall linings.
per rrugated board lti wall sacking						
llpaper						Employed where lime resistance is required
yl coated paper ystyrene (foam)						Contains a fungicide to prevent biodeterioration

Trade source	Trade name or designation	Basic type	No. of compo-nents	Colour	Physical form, consistency or viscosity Ns/m^2	Working life	Method of application	Processing		Service temp. range °C	Main uses
								Curing cycle	Bonding pressure N/cm^2		
160	Samson 4622	Casein	1	Light fawn	Liquid		Brush machine	Dries in air	Contact		Carton gluing, bottle labelling, where high resistance to water and water vapour required
160	Samson 7123	Casein	1	Pale cream	Medium viscosity liquid		Brush Roller coater	Dries in air	Contact		Labelling lacquered tins. Gluing to printed and varnished surfaces
161	I.S.–12*	Cyano-acrylate	1	Trans-parent clear	Liquid 0·09–0·12	Sets under pressure in joint	Dispensed from container or automatic applicator	15 s to 10 min at 20°C Substrate dependent	Contact to 4	−80 to 80	Rapid assembly of metal, glass, plastic, rubber component Light engineering applications; instrumentation
161	I.S.–06*		1		Medium viscosity liquid 0·06			15 s to 1 min at 20°C Substrate dependent		−80 to 80	As for I.S.–12
161	I.S.–04E*		1		Liquid 0·04			1–6 s at 20°C Substrate dependent		−80 to 80	As for I.S.–12 but especially suitable for rubber bonding

Adherends	Test spec.	Test conditions	Test temp. °C	Physical data	Key	Remarks
er dboard ss						Rapid setting adhesive with high tack, giving flexible bonds
plate (lacquered) er						Slow setting adhesive with low tack. Flexible. High resistance to water. May be diluted with water. Slightly alkaline in reaction. Contains rust inhibitor
el—steel	ASTM–D1002–64		20	2 000 N/cm^2	b	Anaerobic adhesives which cure rapidly by monomer polymerisation in the absence of oxygen (air).
minium—steel	,,		20	1 240 N/cm^2	b	Maximum bond strength is
minium—aluminium	,,		20	1 720 N/cm^2	b	obtained after a 12 h storage
nolic—butyl rubber	,,		20	138 N/cm^2	b	period. Optimum performance
yl—butyl	,,	After 3 y at 20°C/45% R.H.	20	44 N/cm^2	b	obtained with thin gluelines generally not exceeding
prene—neoprene	,,	After 3 y at 20°C/45% R.H.	20	59 N/cm^2	b	0·10 mm.
ber—rubber (natural)	,,	After 3 y at 20°C/45% R.H.	20	39 N/cm^2	b	Unsuitable for fluorocarbons, polythene or polypropylene
on—nylon	,,	After 1 y at 20°C/45% R.H.	20	421 N/cm^2	b	without special treatments. Silicone rubbers do not bond
ylic—acrylic	,,	After 1 y at 20°C/45% R.H.	20	279 N/cm^2	b	satisfactorily and rigid nature of cured glueline limits
prene—neoprene	,,		20	54 N/cm^2	b	usefulness of the adhesives for
el—steel	,,	After 6 mth in water, 18°C	20	2 255 N/cm^2	b	flexible materials.
minium—aluminium	,,		20	2 059 N/cm^2	b	Cured joints generally have good resistance to alcohols,
hnical Data Sheet lists 60 joint performances			20	80	l	acetone, and aromatic hydro-carbons; 24 h immersion has
		At 10^6 Hz	20	3·34	n	little effect on initial bond
		At 10^6 Hz	20	2·02%	o	strength. No outgassing of
el—steel	ASTM–D1002–64		20	1 961 N/cm^2	b	volatiles apparent under vacuum of 10^{-9} mm.
	,,		20	2 745 N/cm^2	d	I.S. Activator A.C. available as primer aerosol to extend
	,,					utility to the bonding of
	,,		20	80	l	woods, ceramics and certain
		At 10^6 Hz	20	3·34	n	plastics.
		At 10^6 Hz	20	2·02%	o	Range of applicators available from manufacturer for
el—steel	ASTM–D1002–64		20	1 716 N/cm^2	b	production assembly purposes
	,,		20	2 060 N/cm^2	d	
	,,		20	82	l	
	,,	At 10^6 Hz	20	3·34	n	
		At 10^6 Hz	20	2·02%	o	

Trade source	Trade name or designation	Basic type	No. of compo-nents	Colour	Physical form, consistency or viscosity Ns/m^2	Working life	Method of application	Processing		Service temp. range °C	Main uses
								Curing cycle	Bonding pressure N/cm^2		
161	I.S.–150		1		High viscosity liquid 1·5			15–45 s at 20°C Substrate dependent		−80 to 80	As for I.S.–12 but especially suitable for high impact resistance applications. Preferred product where gap-filling requirement exist (up to 0·25 mm)
161	Loctite 407	Cyano-acrylate +additive	2		Liquid 0·04 +solid additive		Dispensed from container	6 h at 20°C (maxi-mum). Handling strength in 10 min at 20°C and ~80% of ultimate strength in 30 min		−80 to 120	Bonding of metal alloys and heat-resistant plastics required to operate at elevate temperatures
161	I.S.–901	Cyano-acrylate	1		High viscosity liquid 2·5			10 s to 15 min at 20°C		−30 to 80	Designed for gap-filling applications or slightly porous materials. Rubber-to-metal bonding
21	Avdelbond Blue 1	Cyano-acrylate	1	Col-ourless	Liquid 0·002 at 20°C	Sets under pressure, in the joint	Dispensed from applicator or container	3–50 s at 20°C Substrate dependent	Contact		Rapid assembly applications involving metals, thermosetting plastics, rubbers, ceramics, etc., e.g. bonding electronic components
21	Avdelbond Yellow 2	Cyano-acrylate	1	Col-ourless	Liquid 0·075–0·1	Sets under pressure, in the joint	Dispensed from container or applicator	3–150 s at 20°C Substrate dependent	Contact		Metal inserts, sha pins, etc. Glass to metal in instrumentation
21	Avdelbond Green 3	Cyano-acrylate	1	Col-ourless	Liquid 0·015–0·025	Sets under pressure, in the joint	Dispensed from container or applicator	5–180 s at 20°C Substrate dependent	Contact		

Adherends	Test spec.	Test conditions	Test temp. °C	Physical data	Key	Remarks
el—steel	ASTM–D1002–64		20	2 206 N/cm²	b	
	"		20	2 451 N/cm²	d	
	"		20	75	l	
	"	At 10⁶ Hz	20	3·34	n	
		At 10⁶ Hz	20	2·02%	o	
el—steel	ASTM–D1002–65		20	1 825 N/cm²	b	
	"		20	2 308 N/cm²	d	
	ASTM 950–54		20	20·3 J for 25·4 mm	g	
			20	8 × 10¹² Ω/cm	p	
el—steel	ASTM–D1002–65	After 1 000 h at 100°C	20	1 088 N/cm²	b	
minium—aluminium	"	After 300 h at 100°C	20	964 N/cm²	b	
el—steel	"	After 300 h at 120°C	20	689 N/cm²	b	
minium—aluminium	"	After 300 h at 120°C	20	550 N/cm²	b	
ycarbonate—polycarbonate	"	After 200 h at 120°C	20	1 136 N/cm²	b	
el—steel	ASTM–D1002–65	After 24 h at 20°C	20	1 481 N/cm²	b	Optimum bond strength after 24 h storage
ober—rubber (natural)	"	After 0·5 min at 20°C	20	799 N/cm²	b	
ober—rubber (natural)		After 48 h storage	20	316 N/cm²	b	Anaerobic adhesives, which set through the catalytic action of moisture film present on substrate. Unaffected by solvents such as propane, petroleum ether, light oils, alcohols, benzene and aromatic hydrocarbons
oprene—neoprene			20	434 N/cm²	b	
nolic—phenolic			20	1 012 N/cm²	b	
C—PVC (rigid)			20	1 990 N/cm²	b	
oper—copper			20	1 651 N/cm²	b	
el—steel			20	1 508 N/cm²	b	
ober—rubber (natural)		After 48 h storage	20	337 N/cm²	b	Not resistant to water or saturated water vapour conditions. Cured materials are soluble in dimethyl formamide, dimethyl sulphoxide, or nitromethane. Flash point of these products is 83°C and storage life at 5°C is ~1 year
oprene—neoprene			20	391 N/cm²	b	
nolic—phenolic			20	1 026 N/cm²	b	
C—PVC (rigid)			20	1 894 N/cm²	b	
oper—copper			20	1 797 N/cm²	b	
el—steel			20	1 729 N/cm²	b	
ober—rubber (natural)		After 48 h storage	20	241 N/cm²	b	HAZARD! Cyanoacrylates react rapidly with human tissue so that contact with the skin or eyes should be avoided. Prolonged exposure to adhesive vapours in poorly ventilated areas should be avoided
oprene—neoprene			20	372 N/cm²	b	
nolic—phenolic			20	1 019 N/cm²	b	
C—PVC (rigid)			20	1 956 N/cm²	b	
oper—copper			20	2 632 N/cm²	b	
el—steel			20	2 331 N/cm²	b	

Trade source	Trade name or designation	Basic type	No. of compo-nents	Colour	Physical form, consistency or viscosity Ns/m^2	Working life	Method of application	Processing		Service temp. range °C	Main uses
								Curing cycle	Bonding pressure N/cm^2		
21	Avdelbond Brown 4	Cyano-acrylate	1	Col-ourless	Liquid 2–2·4	Sets under pressure, in the joint	Dispensed from container or applicator	10–900 s at 20°C Substrate dependent	Contact		
105	Eastman 910*	Cyano-acrylate with additives	1	Trans-parent	Liquid (1·10 kg/l) p = 1·0959 0·06–0·1 at 25°C	Sets under pressure, within the joint	Appli-cator tube	15 s to 10 min at 20°C Substrate dependent	Contact to 4		Assembly of components in rapid sequence. Precision bonding of delicate or intricate assemblies based on a wide range of materials. Optical components. Subject to dry service conditions. Maximum service temperature = 77°C. Unsuitable for porous adherends
105	Eastman 910 MHT	Cyano-acrylate with additives	1	Trans-parent	Medium viscosity. Liquid 0·002 at 25°C	Sets under pressure, within the joint	Appli-cator tube	15 s to 10 min at 20°C Substrate dependent	Contact to 4		As for Eastman 910 where service temperatures approaching 230°C are envisaged
161	Loctite 309 Flexible	Polyacry-late resin	1	Trans-parent	Liquid to paste	Sets under pressure	Dispensed from container	3 min at 120°C to 45 min at 65°C or 7 d at 20°C	Contact to 7	−55 to 65	Bonding and sealing flexible metal sheet. Deck stiffeners to aluminium skin, flexible containers and aircraft fuselages. Thermo-setting plastics and glass materials

Adherends	Test spec.	Test conditions	Test temp. °C	Physical data	Key	Remarks
bber—rubber (natural) oprene—neoprene enolic—phenolic C—PVC (rigid) pper—copper eel—steel		After 48 h storage	20 20 20 20 20 20	268 N/cm^2 282 N/cm^2 923 N/cm^2 $1\,075 \text{ N/cm}^2$ $1\,780 \text{ N/cm}^2$ $1\,619 \text{ N/cm}^2$	b b b b b b	Leaflet L10 and other data sheets give extensive performance details on a wide range of substrates
ass rrous metals ass lyester enolics pper uminium astomers—SBR, butyl, oprene, natural	ASTM D–150– 54T ASTM D–1002– 53T	 Aluminium— aluminium	 20 20 20	Refractive index, 1·451 7 (Sodium D line at 20°C) 3·34 (at 10^6 Hz) 2·02 (at 10^6 Hz) $7·23 \times 10^{-4}$ $1\,022 \text{ N/cm}^2$ after 10 min $1\,507 \text{ N/cm}^2$ after 1 h $1\,860 \text{ N/cm}^2$ after 48 h	 n o q b b b	Anaerobic adhesive. Curing action is based on rapid polymerisation of the monomer under the influence of basic catalysts. Polymerisation initiated by substrate moisture. Unsuitable for fluorocarbons, polythene, and polypropylene without special pretreatments. Poor gap filling properties. Extensive performance data on metal, plastics and rubber joints with respect to durability towards humidity, chemicals temperature and ageing, are to be found in Technical Bulletin 3-2-910 from trade source
listed for Eastman 910 eel—steel	ASTM D–1002– 53T	After 7 days, at 20°C After 7 days, at 100°C After 7 days, at 150°C After 7 days, at 200°C After 7 days, at 250°C After 7 days, at 275°C	20 20 20 20 20 20	$2\,273 \text{ N/cm}^2$ 999 N/cm^2 668 N/cm^2 737 N/cm^2 296 N/cm^2 21 N/cm^2	b b b b b b	Anaerobic adhesive which cures in a similar manner to Eastman 910. Extensive performance data on thermal ageing properties are presented in Technical Bulletin 3-2-910A. Higher viscosity adhesive, Eastman THT, is available as an alternative (Brookfield at 25°C = 0.06–0·1 NS/m²)
luminium—aluminium	ASTM D1002 —64T ASTM D1002 —64T ASTM D1002 —64T ASTM D1002 —64T ASTM D1002 —64T ASTM D1002 —64T ASTM D1002 —64T	Cured at 25°C Cured at 120°C, 10 min Cured at 120°C and immersed in solvents for 30 d Acetone at 20°C Trichloroethylene at 20°C JP4 at 65°C Glycol/water at 65°C	20 20 −55 25 65 20 20 20 20	825 N/cm^2 965 N/cm^2 965 N/cm^2 965 N/cm^2 412 N/cm^2 152 N/cm^2 86 N/cm^2 575 N/cm^2 435 N/cm^2	b b b b b b b b b	Anaerobic adhesive. Joints conditioned in 100% relative humidity at 80°C (1 mth) lost all strength. Optimum performance obtained with 0·0025—0·0075 cm cured bond line thickness. Maximum glue line thickness is 0·0124 cm

continued

Trade source	Trade name or designation	Basic type	No. of compo-nents	Colour	Physical form, consistency or viscosity Ns/m²	Working life	Method of application	Processing		Service temp. range °C	Main uses
								Curing cycle	Bonding pressure N/cm²		
Loctite 309 Flexible — continued											
161	Loctite 308 Impact	Polyacry-late resin	1	Trans-parent	Liquid to paste	Sets under pressure	Dispensed from container	3 min at 120°C to 45 min at 65°C or 7 d at 20°C	Contact to 7	−55 to 95	Assembly require-ments requiring hi, resistance to impac or shock loading. Metals, glass and thermosetting plastics
161	Loctite 300 series: 306	Modified polyacrylate resins	1	Trans-parent yellow	Liquid 25	Indefinite. Anaerobic adhesives which set under pressure, within the joint	Dispensed from con-tainer vessel or from semi-automatic, automatic, or hand applicator equipment	3–4 h at 20°C with primed substrate* or 3 min at 120°C	Contact to 7		See also 'Remarks' Hot strength adhesive for use from −50°C to +150°C *Loctite Primer T
	307		1	Trans-parent yellow	Liquid 2·5			3–4 h at 20°C with primed substrate* or 3 min at 120°C	Contact to 7	−50 to 125	Low viscosity, hig tensile strength adhesive suitable f production line work *Loctite Primer T

Adherends	Test spec.	Test conditions	Test temp. °C	Physical data	Key	Remarks
	ASTM D1002 —64T	Water at 65°C	20	620 N/cm²	b	
	ASTM D897 —49	Cured at 25°C	20	1 380 N/cm²	d	
	ASTM D1876 —61T	Cured at 25°C	20	23 N/cm	a	
		Cured at 25°C	20	9·5 J for 2·54 cm notch	g	
			20	100%	k	
Aluminium—aluminium	ASTM D1002 —64T	Cured at 25°C	20	1 380 N/cm²	b	Anaerobic adhesive. Joints conditioned in 100% relative humidity at 80°C (720 h) retained 50% of initial strength
		Cured at 120°C, 10 min	20	1 720 N/cm²	b	
	,,	Cured at 120°C and immersed in solvents for 30 d	−55	1 720 N/cm²	b	As for Loctite 309, maximum glue line thickness is 0·0124 cm.
			25	1 720 N/cm²	b	
			95	1 030 N/cm²	b	
	,,	Acetone at 20°C	20	345 N/cm²	b	Impact (g) 16·3J for 2·54 cm.
	,,	Trichloroethylene	20	430 N/cm²	b	Elongation (k) is 15%
	,,	JP4 at 95°C	20	1 500 N/cm²	b	Coverage (0·007 cm thick) is 160 cm²/cc
	,,	Glycol/water at 95°C	20	1 620 N/cm²	b	
	,,	Water at 95°C	20	1 180 N/cm²	b	
	ASTM D897 —49	Cured at 25°C	20	2 750 N/cm²	d	
	ASTM D1876 —61T	Cured at 25°C	20	8·7 N/cm	a	
Copper Steel Ceramics, ferrites Glass Nylon Thermoset plastics Aluminium—aluminium	ASTM– D1002– 64T	Cured at 120°C, 10 min	20	1 240 N/cm²	b	Anaerobic adhesives which cure in the absence of air by rapid polymerisation of the monomer under the influence of basic catalysts. Absorbed moisture on most surfaces suffices to initiate polymerisation and achieves bond formation. Unsuitable for natural rubber, fluorocarbons, polythene or polypropylene without special pretreatments. Unsuitable for plastics such as styrene, polyvinyl chloride, or polycarbonate. Optimum performance is usually obtained with glueline thicknesses not exceeding 0·124 mm.
	ASTM– D897–49		20	1 929 N/cm²	d	
			20	1·35 J for 25·4 mm	g	
	ASTM– D1876– 61T		20	5·25 N/cm	a	
As for Loctite 306 Aluminium—aluminium	ASTM– D1002– 64T	Cured at 120°C, 10 min	20	2 067 N/cm²	b	Automatic application equipment systems are available, or can be designed, to suit production assembly lines.
			20	5 167 N/cm²	d	
			20	10·84 J for 25·4 mm	g	
			20	17·5 N/cm	a	

Trade source	Trade name or designation	Basic type	No. of compo-nents	Colour	Physical form, consistency or viscosity Ns/m^2	Working life	Method of application	Processing		Service temp. range °C	Main uses
								Curing cycle	Bonding pressure N/cm^2		
	308		1	Trans-parent yellow	Liquid 25			3–4 h at 20°C with primed substrate* or 3 min at 120°C	Contact to 7	−50 to 95	Assembly require-ments which need high resistance to impact or shock loading *Loctite Primer T
	310		1	Trans-parent yellow	Liquid 25			3–4 h at 20°C with primed substrate* or 3 min at 120°C	Contact to 7	−50 to 95	High strength bonding on dis-similar base materials *Loctite Primer T
	312		1	Trans-parent yellow	Liquid 1			12 h at 20°C with primed substrate	Contact to 7	−50 to 80	For bonding close-fitting rigid parts primed with Loctite Primer NF
107	Evo-stik 8951	Acrylic copolymer resin in hydrocarbon ester solvent	1	Clear	Liquid 30% w/w solids	6 mth at 5–30°C shelf-life	Knife or bar coater	Contact bonded after drying	Contact to 4		General purpose high performance pressure sensitive adhesive, developed for the printing industry. Maximum service temperature, 60°C
107	Evo-stik	Acrylic resin in aqueous dispersion	1	White to colour-less when dry	Liquid 47% w/w solids	12 mth at 5–30°C shelf-life	Knife or bar coater. Brush Spray	Dried 10–20 min at 20°C to 5–10 min at 60°C	Contact to 4		General purpose high performance pressure sensitive adhesive for plastics rubbers

Adherends	Test spec.	Test conditions	Test temp. °C	Physical data	Key	Remarks
luminium—aluminium	ASTM–D1002–64T	Cured at 25°C	20	1 380 N/cm²	b	Cured gluelines are resistant to oil, water, fuels, natural and synthetic oils, hydraulic fluids. Hot concentrated mineral acids and alkalis may slowly attack cured adhesives
		Cured at 120°C, 10 min	20	1 720 N/cm²	b	
			−55	1 720 N/cm²	b	
			95	1 030 N/cm²	b	
	„	After 30 d immersion in acetone at 20°C	20	345 N/cm²	b	
	„	glycol/water at 95°C	20	1 620 N/cm²	b	
	„	water at 95°C	20	1 180 N/cm²	b	
	ASTM–D897–49	Cured at 25°C	20	2 750 N/cm²	d	
	ASTM–D1876–61T	Cured at 25°C	20	8·7 N/cm	a	
			20	16·3 J for 25·4 mm	g	
			20	15%	k	
s for Loctite 306 luminium—aluminium	ASTM–D1002–64T	Cured at 120°C, 10 min	20	2 411 N/cm²	b	
			20	3 789 N/cm²	d	
			20	16·3 J for 25·4 mm	g	
			20	21 N/cm	a	
s for Loctite 306 luminium—aluminium	ASTM–D1002–64T	Cured at 120°C, 10 min	20	2 239 N/cm²	b	
			20	3 445 N/cm²	d	
			20	13·5 J for 25·4 mm	g	
			20	21 N/cm	a	
VC—aluminium		Initial	25	5·95 N/cm	a	Suitable for screen printing or roller coating. Alternative product 8950 (40% w/w solids) available. Coverage 7·5 m²/l. Flash Pt. below 10°C
		After 48–72 h	25	6·65 N/cm	a	
		After 24 h in water	25	7 N/cm	a	
		40° peel data				
VC—aluminium		150 mm peel front	25	24 N/cm	a	May be diluted with water (2x) for spray application. Coverage 13 m²/l. Flash Pt. none
		50 × 50 mm² area	25	69–83 N/cm²	b	

Trade source	Trade name or designation	Basic type	No. of compo-nents	Colour	Physical form, consistency or viscosity Ns/m^2	Working life	Method of application	Processing		Service temp. range °C	Main uses
								Curing cycle	Bonding pressure N/cm^2		
161	Loctite 353*	Modified acrylic ester	1	Light amber	Liquid 3 **Specific** gravity 1·1–1·2	Indefinite	Spatula Dispenser Brush	Cures on exposure to ultra-violet light 3650 Å in seconds to minutes	Contact to 4	−55 to 100	Glass-to-metal bonding in auto-motive and other industries. Optical component
175	Lion glue	Dextrine	1	Brown	Semi-liquid	Indefinite	Brush	Air drying	Contact		General purpose glue for absorbent materials
88	Croid 3102	Dextrine	1	Brown clear	Medium viscosity aqueous solution	Indefinite shelf life of 2 years in container	Brush Roller	Sets by water loss at room tempera-ture	Contact to +10		For machine labelling, carton sealing and paper bonding, in the packaging industry. Paper and board lamination
160	Samson 5603	Dextrine	1	White	Thixo-tropic paste	Indefinite	Brush Machine	Air drying	Contact		Bag sealing and board lining. Bookbinding
160	Samson 1704	Dextrine	1	Buff	Thick paste	Indefinite	Brush Machine	Air drying	Contact		Labelling silicone treated bottles in Pharmaceutical industry
264	Swift K1552	Dextrine-starch blend	1	Light tan	Liquid 1·2–1·6 at 20°C	Indefinite	Brush Roller coater	Applied above 15°C Air drying	Contact	48	Labelling, carton sealing. Spiral tube winding
50	Bakelite resin R18774/1 *	Epoxy resin +catalyst Q19027/1	2	Amber	4·6 at 25°C	35 min at 25°C 60 g mass	Spatula Roller Extruder	16 h to 9 d at 20–22°C to 1 h at 100°C	Contact to 20		Metals, plastic laminates and ceramic mouldings
47	Bostik 2001	Epoxy resin +catalyst	2	Grey	Medium thick liquid	2–2½ h at 25°C	Spatula	3 d at 18°C to 1 h at 50°C	Contact to 20	100	Metals

Adherends	Test spec.	Test conditions	Test temp. °C	Physical data	Key	Remarks
Glass						Refractive index = 1·470. Additional information on environmental performance of glass-to-steel joints re humidity, temperature, vibration available from Sira Institute. See also manufacturer's Data Sheet for other performance data. Loctite 355 available as alternative product (Viscosity = 50 N/cm^2)
Paper, Cardboard, Leather, Wood, Pottery						Medium drying period of 2–3 h
Paper, board materials						May be diluted as required with water but is used neat where high tack properties required. Poor resistance to water and humidity
Paper						Has low water content for rapid drying. May be diluted with water, medium tack
Glass (treated) Paper						Rapid setting material with high tack property. Moderately flexible when cured. Acidic material
Cellulosic materials Cardboard Paper						Fast setting adhesive. May be diluted with water. Cures to a nearly colourless film
Polyester—glass fibre (sheet) Epoxy—glass fibre (sheet) Phenolic—(laminate) Glass Wood Aluminium (alloy)—aluminium	BS 1470	Cured 16 h at 22°C 9 d at 22°C 2 h at 60°C 1 h at 100°C	22 22 22 22	535 N/cm^2 1 240 N/cm^2 1 570 N/cm^2 2 020 N/cm^2	b b b b	Recommended filler materials are whiting (100% w/w) and china clay (75% w/w) Good retention of bond strength on immersion in water for long periods
Steel—steel Aluminium—aluminium Wood—wood Glass—glass Tile (unglazed)—tile Aluminium—aluminium			20 20 20 20 20 100	620 N/cm^2 480 N/cm^2 480 N/cm^2 1 510 N/cm^2 760 N/cm^2 310 N/cm^2	b b b b b b	Fairly good resistance to esters, ketones and carbon tetrachloride, petrol, oil, but not trichloroethylene and organic acids. Non-staining and withstands u. violet and humidity. Coverage, 4 m^2/l

Trade source	Trade name or designation	Basic type	No. of compo- nents	Colour	Physical form, consistency or viscosity Ns/m^2	Working life	Method of application	Processing		Service temp. range °C	Main uses
								Curing cycle	Bonding pressure N/cm^2		
107	Evostik E20	Epoxy resin +catalyst	2		Paste	8 h at 20°C	Spatula	90 min at 50°C to 30 min at 70°C	Contact to 20	100	Rigid materials such as metals, plastics and ceramics
See re- marks	Budd GA-60	Epoxy resin +catalyst	2		Solid at 20°C	3 d at 0°C	Spatula	2 h at 180°C+ 1 h at 180– 315°C de- pending on maximum service tempera- ture	Contact	300	Strain gauge adhesive for high temperature use
286	M-Bond 43-B	Epoxy resin +catalyst	1	Yellow	Liquid 25% solids in xylene	Open time 15 min at 24°C	Brush	1 h at 176°C or 2 h at 190°C	10–70 or 27–35 for trans- ducers	150 (static) to 176 (dy- namic)	Strain gauge adhesive. Protective coating for transducers
184	Chemgrip HT	Epoxy resin +catalyst	2	Amber	Paste	16 h at 20°C	Spatula	8 h at 121°C to 1 h at 204°C	Contact to 20	−54 to 260 For short expos- ures 370	High strength adhesive for elevated temperature use. Formulated for bonding fluoro- carbons to various materials
184	Chemgrip	Epoxy resin +catalyst	2	Orange	Thixotropic paste	45 min at 20°C	Spatula	24–48 h at 20°C to 45 min at 93°C to 10 min at 150°C	Contact to 20	−40 to 121	Bonding of fluoro- carbon materials
193	Scotchweld EC1614B/A *	Epoxy resin +catalyst	2	Tan	Paste	45 min at 23°C (450 g mass)	Spatula	24–48 h at 20°C to 20 min at 120°C	Contact to 7	100	General purpose structural adhesive

Adherends	Test spec.	Test conditions	Test temp. °C	Physical data	Key	Remarks
luminium lass olymethyl methacrylate olyester laminates ood						Maximum bond strength in 48 h when cured at 20°C. Useful pot life with optional fast curing for assembly work
ickel-chrome alloys						Hardener is in powder form which is added to resin at 40°C. Long storage life when refrigerated. Developed by Westland Aircraft Ltd, Saunders Roe, Div. East Cowes, I.O.W.
etals ansducer materials			24 −233	±4% max ±1% max	k k	Post cure for 2 h at 205°C recommended for transducers. Excellent resistance to chemicals and moisture. Shelf life is 9 mth at 24°C
olytetrafluorethylene ood eramics eel lass luminium—aluminium			25 204 25 204 25 204 25 20	220 kV/cm 195 kV/cm 3·6 (at 10^3 Hz) 3·83 (at 10^3 Hz) 0·0090 (at 10^3 Hz) 0·0054 (at 10^3 Hz) 10^{16} Ω/cm 2 340–2 600 N/cm²	m m n n o o p b	Resistant to creep, impact, vibration and thermal shock for fluorocarbon assemblies. Peel strengths approach 44 N/cm. Pretreatment of fluorocarbons with etchant chemgrip is recommended. Adhesive has slow burning flammability. Coverage, 17 m²/kg
olytetrafluorethylene ood eel lass eramics luminium—aluminium			−54 204	1 240 N/cm² 690 N/cm²	b b	Non-inflammable. Coverage, 6·6 m²/l
eel lass olyester—glass fibre composite luminium—aluminium		Etched in chromic-sulphuric acids	20	1 370 N/cm²	b	Wide range of curing cycles, e.g. may be gelled at 150°C in 2 min and allowed to complete cure at room temperature

Trade source	Trade name or designation	Basic type	No. of components	Colour	Physical form, consistency or viscosity Ns/m^2	Working life	Method of application	Processing		Service temp. range °C	Main uses
								Curing cycle	Bonding pressure N/cm^2		
193	Scotchweld EC1838B/A *	Epoxy resin + catalyst	2	Green	Paste	45 min at 20°C (450 g mass) or 3 wk at 37°C	Spatula Trowel Extruder gun	8 h at 24°C to 2 h at 66°C to 45 min at 121°C	Contact to 7	65	Bonding of metals, glass, ceramics and plastic composites
73	Hidux 1033M	Modified epoxy with aluminium powder filler	2	Brown	Paste	2 d at 20°C	Brush	Dried for 30 min at 60°C. Cured for 30 min at 150°C under pressure. Pressure released, on cooling, at 50°C	69	−40 to 150	Metal bonding for structural assembli at elevated temperatures
73	Hidux 1233	Modified epoxy + aluminium powder filler	3	Grey	Paste	2 d at 20°C	Brush priming with adhesive	Dried for 30 min at 60°C and heat cured. 30 min at 150°C. Pressure released, on cooling, at 50°C	69	−40 to 150	Metal bonding for structural assembl at elevated temperatures

Adherends	Test spec.	Test conditions	Test temp. °C	Physical data	Key	Remarks
Steel Copper and alloys Zinc Silicon carbide Wood Masonry Polyester—glass fibre composite Aluminium—aluminium	ASTM D1002–53T	Cured 8 h at 24°C	−55 24 60	690 N/cm² 1 720 N/cm² 482 N/cm²	b b b	Maximum bond strength in 24–48 h for 20°C cure. Cures to strong durable bond without applied bonding pressure. No tendency to sag during cure. Similar to EC 2216 B/A but has greater impact strength and resistance to oil
		Cured 2 h at 66°C	−55 24 60	1 380 N/cm² 2 475 N/cm² 1 380 N/cm²	b b b	
		Cured 7 d at 23°C and exposed 2 wk to, JP4 Fuel at 60°C 20% NaCl salt spray at 35°C 100% R.H. at 60°C	27 27 27 27	2 140 N/cm² (control) 1 350 N/cm² 1 500 N/cm² 1 820 N/cm²	 b b b	
			20 100	3 250 N/cm² 0	e e	
Aluminium (alloy)—aluminium to DTD 746	ASTM D1002–53T	Etched in chromic sulphuric acid to DTD 915B	20 125 200 225	2 079 N/cm² 1 450 N/cm² 1 160 N/cm² 1 063 N/cm²	b b b b	Bond strength adequate for short periods up to 150°C. Bond strength deterioration for 1 000 h immersion in fluids at 130°C is comparable with that for thermal ageing in air. Data provided by B.A.C. (Operating) Ltd., Filton, Bristol
		Aged at 120°C, 1 000 h 10 000 h 30 000 h 1 000 h 10 000 h 30 000 h	−40 −40 −40 120 120 120	2 450 N/cm² 1 710 N/cm² 2 010 N/cm² 1 490 N/cm² 1 410 N/cm² 1 460 N/cm²	b b b b b b	
		Aged at 130°C, 1 000 h 10 000 h 20 000 h 1 000 h 10 000 h 20 000 h	−40 −40 −40 130 130 130	2 260 N/cm² 1 940 N/cm² 1 960 N/cm² 1 350 N/cm² 1 410 N/cm² 1 190 N/cm²	b b b b b	
		Aged at 150°C, 1 000 h 3 000 h 1 000 h 3 000 h	−40 −40 150 150	2 040 N/cm² 1 930 N/cm² 1 430 N/cm² 1 470 N/cm²	b b b b	
Aluminium (alloy)—aluminium to DTD 746	ASTM D1002–53T	Etched in chromic sulphuric acid to DTD 915B	20 125 200 225 250	2 490 N/cm² 1 498 N/cm² 1 353 N/cm² 1 257 N/cm² 1 208 N/cm²	b b b b b	Good bond strength retention for long periods, up to 150°C. Bond strength deterioration for 1 000 h immersion in fluids at 130°C is comparable with that for thermal ageing in air, e.g. Kerosene (Spec. DERD 2494) Hydraulic oil (Spec. DTD 585) Chlorinated Silicone oil (Spec. DTD 900/4725) Synthetic ester lubricant (Spec. DERD 2487) Not subject to creep at 100°C, 150°C, for sustained loads (40% ultimate tensile shear strength).
		Aged at 120°C, 1 000 h 10 000 h 30 000 h 1 000 h 10 000 h 30 000 h	−40 −40 −40 120 120 120	1 880 N/cm² 1 790 N/cm² 1 900 N/cm² 1 520 N/cm² 1 490 N/cm² 1 350 N/cm²	b b b b b b	
		Aged at 130°C, 1 000 h 10 000 h 20 000 h 1 000 h 10 000 h 20 000 h	−40 −40 −40 120 120 120	2 140 N/cm² 1 830 N/cm² 2 210 N/cm² 1 460 N/cm² 1 490 N/cm² 1 190 N/cm²	b b b b b b	

continued

Trade source	Trade name or designation	Basic type	No. of compo- nents	Colour	Physical form, consistency or viscosity Ns/m^2	Working life	Method of application	Processing		Service temp. range $°C$	Main uses
								Curing cycle	Bonding pressure N/cm^2		
Hidux 1233—continued											
91	Blooming- dale FM 96U with BR227 primer (20% w/w solids)	Modified epoxy	1	Brown	Unsup- ported film (0·32 mm thick)		Brushed primer + press curing of film	Primer air dried 30 min at 20°C + 60 min at 100°C. Film cured 60 min at 177°C under pressure. Pressure released, on cooling, at 50°C	28	−40 to 150	Metal bonding for structural assemblies at elevated temperatures
91	Blooming- dale FM 61 with BR227A primer (20% w/w solids)	Elasto- meric modified epoxy	1		Film supported on nylon carrier		Brushed primer + press curing of film	Primer air dried, 30 min at 20°C + 60 min at 100°C. Film cured 60 min at 177°C under pressure. Pressure released, on cooling, at 50°C	17	−40 to 125	Metal bonding for structural assemblies at elevated temperatures
266	Kollerbond 22	Epoxy	2	Grey	Putty	1 h at 21°C	Spatula	4·6 h at 20°C to 30 min at 65°C to 15 min at 100°C	Contact		General purpose bonding agent. Defect repairs in sheet metals, castings, tanks, concrete, building applications

Adherends	Test spec.	Test conditions	Test temp. °C	Physical data	Key	Remarks
		Aged at 150°C, 1 000 h	−40	1 980 N/cm²	b	Fatigue strength exceeds
		10 000 h	−40	1 660 N/cm²	b	25% ± 10%. Ultimate shear
		20 000 h	−40	1 590 N/cm²	b	strength at 10⁷ cycles (50Hz)
		1 000 h	150	1 460 N/cm²	b	at 120°C. Fatigue failure
		10 000 h	150	1 570 N/cm²	b	occurs at 150°C
		20 000 h	150	1 430 N/cm²	b	Data provided by B.A.C.
		Sustained loading at test temperature	150	Withstands 860 N/cm² (+500 h)	b	(Operating) Ltd., Filton, Bristol
luminium (alloy)—aluminium to DTD 746	ASTM D1002–53T	Etched in chromic sulphuric acid to DTD 915B	20	3 481 N/cm²	b	Good bond strength retention,
			125	2 828 N/cm²	b	for long periods, up to 150°C.
			150	1 590 N/cm²		Not subject to creep at 100°C,
		Aged at 120°C, 1 000 h	−40	3 310 N/cm²		150°C for sustained loads
		10 000 h	−40	3 080 N/cm²	b	(40% ultimate tensile shear
		20 000 h	−40	3 240 N/cm²	b	strength).
		1 000 h	120	3 760 N/cm²		Fatigue strength exceeds 25%
		10 000 h	120	3 660 N/cm²	b	±10%.
		20 000 h	120	3 380 N/cm²	b	Ultimate shear strength at 10⁷
		Aged at 150°C, 1 000 h	−40	3 040 N/cm²	b	cycles (50 Hz) at 20°C.
		10 000 h	−40	1 900 N/cm²	b	Fatigue failure occurs at 150°C
		20 000 h	−40	2 830 N/cm²		Data provided by B.A.C.
		1 000 h	150	2 340 N/cm²	b	(Operating) Ltd., Filton,
		10 000 h	150	2 860 N/cm²		Bristol
		20 000 h	150	2 900 N/cm²		
		Sustained loading at test temperature	100	Withstands 1 940 N/cm² (+500 h)	b	
				Ruptured by 324 N/cm² (85 h)	b	
uminium (alloy)—aluminium to DTD 746	ASTM D1002–53T	Etched in chromic sulphuric acid to DTD 915B	20	1 716 N/cm²	b	Good bond strength retention,
			125	1 015 N/cm²	b	for long periods, up to 150°C.
		Aged at 130°C, 1 000 h	−40	1 870 N/cm²		Subject to creep at 100°C,
		10 000 h	−40	1 490 N/cm²	b	150°C for sustained loading.
		20 000 h	−40	1 660 N/cm²	b	(40% ultimate tensile shear
		1 000 h	130	1 080 N/cm²	b	strength produces 0·3 mm
		10 000 h	130	1 100 N/cm²		deformation, with standard lap
		20 000 h	130	1 350 N/cm²	b	joint, in 500 h).
		Aged at 150°C, 1 000 h	−40	1 900 N/cm²		Fatigue strength exceeds 25%
		3 000 h	−40	1 840 N/cm²	b	±10%.
		1 000 h	150	1 100 N/cm²		Ultimate shear strength at 10⁷
		3 000 h	150	1 250 N/cm²		cycles (50 Hz) at 20°C.
		Sustained loading at test temperature	150	Withstands 344 N/cm² (+500 h)	b	Data supplied by B.A.C. (Operating) Ltd., Filton, Bristol
ass			20	5 500–6 900 N/cm²	c	Kollerbond 24 also available
ramics			20	4 120–5 500 N/cm²	d	with thicker consistency than
od			20	70–85 (Barcol, GYZJ 935)	1	22.
el—steel						Hardener SL recommended for
minium—aluminium			20	1 380 N/cm²	b	maximum adhesion. Hardener
ncrete—concrete			20	965 N/cm²	b	S for general purpose bonding.
			20	Cohesive failure of concrete	b	Hardener FX for rapid setting. Optimum strength after 48 h and cured materials can be machined or filed. Resistant to oils, greases, hydrocarbon solvents

Trade source	Trade name or designation	Basic type	No. of compo-nents	Colour	Physical form, consistency or viscosity Ns/m^2	Working life	Method of application	Processing		Service temp. range °C	Main uses
								Curing cycle	Bonding pressure N/cm^2		
286	M–Bond 43–B	Epoxy with solvent	1	Yellow	Low viscosity liquid. **Solids** content 25% w/w	9 mth at 24°C to 18 mth at 5°C	Spatula Brush	1–2 h at 175°C Post cure for 2 h at 205°C recom-mended	10–69 27–34 (opti-mum)		As for M–Bond 610 where service temperature range is −270°C to +120°C
286	M–Bond AE–10	Epoxy	2	Yellow	Medium viscosity liquid	15–20 m at 24°C	Spatula	24–28 h at 24°C +2 h at 15°C above maximum service temp. or 2 h at 50°C to 0·5 h at 100°C	3–14		Strain gauge bonding for general purpose stress analysis from −200°C to +100°C
286	M–Bond GA–2	Epoxy with filler material	2	Yellow	High viscosity liquid	15 m at 24°C	Spatula	40 h at 24°C +2 h at 15°C above maximum service temp. or 2 h at 50°C to 0·5 h at 100°C	3–14		As for M–Bond AE–10
101	Dupoxy EP-018	Epoxy	2	Green	Thick paste	30 min at 21°C	Trowel	24 h at 21°C	Contact	50	Civil engineering applications such as building, road repairs, bridgework and other structural assemblies
101	Dupoxy EP-010/3A	Epoxy	2		Thick paste	40–60 min at 21°C	Special hand tool	48 h at 21°C	Contact	−5 to 50	Road repair, bonding of kerb-stones to concrete
70	Nitoflor R251	Epoxy	2	White	Thick paste Non-sagging	1 h at 18°C	Trowel	24 h at 21°C	Contact		Building industry, repair of spalled concrete structures or grouting of steelwork in rein-forced concrete

Adherends	Test spec.	Test conditions	Test temp. °C	Physical data	Key	Remarks
for M–Bond 0			−269 24 150	1% 4% 2%	k k k	Capable of forming very thin gluelines similar to M–610 (0·005 mm). Highly resistant to moisture and chemicals. Bulletin B–130 details installation procedures
el uminium rain gauge foil terials, e.g. oxy, polyimide			−196 24 +95	1% 6–10% 15%	k k k	Essentially creep-free data are obtained with cure periods as low as 6 h at 24°C. Maximum stability is given by elevated temperature curing. Post cured assemblies are highly resistant to moisture and chemicals
			−196 24 +95	2% 10–15% 15%	k k k	Alternative product AE–15 available for more critical applications. Suitable for multiple gauge assemblies by virtue of longer pot life. Installation procedures for AE–10 and AE–15 are given in Bulletin B–137
for M–Bond -10		*After 6 h cure at 24°C with post curing	−196 24 +95	4% 10–15%* 15–20%	k k k	Optimum performances are obtained with elevated temperature cure. Suitable for irregular surfaces; uneven gluelines readily detectable by non-uniformity of bond colour. Installation procedures are detailed in Bulletin B–137
el ass ncrete amics minium—aluminium			20	1 030 N/cm²	b	
ncrete—concrete			20	690 N/cm²	b	One of a range of bonding materials for outdoor usage. Good weathering resistance. Application at 2–3°C feasible
ncrete nework amic tiles el od ass		Intended for outdoor exposure				Absorbent surfaces may require adhesive priming before bonding. Good resistance to abrasion, impact, sea water and the corrosive action of chemical fumes when bonded structure is treated with Nitowall epoxy coating

Trade source	Trade name or designation	Basic type	No. of compo-nents	Colour	Physical form, consistency or viscosity Ns/m²	Working life	Method of application	Processing		Service temp. range °C	Main uses
								Curing cycle	Bonding pressure N/cm²		
252	Epikote 828*	Epoxy + amine catalyst (Ancamine LT)	2	Pink	Liquid 1·2–1·4	75 min at 25°C (25 g mass)	Spatula	2–7 d at 20°C for 33% w/w catalyst content	Contact	−5 to 60	Repair of concrete, roads and stone surfaces
157	Delta Bond 152	Epoxy + catalyst ETD	2	Amber	Liquid 20	24 h at 25°C	Spatula Extruder	3 h at 25°C to 45 min at at 121°C Post cure for 8 h at 176°C	Contact to 15	−190 to 250 or in-termit-tently at 290	As for Deltabond 152 with ETC catalyst. Adhesive has low viscosity and longer pot life which may favour processing
117	Epoxi-Patch Kit 3X	Epoxy + catalyst	2	Dark Grey	Paste	60–90 min at 25°C for 100 g mass	Spatula	1 h at 60°C to 15 min at 100°C	Contact to 15	−51 to 107	Strain gauges, printed circuits repair. Bonding of metals to ceramics and glass where resistance to therm and mechanical shock is needed. Plastic laminate bonding. Rubber hosing to steel tubes
117	Epoxi-Patch Kit 0151	Epoxy + catalyst	2	Pale Amber	Liquid	90–120 min at 25°C for 100 g mass	Spatula	24 h at 25°C to 2 h at 60°C	Contact to 15	−51 to 107	Resilient adhesive for bonding elasto meric to rigid substrates. Glass bonding where transparent glue-li is required. Optica cement

Adherends	Test spec.	Test conditions	Test temp. °C	Physical data	Key	Remarks
Concrete Stonework		Cured 3 d at 20°C Cured 2 h at 20°C +2 h at 100°C		43°C heat distortion temp. 66°C heat distortion temp 9 000 N/cm² 3–5% 1·6 refractive index	 f k	Applicable at low temperatures (−5°C). Has excellent pigment-wetting properties. Effective under water and suited to application under adverse wet or cold conditions. Good chemical resistance to dilute acetic acid and aqueous media. Ancamine LT curing agent from Trade Source 9
As for Delta Bond 152 with ETC catalyst Aluminium—aluminium		Data refers to resin: catalyst ratio of 77 : 23 Electrical properties dependent on catalyst content After 30 d in water	25 93 160 25 160 25 93 25 100 25 25	6·0 (at 10⁶Hz) 7·0 (at 10⁶Hz) 8·0 (at 10⁶ Hz) 0·03−0·02 (at 60−10⁷ Hz) 0·075 (at 10³ Hz) 156 kV/cm 1·2 × 10¹⁵ Ω/cm 1·2 × 10¹⁴ Ω/cm 0·72 Jm/m²/s/°C 36 × 10⁻⁶ cm/cm/°C 3·25 J for 2·54 cm notch 2 670 N/cm² 482 N/cm² 2 270 N/cm² 9 800 N/cm²	n n n o o m p p q r g b b b c	Similar to ETC cured system above but has better heat resistance. ETD catalyst: resin ratio can be altered to change hardness of cured resin; 15–55% w/w catalyst is recommended. Flexibility is directly proportional to an increase in hardener
Steel (stainless) Epoxy (laminate) Phenolic (laminate) Glass and fibre glass Pyroceram to pyrex Glass to aluminium Polyurethane (foam) Aluminium—aluminium		Thickness 9·17 mm After 70 d at 66°C, 95% RH After 168 h in JP4 fuel	 25 25 25 −45 25 25 82 25	184 kV/cm 3·5 × 10³ (at 100 Hz) 0·03 × 10³ (at 100 Hz) 1·5 × 10⁻⁴ Ω/cm 2·8 × 10⁻⁵ Jm/m²/°C 11·7 × 10⁻⁶ cm/cm/°C 8 900 N/cm² 14 100 N/cm² 1 310 N/cm² 1 380 N/cm² 1 550 N/cm² 275 N/cm² 970 N/cm²	m n o p q r f c b b b b b	Post cure for 12 h at 21°C recommended after 100°C heat cure. Linear shrinkage on curing at 25°C is 0·5%
Steel Glass Ceramics Aluminium—aluminium		Thickness 0·127 mm After 70 d at 66°C, 95% RH	 25 25 25 −45 25 25 82	780 kV/cm 3·35 (at 10⁶ Hz) 0·03 1·0 × 10¹⁴ Ω/cm 2·1 × 10⁻⁵ Jm/m²/s/°C 12·6 × 10⁻⁶ cm/cm/°C 7 580 N/cm² 13 100 N/cm² 1 790 N/cm² 1 340 N/cm² 1 650 N/cm² 275 N/cm²	m n o p q r f c b b b b	Refractive index for film cured 2 h at 60°C is 1·531. Linear shrinkage on curing at 25°C is 0·67%

Trade source	Trade name or designation	Basic type	No. of components	Colour	Physical form, consistency or viscosity Ns/m²	Working life	Method of application	Processing		Service temp. range °C	Main uses
								Curing cycle	Bonding pressure N/cm²		
117	Epoxi-Patch Kit 615	Epoxy + catalyst	2	Pale Blue	Paste at 64	2 min after mixing	Spatula	10 min at 25°C Optional post cure, 10 min at 120°C	Contact to 15	−51 to 107	Rapid bonding adhesive where heat for curing is not generally available. Assembly work with metals and thermo-setting plastics
117	Epoxi-Patch Kit 0266	Epoxy + catalyst	2	Brown	Paste 30% w/w solids	90 min at 25°C for 100 g mass	Spatula	24–36 h at 25°C	Contact to 15	−45 to 93	Metal to metal bonding. Aluminium hangers to honey-comb structures. Epoxy inserts to silver plated brass tubing. Fluorocarbons to steel. High peel strength, structural adhesive bonding
82	Mereco X–305	Epoxy	2	Brown	Liquid	30 s at 20°C	Spatula Dual spray gun	45 s at 20°C	Contact		Rapid assembly of electronic com-ponents instrument parts, printed circuits. Stone setting in jewellery, and as alternative to soldering
82	Meta-bond 315	Epoxy	2	Amber	Liquid 0·45 at 25°C	2–3 d at 20°C	Brush Dip Roller Spray	Dried in air or at 60°C for 30 min + heat cure for 3 h at 60°C to ½ h at 121°C	Contact		Bonding of metals, ceramics, phenolic plastics. Used as a primer coating for encapsulating resin systems and as a protective lacquer

Adherends	Test spec.	Test conditions	Test temp. °C	Physical data	Key	Remarks
eel—steel ·lyester (laminate)—polyester ·lyester (laminate)—polyester uminium—aluminium ·lyester (laminate)—aluminium		 Primed with XP-F861 Primed with XP-F861	30 30 30 25 25 25 25 25	$2 \cdot 7$ (at 10^5 Hz) $0 \cdot 0001$ (at 10^5 Hz) 4×10^4 Ω/cm $3 \cdot 6$ cm/cm/°C 2 040 N/cm^2 830 N/cm^2 1 375 N/cm^2 2 250 N/cm^2 620 N/cm^2	n o p r b b b b b	50% of ultimate bond strength develops in 10 min. Should be mixed within 30 s. Primer XP-F861 available for pre-treating polyester adherends. Linear shrinkage on curing is $1 \cdot 8\%$
eel lver ·ass lass—P.T.F.E. (treated) uminium – aluminium	ASTM D1781–60T	 After 70 d at 66°C, 95% RH	30 30 −45 25 25 82 25	$4 \cdot 3$ (at 100 Hz) $0 \cdot 06$ (at 100 Hz) $2 \cdot 8 \times 10^{-5}$ Jm/m^2/s/°C 50×10^{-6} cm/cm/°C 4 130 N/cm^2 8 250 N/cm^2 2 030 N/cm^2 2 260 N/cm^2 2 340 N/cm^2 290 N/cm^2 $87 \cdot 5$ N/cm	n o q r f c b b b b a	Used to bond brass to glass for use and storage at -184°C Linear shrinkage on curing is $0 \cdot 25\%$
em stones lass eel luminium – aluminium		For glue line thickness, 0·050–0·10 mm, and, Setting time, 5 min Setting time, 120 min Setting time, 240 min Setting time, 7 d Resin : catalyst ratio 10 : 1 (weight or volume) 3 : 1 2 : 1	 20 20 20 20 20 20 20	 1 380 N/cm^2 1 580 N/cm^2 1 650 N/cm^2 3 500 N/cm^2 620 N/cm^2 1 680 N/cm^2 1 820 N/cm^2	 b b b b b b b	Non critical mixing ratio of resin : catalyst. Mixing must be thorough or unreacted base or catalyst functions as a plasticiser for cured adhesive. Allow 10 min before stress loading the cured joint. Bonding surfaces must be free of alkaline substances
eel luminium lumina lass henolic (laminate)	ASTM D257 ASTM D149 ASTM D150 ASTM D150 ASTM D695 ASTM D1002			$1 \cdot 5 \times 10^{14}$ Ω/cm 187 kV/cm $4 \cdot 7$ (at 10^3 Hz) $0 \cdot 07$ (at 10^3 Hz) 4 000 N/cm^2 10–14 N/cm 1 380 N/cm^2	p m n o c a b	Cures to a resilient bond with high peel strength

Trade source	Trade name or designation	Basic type	No. of compo-nents	Colour	Physical form, consistency or viscosity Ns/m^2	Working life	Method of application	Processing		Service temp. range °C	Main uses
								Curing cycle	Bonding pressure N/cm^2		
82	Mereco	Epoxy	1	Dark-red	Paste	Stable at 20°C	Spatula	$2\frac{1}{2}$ h at 121°C	Contact	−70 to 125	Useful adhesive for bonding materials with a wide range of expansion co-efficients
82	Mereco 303	Epoxy	2	Amber	Liquid	1 h at 20°C	Brush	Cured for 5 h at 20°C to 15 min at 80°C (infra-red lamp)	Contact	−60 to 140	
193	EC 2214	Modified epoxy resin	1	Alu-min-ium	Paste 100% w/w solids		Spatula Trowel	40 min at 121°C 7 min at 177°C May be cured up to 204°C	Contact to 17	−56 to 120	One component adhesive for metal bonding
193	EC 2216 B/A	Epoxy resin + amine catalyst	2	Grey	Paste 100% w/w solids	2 h at 24°C (100 g mass)	Spatula	3–7 d at 24°C to 2 h at 60°C to 30–60 min at 93°C	Contact to 17	−69 to 177	Bonding of aluminium to polyester materials

Adherends	Test spec.	Test conditions	Test temp. °C	Physical data	Key	Remarks
Steel Ceramics Phenolics Polyesters Copper				2×10^{16} Ω/cm 156 kV/cm 4·41 (at 10^3 Hz) 0·009 (at 10^3 Hz)	p m o	Formerly called Meta-bond 331. Thixotropic paste curing to a moderately resilient, glossy solid. De-aeration of compound prior to use, for large scale assembly applications is recommended. Resistant to acids, alkalies and many hydrocarbon solvents
Aluminium − aluminium			− 70 25 120	1 950 N/cm² 2 200 N/cm² 1 240 N/cm²	b b b	
Brass − brass			25	1 270 N/cm²	b	
Glass − glass			25	Cohesive failure of glass	b	
Iron Lead Zinc Ceramics	ASTM D257−52T ASTM D149−44 ASTM D150 ASTM D150 ASTM D796−49 ASTM D638−52T			$1·2 \times 10^{14}$ Ω/cm 172 kV/cm 3·1 (at 10^3 Hz) 0·01 (at 10^3 Hz) $8·65 \times 10^{-5}$ cm/cm/°C 34 500 N/cm² 22 000 N/cm²	p m n o r f d	Adhesive has good 'wetting action' and can be applied to moist bonding parts. Resistant to water and chemicals
Aluminium − aluminium			25 121	2 200 N/cm² 412 N/cm²	b b	
Brass − brass			25 82	1 940 N/cm² 790 N/cm²	b b	
Steel (cold rolled) − steel			25 82	1 930 N/cm² 1 030 N/cm²	b b	
Copper − copper			25 82	962 N/cm² 550 N/cm²	b b	
Aluminium − aluminium	ASTM D1002−53T	Cured 40 min at 121°C	− 40 23 82 121 177	2 070 N/cm² 3 450 N/cm² 3 450 N/cm² 1 370 N/cm² 345 N/cm²	b b b b b	High impact and peel strength properties. Bonding pressures required are those sufficient to ensure contact between mating surfaces. Similar to EC 1838 but slightly less resistant to oil. Loses ability to resist flowing or sagging during oven curing if storage exceeds 3 mth at 27°C. Non-flow properties extended, if stored at 4°C, to 8 mth
Steel − steel		Cured 40 min at 121°C	− 40 23 82 121 177	2 070 N/cm² 1 720 N/cm² 1 370 N/cm² 550 N/cm² 137 N/cm²	b b b b b	
Steel−steel		20 gauge metal sheet	23	88 N/cm	a	
Polyester − glass fibre composite Steel − steel			20 100 150	2 350 N/cm²* 2 200 N/cm²* 0*	e e e	Cures to a flexible material. Resistant to hydrocarbon fuels, water brine, anti-freeze solutions, and hydraulic oils
Aluminium − aluminium			20	2 050 N/cm²*	e	

continued

Trade source	Trade name or designation	Basic type	No. of compo-nents	Colour	Physical form, consistency or viscosity Ns/m^2	Working life	Method of application	Processing		Service temp. range °C	Main uses
								Curing cycle	Bonding pressure N/cm^2		
EC 2216 B/A—continued											
133	Double-bond Non-metallic	Epoxy resin + catalyst	2	White	Cream	2–3 h at 22°C (220 g mass)	Spatula	100 min at 35°C to 10 min at 90°C	Contact to 17	120	Metal bonding and sealing adhesive for use where good electrical insulation properties are required
133	Double-Bond Metallic Iron	Epoxy resin + catalyst with iron powder filler	2	Grey	Cream	2–3 h at 22°C (220 g mass)	Spatula	100 min at 35°C to 10 min at 90°C	Contact to 17	120	Metal bonding and sealing adhesive
133	Double-Bond Metallic cream	Epoxy resin + catalyst with aluminium powder filler	2	Grey	Cream	2–3 h at 22°C (220 g mass)	Spatula	12 h at 25°C to 10 min at 90°C	Contact to 17	120	Metal bonding and sealing adhesives
271	DS cement	Epoxy	1	Amber	Liquid		Brush	1 h at 93°C to $\frac{1}{2}$ h at 121°C	10–14	−130 to 316	Strain gauges

Adherends	Test spec.	Test conditions	Test temp. °C	Physical data	Key	Remarks
	ASTM D1002–53T	Cured 7 d at 24°C	−55	1 380 N/cm²	b	
			24	2 050 N/cm²	b	
			82	345 N/cm²	b	
			121	240 N/cm²	b	
			−55	5 N/cm	a	
			24	36 N/cm	a	
		After,	82	5 N/cm	a	
		28 d in air at 71°C	24	3 250 N/cm²	b	
		28 d in air at 71°C	24	42 N/cm	a	
		7 d in JP4 fuel	24	2 270 N/cm²	b	
		100% RH at 49°C	24	1 860 N/cm²	b	
				*Sira evaluation		
Metals Ceramics				Conforms to AID Spec. DTD 900/4572 10 kV/cm	m	Maximum bond strength, for room temperature cure, in 24 h 50% decrease in strength when exposed at 150°C for 1 h. Resistant to oils, water, mild acids and strong alkalies. Trichloroethylene softens the cured material. Not suitable for polyolefine plastics, fluorocarbons or plasticised vinyl polymers. Available in other colours. (Yellow, blue, black, green)
				$3 \cdot 5 \times 10^6$ Ω/cm at 500 volts	p	
				$3 \cdot 3 \times 10^{-5}$ cm/cm/°C	r	
				6 600 N/cm²	c	
				2 320 N/cm²	d	
				0·68J per 2·54 cm notch	g	
Aluminium – aluminium		Catalyst : resin ratio, 1 : 1		85 Shore D	h	
Steel (mild) – steel		Cured 1 h at 100°C	20	1 720 N/cm²	b	
			20	1 500 N/cm²	b	
Metals Ceramics				6 100 N/cm²	c	Non-conductive material with 20 MΩ surface resistivity. Similar resistance properties to Double Bond Non-metallic grade. Available in putty and pourable consistency grades
				0·68J per 2·54 cm notch	g	
Aluminium – aluminium		Catalyst : resin		3 300 N/cm²	d	
		ratio, 1 : 1	20	1 800 N/cm²	b	
Steel (mild) – steel		Cured 1 h at 100°C	20	1 850 N/cm²	b	
Metals Ceramics				Conforms to AID Spec. DTD 900/4572		Available with alternative metal fillers (iron, lead, steel, zinc, copper). Resistance properties are similar to the Non-metallic grade. Available in putty and pourable consistency grades
				$3 \cdot 3 \times 10^{-5}$ cm/cm/°C	r	
				5 150 N/cm²	c	
				2 960 N/cm²	d	
				0·68J per 2·54 cm notch	g	
Aluminium – aluminium		Catalyst : resin ratio of 2 : 3 Cured 10 min at 100°C	20	1 580 N/cm²	b	
Steel (mild) – steel			20	1 650 N/cm²	b	
Metals						Should be stored in refrigerator. Above 94°C the assembly should be held for 1 h before applying the load. Creep takes place above 260°C for long term load cycles

Trade source	Trade name or designation	Basic type	No. of compo-nents	Colour	Physical form, consistency or viscosity Ns/m^2	Working life	Method of application	Processing		Service temp. range °C	Main uses
								Curing cycle	Bonding pressure N/cm^2		
271	Denex	Epoxy with an-hydride-dianhydride catalyst	2	Amber	Liquid	30 min at 71°C to 5 h at 20°C to 7 d at 26°C	Brush	1 h at 107°C+ 1 h at 177°C	10–15	204	Strain gauges
271	Denex No. 3	Epoxy in a solvent	1	Amber	Liquid	20 min at 20°C	Brush	Air dried 20 min at 20°C and cure for, 1 h at 121°C+ 2 h at 177°C followed by post cure 1 h above service temp.	10–15	260	Strain gauges for operation up to 260°C
271	Denex No. 4		2	Amber	Liquid	3 min at 20°C	Brush	Air dried 3 min at 20°C and cured for 1 h at 150°C	10		Strain gauges with high elongation properties
286	BR–600	Epoxy resin in solvent + catalyst	3	Yellow	Liquid	2 wk at 24°C or 6 mth at 32°C	Brush	8 h at 52°C to $\frac{1}{2}$ h at 121°C	4–40	−270 to 371	Strain gauges for cryogenic and elevated tempera-ture use. Micro-measurement strain gauges
286	BR–610	Epoxy resin in solvent + catalyst	2	Yellow	Liquid	4 wk at 24°C or 1 yr at 32°C	Brush	5 h at 112°C to $\frac{1}{2}$ h at 224°C	4–40	−270 to 371	Strain gauges as for BR 600 but more suited to transducer work requiring thinner glue lines

Adherends	Test spec.	Test conditions	Test temp. °C	Physical data	Key	Remarks
Metals						Bond is suitable for static strain measurements up to 150°C and dynamic ones up to 260°C. Past curing at 204°C for 1½ h raises static strain measurement temperature to 204°C; an above 204°C service temperature requires a post cure at 316°C
Metals						Has lower creep at elevated temperatures than DS cement grade. For transducer work where only negligible creep can be tolerated, post cure is extended to several hours
Metals						High elongation cement
Aluminium Magnesium } alloys for elevated temperature service			24 −268	Exceeds 5% 2%	k k	Three component kit is mixed on receipt and stored in bottle. Cured material resists outgassing in high vacuum. Transducer applications require post cure for 2 h at 38–56°C above maximum operating temperature
Aluminium alloys Magnesium alloys						Two component kit is mixed on receipt and stored in bottle. Similar to BR 600 but slower in reaction rate; requires longer cure cycles. Easier to handle and less sensitive to conditions of use than BR 600 for transducer work. Transducer applications require post curing at 38–56°C above maximum service temperature

Trade source	Trade name or designation	Basic type	No. of compo-nents	Colour	Physical form, consistency or viscosity Ns/m^2	Working life	Method of application	Processing		Service temp. range °C	Main uses
								Curing cycle	Bonding pressure N/cm^2		
105	Eccobond 285/24 HV	Epoxy	2	Grey	Paste	45 min at 21°C	Spatula	Sets in 6 h at 20°C with full cure in 48 h or 3 h at 50°C	Contact to 15	−198 to 149	Metal and plastic bonding for pipe and tube applications. Building and plumbing work. Low temperature structural adhesive
105	Eccobond 45 LV	Epoxy	2	Black	Liquid	3 h	Brush Spatula Roller	24 h at 20°C to 15 min at 100°C	Contact to 7	−57 to 150	Metals, glass ceramics, plastics
105	Eccobond 55	Epoxy	2	Off White	Liquid 8	4 h	Brush Spatula Roller	6 h at 20°C to $\frac{1}{2}$ h at 150°C	Contact to 7	−57 to 150	General purpose adhesive for metals, glass, ceramics
105	Eccobond PDQ	Epoxy	2		Liquid Density of 1·6 g/cc	1 min for 20 g mass	Spatula	2 min at 20°C	Contact to 7		Fast curing adhesive for production line use metals, glass, ceramics

Adherends	Test spec.	Test conditions	Test temp. °C	Physical data	Key	Remarks
opper				$1·55$ Jm/m^2/s/°C	q	Good resistance to vibration at 4 kHz. Pipework joints withstand water pressure of 240 N/cm^2 at 93°C when cycled 3×10^6 times. Withstands immersion in water for 30 d without loss of strength.
rass				27×10^{-6} cm/cm/°C	r	
agnesium			20	10^{16} Ω/cm	p	
henolic resin			93	10^{13} Ω/cm	p	
olyacetals				10 300 N/cm^2	f	
olycarbonates			−57	1 240 N/cm^2	b	
olychloroprene			20	1 440 N/cm^2	b	
lyvinyl chloride			93	965 N/cm^2	b	
etal (unspecified) — metal						
teel (stainless) — platinum			−195	1 060 N/cm^2	b	Cryogenic adhesive for transducer, strain gauge and structural assembly work
			25	520 N/cm^2	b	
		After 4 cycles, −195 to 25°C	25	460 N/cm^2	b	
Aluminium — platinum			−195	770 N/cm^2	b	
			25	282 N/cm^2	b	
		After 4 cycles, −195 to 25°C	25	185 N/cm^2	b	
steel (stainless)—platinum			−195	6 150 N/cm^2	d	
			25	880 N/cm^2	d	
		After 4 cycles, −195 to 25°C	25	810 N/cm^2	d	
ira evaluation						
eoprene				160 kV/cm	m	Flexibility controlled by catalyst content. Less viscous than Eccobond 45. Designed for use where shock and peel resistance required. Available in other colour grades. Remove surplus adhesive with toluene or trichloroethylene
olyamides				3×10^{13} Ω/cm	p	
uminium				$3·2-2·9$ (at 10^2-10^9 Hz)	n	
olystyrene				$0·03-0·04$ (at 10^2-10^9 Hz)	o	
olythene	ASTM C293–57	Data on cured adhesive 1:1 ratio of resin: catalyst		3 800 N/cm^2	f	
				5·4J/2·54 cm notch	g	
	ASTM D1706–61			Shore D 40	l	
			20	2 200 N/cm^2	b	
uminium — aluminium		After 30 d in water	20	1 940 N/cm^2	b	
				172 kV/cm	m	Rigid adhesive. Available in other colour grades. Remove surplus material with trichloroethylene
			20	5×10^{14} Ω/cm	p	
			93	3×10^{11} Ω/cm	p	
				59×10^{-6} cm/cm/°C	r	
				$2·9-3·3$ (at 10^2-10^{10} Hz)	n	
				$0·01-0·02$ (at 10^2-10^{10} Hz)	o	
	ASTM C293–57			10 000 N/cm^2	f	
uminium — aluminium		After 30 d in water	20	1 380 N/cm^2	b	
	ASTM D696–44			11×10^{-5} cm/cm/°C	r	Possible to apply resin to one adherend and catalyst to other for bonding procedure., i.e. where limited pot life is a disadvantage
uminium — aluminium			20	620 N/cm^2	b	

Trade source	Trade name or designation	Basic type	No. of compo-nents	Colour	Physical form, consistency or viscosity Ns/m^2	Working life	Method of application	Processing		Service temp. range °C	Main uses
								Curing cycle	Bonding pressure N/cm^2		
105	Eccobond 32F	Epoxy	2	Black	Heavy cream	1 h for 50 g mass	Spatula	2 d at 0°C to 8 h at 20°C	Contact to 7		Low temperature curing adhesive intended for outdoor and construction work
105	Eccobond 88	Epoxy	1	Black	Thick paste	6 mth	Spatula	2 h at 150°C to 15 min at 260°C	Contact to 7	−57 to 204	Metals Glass Ceramics Plastics
105	Epoxy 98	Epoxy	1	Amber	Powder	6 mth	Dusted or spread on surface	40 min at 177°C to 3 h at 121°C	Contact to 7	−57 to 204	Metal Glass Ceramics Plastics
105	Eccobond Solder 56C	Epoxy + Silver filler	1	Silver	Smooth paste	Up to 8 h	Spatula	2 h at 50°C to 3 min at 93°C	Contact to 7	−53 to 180	Plastic cement with low electrical resistance. Electrical connection where hot soldering is impractical. Metal Ceramics.
105	Eccobond 60L	Epoxy + filler	2	Black	Smooth paste	2 h	Spatula	4 h at 20°C to 15 min at 150°C	Contact to 7	−53 to 180	General purpose bonding where electrical conductivity must be maintained. Waveguide applications Electronic circuitry
105	Eccobond solder 59	Epoxy + Silver filler + Silicone rubber	1	Silver	Liquid	20 min to tacky state	Brush Spatula	6 h at 150°C	Contact to 7	260°C	Applications requiring highly conductive, flexible bonding

Adherends	Test spec.	Test conditions	Test temp. °C	Physical data	Key	Remarks
...l ...ss ...od ...crete ...minium – aluminium	ASTM D1706–61		 20	 Shore D 85 690 N/cm²	 l b	Cures in the presence of moisture (rain)
...minium – aluminium	ASTM C293–57	 After 30 d in water	 93 −57 93 20	160 kV/cm 10¹² Ω/cm 4·5–4·2 (at 10²–10¹⁰ Hz) 11 000 N/cm² 1 920 N/cm² 1 500 N/cm² 2 500 N/cm²	m p n f b b b	One part adhesive with non-flow properties. Shelf life extended at 0°C. Shear strength retained after immersion in water for 1 mth
...minium – aluminium	ASTM C293–57	 After 30 d in water	 −57 93 20	168 kV/cm 10¹⁶ Ω/cm 4·5–4·2 (at 10²–10¹⁰ Hz) 54 × 10⁻⁶ cm/cm/°C 10 500 N/cm² 1 920 N/cm² 1 450 N/cm² 3 100 N/cm²	m p n r f b b b	One part adhesive which melts on moderate heating. Coat thickness can be built up
...minium ...hrome wire ...ductive plastics ...specified metal – metal			 20	2·2 × 10⁴ Ω/cm 8 400 N/cm² 0·15 Jm/m²/s/°C 34 × 10⁻⁶ cm/cm/°C 310 N/cm²	p f q r b	May be thinned with toluene 20% w/w for thin film application. Curing at 65°C preferred for low resistivity. Good resistance to water and chemicals
...l ...hrome wire ...ductive materials ...minium – aluminium		 After 30 d in water	20 93 −53 93 20	50 Ω/cm 50 Ω/cm 4 100 N/cm² 0·007 Jm/m²/s/°C 49 × 10⁻⁶ cm/cm/°C 5000 × 10⁶ Hz 40 × 10¹⁰ Hz 15 × 10⁶ Hz 0·6 × 10¹⁰ 1 580 N/cm² 760 N/cm 2 100 N/cm²	p p f q r n n o o b a b	Not recommended where precise resistive properties are required. Toluene can be used to thin the material. Cure volumes exceeding 6·5 cm³ at 20°C are not recommended. Bulk resistivity is higher than Eccobond 56C grade. Good resistance to water and chemicals
...specified metal – metal			 20	0·001 Ω/cm 0·06 Jm/m²/s/°C 63 × 10⁻⁶ cm/cm/°C 210 N/cm²	p q r b	Evaporation of solvent leaves permanently tacky adhesive. Material may be catalytically cured to a rigid state (add 2% w/w catalyst 59) by heat curing. Uncatalysed bonds can be resealed. Thin with toluene if required

Trade source	Trade name or designation	Basic type	No. of components	Colour	Physical form, consistency or viscosity Ns/m^2	Working life	Method of application	Processing		Service temp. range °C	Main uses
								Curing cycle	Bonding pressure N/cm^2		
151	Ablebond 20–1	Epoxy resin with silver filler	1	Grey	100% solids smooth paste. Sp. gr. = 2·4 g/ml	Indefinite	Syringe, silk screening, or automatic dispenser	1 h at 121°C to 0·5 h at 150°C	Contact to 4		Microelectronics chip bonding and other applications requiring electric conductive gluel
151	Ablebond 58–1	Epoxy resin with gold filler	1	Black	100% solids, smooth paste. Sp. gr. = 2·3 g/ml	Indefinite	Syringe, silk screen, or automatic dispenser	1 h at 125°C to 0·5 h at 150°C	Contact to 4		Designed specifically for microelectronics chip bonding
151	Abletherm 8–2	Modified epoxy resin with amine hardener	2	Blue	Medium paste. Sp. gr. = 2·3 g/ml	Exceeds 4 h at 25°C	Spatula	45 min at 120°C or 4 h at 73°C or 24 h at 60°C	Contact to 4		For instrument applications where high thermal conductivity is required. Bonding of noble and other metals
105	Eccobond 76	Epoxy	2	Black	Thick paste	1 wk	Spatula	3 h at 105°C + 3 h at 150°C	Contact to 70	−53 to 260 Short period at 315°	Metal structural sealant for aircraft industry. Ceramic and high temperature plastic composite bonding

Adherends	Test spec.	Test conditions	Test temp. °C	Physical data	Key	Remarks
minium—aluminium	MMM–A–132		20	826 N/cm²	b	Thixotropic paste. Non-corrosive towards metals, and low out-gassing under reduced pressure are features of the cured adhesive
d—gold	"		20	1 205 N/cm²	b	
er—silver	"		20	757 N/cm²	b	
ss—brass	"		20	999 N/cm²	b	
pper—copper	"		20	827 N/cm²	b	
		After 1 000 h in water	20	0·0003 Ω/cm	p	
			20	0·0002 Ω/cm	p	
			20	0·0001 Ω/cm	p	
			205	0·0002 Ω/cm	p	
			315	0·0004 Ω/cm	p	
			400	0·0017 Ω/cm	p	
			120	$0·173$ J mm^{-2}s^{-1} $1°C^{-1}$	q	
minium						Non-corrosive towards aluminium
er						
pper	MMM–A–132		20	1 653 N/cm²	b	
d—gold			150	0·0008 Ω/cm	p	
			200	0·0008 Ω/cm	p	Storage life is 2 weeks at 25°C. Available in 2–14 g containers
			300	0·0005 Ω/cm	p	
			400	0·0005 Ω/cm	p	
		After heating to 400°C	25	0·0001 Ω/cm	p	
d—gold (plated steel)	MMM–A–132		25	2 273 N/cm²	b	Cured bonds withstand continuous exposure to 150°C and intermittent exposure to 200°C. Unusually high thermal conductivity although adhesive contains no toxic beryllia filler. Storage life is 1 year at 25°C. Available as premixed, frozen product ($-5°C$) in 2·5–68 g containers
er—silver (plated brass)			25	1 894 N/cm²	b	
ss—aluminium			25	1 894 N/cm²	b	
pper—copper			25	1 446 N/cm²	b	
der—solder (% Sn + 37% Pb)			25	1 688 N/cm²	b	
minium—aluminium			25	1 805 N/cm²	b	
minium—aluminium			65	1 908 N/cm²	b	
minium—aluminium			120	668 N/cm²	b	
minium—aluminium			205	296 N/cm²	b	
	ASTM D257–61		25	3×10^{14} Ω/cm	p	
	ASTM D150–59T	At 1 000 Hz	25	6·07	n	
		At 1 000 Hz	25	0·013	o	
			−20 to 25			
			25 to 94			
			121	$0·151$ J mm^{-2}s^{-1} $1°C^{-1}$	q	
			20	10^{16} Ω/cm	p	Power mixing of paste preferred. Stainless steel bonding improved by a post cure of 5 h at 260°C. Better flow of adhesive on hot surfaces. Good resistance to chlorinated hydraulic fluids
			260	10^{12} Ω/cm	p	
el (stainless)				156 kV/cm	m	
uminium				4·4–4·1 (at 10^2–10^{10} Hz)	n	
umina				<0·02 (at 10^2–10^{10} Hz)	o	
				9 600 N/cm²	b	
			260	1050 N/cm²	b	
uminium—aluminium		After 30 d in water	20	2 950 N/cm²	b	

Trade source	Trade name or designation	Basic type	No. of components	Colour	Physical form, consistency or viscosity Ns/m^2	Working life	Method of application	Processing		Service temp. range °C	Main uses
								Curing cycle	Bonding pressure N/cm^2		
105	Eccobond 104	Epoxy	2	Black	Medium thick paste. Density is 1·5 g/cc	24 h	Spatula	6 h at 120°C to 1 h at 200°C	Contact to 70	240° Short period at 290°	High strength bonding for porous and non-porous materials at high temperature
162	Kelseal 12·091	Epoxy	2	Light Brown	Paste	4 h	Trowel	Cold curing in 24 h	Contact to 7		Building applications
162	Kelseal* 18·188	Epoxy	2		Filled resin Density 14·2 kg/l		Cartridge gun, Machine	20 min at 155°C	Contact to 7		Metals with clean or oily surfaces
173	Trinasco J resin	Epoxy	2	Green on mixing	Pourable slurry. May be sand filled (BS25/200 grade)	Initial set in 4 h after mixing	Trowel	12 h at 20–25°C 24 h at 10–15°C	Contact to 7		Pile jointing and repair work under outdoor conditions
133	Trinasco A resin	Epoxy	2	Grey on mixing	Paste	30 min	Trowel	24 h at 20–25°C	Contact to 7		Building and civil engineering work
105	Eccobond 45 (clear)	Epoxy	2	None	Liquid 4	2–3 h	Spatula Brush Roller	8 h at 20°C 45 min at 70°C 15 min at 100°C	Optional	−57 to 121	Metals, glass ceramics and many plastics where flexible bonding required

Adherends	Test spec.	Test conditions	Test temp. °C	Physical data	Key	Remarks
minium el (stainless) ss ss mina rmoset plastics minium – aluminium			24 177 24 150 230 290	10^{15} Ω/cm 10^{13} Ω/cm 153 kV/cm 49×10^{-6} cm/cm/°C 1 730 N/cm² 1 240 N/cm² 689 N/cm² 34 N/cm²	p p m r b b b b	Post cure at 260°C for 1 h for optimum properties. Excellent resistance to acids, alkalies, organic solvents and fuels
crete sonry		Outdoor exposure	20	1 380 N/cm²	d	Fully cured in 3 d. Flexible, gap sealant
minium – aluminium l (mild) – steel		0·08 cm glueline 0·16 cm glueline	20 20	21 N/cm² 16 N/cm²	b b	Resistant to water, dilute acids, alkalies. Shelf life 3 mth. 2% volume shrinkage on curing. Withstands vibration
crete – concrete		Underground exposure	15	4 470 N/cm² 5% elongation at yield 8 700 N/cm² When sand filled (20% w/w), system gives 2 750 N/cm² 6 850 N/cm²	d k f d c	Employs non-dermatitic hardener. Strength data shown obtained after 7 d cure at 15°C. Coverage 0·22 m² for 9·5 mm glueline
crete ks s od	ASTM C307–55 ASTM C293–57 ASTM D790–61 BS2782–303A ASTM D1706–61	Strength data on cured adhesive material	 20 0 20 0 20 0	60×10^{-6} cm/cm/°C 965 N/cm² 1 210 N/cm² 2 620 N/cm² 3 800 N/cm² 350 N/cm² 6 500 N/cm² 6 800 N/cm² Shore D90, Barcol 80	r d d f f j c c l	Special hardeners available for 2–4 h curing at 15°C. Deflection temperature is 65°C (BS 2782, 102G). Resistant to acids, oils and water. Coverage is 0·6 kg/m²
(mild) – steel ninium – aluminium			20 20	345 N/cm² 270 N/cm²	b b	
ninium steel (stainless) nesium styrene thene tetrafluorethylene acetal prene ninium – aluminium	 ASTM D1706–61	Data on cured adhesive 1 : 1 resin to catalyst ratio	 20	72×10^{-6} cm/cm/°C 121 kV/cm 10^{13}–10^{14} Ω/cm 3·2–2·9 (at 10^2–10^{10} Hz) 0·03–0·06 (at 10^2–10^{10} Hz) Shore D 80 1 650 N/cm²	r m p n o l b	Flexibility of system depends on catalyst content. Useful for bonding two materials with different thermal expansions. Good vibration damper and sustains high G forces. Remove surplus adhesive with toluene, acetone

Trade source	Trade name or designation	Basic type	No. of compo-nents	Colour	Physical form, consistency or viscosity Ns/m^2	Working life	Method of application	Processing		Service temp. range °C	Main uses
								Curing cycle	Bonding pressure N/cm^2		
45	Epophen EL-5	Epoxy + catalyst EHT-24	2	Amber	Thick liquid 7·5–17·5	1–1$\frac{1}{2}$ h at 21°C for 120–1 000 g batch	Spatula	16 h at 25°C + optional post cure 8 h at 60°C	Contact to 15	80	Designed specific for bonding glass room temperature Decorative glass panels, double glazing window c struction. Glass t metal or rubber assemblies
45	Epophen EL-5	Epoxy + catalyst EHT-3	2	Amber	Liquid 10 at 25°C	1$\frac{3}{4}$ h at 24°C for 115–230 g batch	Spatula	16 h at 25°C to $\frac{1}{2}$ h at 120°C	Contact to 15		Bonding of porou and non-porous materials based o glass, rubber, ceramics, phenoli laminates
157	Delta Bond 152	Epoxy + catalyst ETC	2	Amber	Paste of high thixo-tropy	4 h at 25°C	Spatula Machine dispenser	36 h at 25°C to 2 h at 121°C. Post cure for 16 h at 176°C	Contact to 15	−190 to 190 or in-termit-tently at 230	Thermally condu tive epoxy resin system for semi-conductors, heat sinks and bondin of assemblies re-quiring high hea conductivity. Vacuum sealing agent. Cryogenic adhesive

Adherends	Test spec.	Test conditions	Test temp. °C	Physical data	Key	Remarks
— glass		Cured 16 h at 25°C and then immersed in water at 25°C for				Resin: catalyst ratio is 100 : 60 ET-2A is an alternative resin to EL-5 where thixotropic gap filling mix is required.
		3 d	25	2 070 N/cm²*	b	
		14 d	25	2 960 N/cm²*	b	Post curing for 8 h at 60°C
		50 d	25	1 180 N/cm²*	b	improves bond strength.
		Immersed in water at 99°C for				Performance data on glass specimens 6 mm thick over-
		1 h	25	1 310 N/cm²*	b	lapped for an area 6·4 cm².
		7 h	25	1 460 N/cm²*	b	Glue-line thickness of 0·1 mm
		3 wk (thermal cycling 8 h at 100°C + 16 h at 25°C)	25	330 N/cm² *cohesive failure of glass	b	
ls	BS 903		20	4·12–4·07 (at 10^3– 10^4 Hz)	n	Resin: catalyst ratio is 100 : 65. Cured resin has good resistance
nics			60	4·2–4·1 (at 10^3–10^4 Hz)	n	to moisture
	BS 903		20	0·009–0·005 (at 10^3–10^4 Hz)	o	
			60	0·01–0·01 (at 10^3–10^4 Hz)	o	
	BS 2782/ 202A		20	$2·4 \times 10^{15}$ Ω/cm	p	
				56×10^{-6} cm/cm/°C	r	
				0·188 Jm/m²/s/°C	q	
	BS 2782		20	6 500 N/cm²	d	
	BS 2787/ 304B			10 300 N/cm²	f	
	BS 2787/ 303A			13 800 N/cm²	c	
(mild) — steel		Cured 16 h at 25°C	20	690 N/cm²	b	
			20	123 N/cm	a	
inium er		Data refers to a resin: catalyst ratio of 95·5 : 4·5	25	6·0 (at 10^6 Hz)	n	Good thermal shock performance between −55°C and
			93	7·0 (at 10^6 Hz)	n	155°C. A permanent copper—
			160	8·0 (at 10^6 Hz)	n	copper seal is maintained at
			25	0·03–0·02 (at 60–10^7 Hz)	o	liquid helium temperatures
nics			160	0·075 (at 10^3 Hz)	o	(−198·5°C). Withstands high
y laminates						vacuum conditions without
olic laminates		Thickness, 3·17 mm		156 kV/cm	m	outgassing and shows good
d			25	$2·4 \times 10^{15}$ Ω/cm	p	resistance to acids and alkalies
			93	$4·0 \times 10^{14}$ Ω/cm	p	at 25°C (25 % w/w of reagent),
				0·91 Jm/m²/s/°C	q	hot fuel oils, trichloroethylene
				18×10^{-6} cm/cm°C	r	ammonia and alcohols. Post
				0·325 J for 2·54 cm notch	g	cure at 176°C recommended to realise optimum performance.
inium — aluminium			25	1 650 N/cm²	b	Catalyst requires warming at
			100	1 030 N/cm²	b	37°C before use
		After 30 d in water	25	1 650 N/cm²	b	
			25	11 400 N/cm²	c	

Trade source	Trade name or designation	Basic type	No. of components	Colour	Physical form, consistency or viscosity Ns/m^2	Working life	Method of application	Processing		Service temp. range °C	Main uses
								Curing cycle	Bonding pressure N/cm^2		
117	Mhobond M20	Epoxy + catalyst	2	Silver	Paste	10–20 min at 25°C	Spatula	24 h at 20°C to 2 h at 60°C	Contact	to 150	High electrical conductivity adhesive. Attachment of cathodic protect systems to burie metal pipes
222	Quentbond	Epoxy + catalyst	2	Trans-lucent	Liquid	4 h at 25°C	Spatula Brush	12 h at 20°C to ½ h at 100°C	Contact to 15	100	Civil engineerin and general pur structural adhes Aircraft and au motive industry Sealing of electr components. Concrete repair
47	Bostik 2000	Epoxy	2	Brown	Heavy liquid 100% solids	1–2 h at room temp. for mixture	Brush Pressure nozzle	16 h at 25°C to 20 min at 70°C	Contact to 4		Domestic and industrial bondi of metallic and non-metallic materials (plasti ceramics, rigid plastics). Applications for cutlery, shears, garden equipme etc.
47	Bostik 2001	Epoxy	2	Grey	Medium thick liquid 100% solids	2–2·5 h at 20°C	Knife blade	24 h at 20°C to 1 h at 50°C	Contact to 4		Engineering applications in aircraft, electron and domestic appliance indus
47	Bostik 2003	Epoxy polyamide	2	Grey	Thick liquid 100% solids. Sp. gr. 1·6	2 h at 20°C	Trowel blade	24 h at 20°C	Contact	−30 to 120	Bonding applica tions in civil engineering; me window frames, sandwich panels tiles in exterior cladding

Adherends	Test spec.	Test conditions	Test temp. °C	Physical data	Key	Remarks
		Cured 24 h at 20°C	22	0·013 Ω/cm	p	High conductivity is main-
		Cured 2 h at 60°C	22	0·002 Ω/cm	p	tained at elevated temperatures.
		Cured 2 h at 60°C	90	0·002 Ω/cm	p	Optimum performance when
		Cured 2 h at 60°C	132	0·010 Ω/cm	p	heat cured
		Cured 2 h at 60°C	140	0·100 Ω/cm	p	
		Aged 8 d at 150°C	22	0·0007 Ω/cm	p	
		Aged 8 d at 150°C	150	0·0009 Ω/cm	p	
				8.4×10^{-5} Jm/m^2/s/°C	q	
␣minium – aluminium		Cured 2 h at 60°C	22	620 – 825 N/cm^2	b	
␣estos				168 kV/cm	m	At 100°C, 30% of full strength
␣amics				1.1×10 Ω/cm	p	is realised. Resistant to weak
␣ncrete				5 420 N/cm^2	d	acids and alkalies, detergents,
␣ss				10 300 N/cm^2	f	hydrocarbon solvents, petrol
␣phite				6 720 N/cm^2	c	and hydraulic oils
				8×10^{-5} cm/cm/°C	r	
␣minium – aluminium		After ½ h cure at 100°C	22	1 720 N/cm^2	b	
␣ss – aluminium		After ½ h cure at 100°C	22	860 N/cm^2	b	
␣romium – aluminium		After ½ h cure at 100°C	22	1 370 N/cm^2	b	
␣d/Tin solder – aluminium		After ½ h cure at 100°C	22	515 N/cm^2	b	
␣kel – aluminium		After ½ h cure at 100°C	22	1 370 N/cm	b	
␣gnesium – aluminium		After ½ h cure at 100°C	22	1 370 N/cm^2	b	
␣er – aluminium		After ½ h cure at 100°C	22	1 370 N/cm^2	b	
␣el (stainless) – aluminium		After ½ h cure at 100°C	22	1 720 N/cm^2	b	
␣c – aluminium		After ½ h cure at 100°C	22	412 N/cm^2	b	
␣pper – aluminium		After ½ h cure at 100°C	22	860 N/cm^2	b	
␣lyethylene terephthalate – ␣luminium		After ½ h cure at 100°C	22	Cohesive failure of polymer	b	
␣rylic (rigid) – aluminium		After ½ h cure at 100°C	22	550 N/cm^2	b	
␣ythene – aluminium		After ½ h cure at 100°C	22	685 N/cm^2	b	
␣gsten carbide – aluminium		After ½ h cure at 100°C	22	1 210 N/cm^2	b	
␣el – steel		Shear tested on 25·4 ×	20	620 N/cm^2	b	Good resistance to water, oil,
␣minium – aluminium		25·4 mm^2 overlap	20	480 N/cm^2	b	fuels, and humidity. Moderate
␣od – wood		joint at 25·4 mm/min	20	480 N/cm^2*	b	resistance to acids and solvents.
␣ss – glass		rate	20	1 515 N/cm^2*	b	Brushable version of Bostik
␣amics				* exceeds material		2001.
␣stics (rigid)				strength		Coverage 1 000 cm^3/l. Applied as a ribbon
␣for Bostik 2000		As for Bostik 2000		Similar performance to Bostik 2000		25·4 mm wide and 0·25 mm thick by 154 m length is yielded by 1 l of material
␣amics						Good resistance to oil, organic
␣minium						solvents, dilute acids and
␣ncrete						alkalies. Bostik Thinner M574 or 6020 available. Coverage 1·8 m^2/kg

Trade source	Trade name or designation	Basic type	No. of components	Colour	Physical form, consistency or viscosity Ns/m^2	Working life	Method of application	Processing		Service temp. range °C	Main uses
								Curing cycle	Bonding pressure N/cm^2		
172	UHU Plus	Epoxy resin + Polyamide hardener	2	Amber	Liquid 35 100% w/w solids	1–1·5 h at 20°C after mixing	Spatula	45 min at 70°C to 10 min at 100°C to 5 min at 150–180°C	Contact to 4		For structural bonding of metals and thermosetting plastics, ceramics, and glass in domestic and industrial applications
47	Bostik 2024	Epoxy-polysulphide	2	Grey	Non-flow thixotropic paste	1·5–2 h at 20°C	Pressure gun Spatula	16 h at 20°C Optimum cure after 24 h	Contact	−30 to 75	Cold curing adhesive sealant for interior and exterior usage
18	Roof Bow 3J6	Epoxy polysulphide	2	Light grey	Paste. Sp. gr. 1·45	25 min for 454 g mixture	Spatula Trowel Automatic equipment	Sets at room temp. Handling strength 4h. Full strength 3–7 d	Contact	−35 to 93	Adhesive-sealant for caulking joints based on metals, masonry insulation material. Roof bow assembly aluminium and wood channelling and frameworks. Double-glaze window manufacture
50	Bakelite Cement J11185	Phenolic resin modified with butyral	1	Orange to Brown	Liquid	Indefinite	Brush	Dried for $\frac{1}{2}$–2 h at 60–80°C then cured for 18 h at 100°C to 20 min at 149°C under pressure	40–70	100	Phenolic stoving cement for metals, plastic laminates and inorganic substrate. Metal foil to plastic laminates as in printed circuit boards
73	Araldite AY 105 + HY 953F	Epoxy polyamide	2	Light Brown	Medium viscosity liquid 120 at 21°C	$2\frac{1}{2}$ h at 20°C	Spatula	24 h at 20°C to 3 h at 60°C to 20 min at 100°C	Contact to 140	−55 to 70	Bonding metals, ceramics, rubber and reinforced plastics

Adherends	Test spec.	Test conditions	Test temp. °C	Physical data	Key	Remarks
ss ramics enolic lamine lyester oxy resins minium—aluminium	DIN 53454 DIN 52612 —	On 10 mm cube At 100 V Overlap joints 25 × 1·5 mm with 10 mm overlap. 1:1 volume ratio of resin: hardener	20 28·3 21 10 30 50 80 100	Up to 67 N/mm² 0·248 J mm⁻²s⁻¹ 1°C⁻¹ 56–58 Ω/cm 2745 N/cm² 1716 N/cm² 981 N/cm² 392 N/cm² ∼340 N/cm²	c q p b b b b b	Unsuitable for thermoplastics such as PVC, polystyrene, polyethylene. Preparation of hardener varied to produce harder or softer cured products. Resin: hardener ratio variable to enhance rigidity or flexibility of glueline. Flash Pt. above 210°C
ss tals ncrete od			20	90° (BS Hardness)		Non-slump, low shrinkage sealant which is resistant to oil, fuels, and mild acids and alkalies. Low moisture permeability
od ss amics minium—aluminium	FSTM 79 FSTM 90b " " "	After 50 min at test/cure temperature specified	 20 60 70 82	50% 70–75 (Shore A) 516 N/cm² 165 N/cm² 137 N/cm² 113 N/cm²	k l d d d d	Optional heat cure at 80°C for 15 min. Cures without shrinkage to give resilient joints with good resistance to solvents, oils, mineral acids
od phite estos nolic (laminate) pper l—steel l—steel l—steel minium (alloy)—aluminium				 4 270 N/cm² 1 290 N/cm² 1 510 N/cm² 2·1 J for 2·54 cm notch 3 470 N/cm²	 d b b g d	Bonding pressures up to 1 000 N/cm² may be needed for joining copper foil to laminate sheets. Adhesive may be thinned with methylated spirits in ratio 1 : 1
estos (rigid) amics ss fibre composites (polyester) ss fibre composites (epoxy) bon (+graphite) ytetrafluorethylene yacetals (treated) vester (treated) ystyrene (treated) volefines (treated) ber (treated) pper (treated) gsten carbide omium (treated) d, Tin, Solder gnesium (alloys)		Aluminium alloy BS 2L73 cured 1 h at 120°C, 138 N/cm² After, 1 d in water, 22°C 7 d 3 mth After, 4 d in water, 100°C 7 d After, weathering 3 mth	 −55 22 40 60 80 100 22 22 22 22 22 22	Complies with DTD900/4713 D.C.I.AFS/16 2 475 N/cm² 3 075 N/cm² 2 730 N/cm² 1 460 N/cm² 580 N/cm² 235 N/cm² 3 400 N/cm² 3 020 N/cm² 2 500 N/cm² 1 600 N/cm² 1 630 N/cm² 2 340 N/cm²	 b b b b b b b b b b b b	Cold or hot curing adhesive. Alternative catalyst HY 951 (triethylene tetramine) available to give less viscous adhesive (5 Ns/m² at 21°C) with slightly higher service temperature (105°C). AY105/HY953F is more suitable for copper and nylon materials and has longer potlife than AY105/HY951 (½–1 h at 20°C) which requires heat curing at 60°C. Similar to the more viscous AV100/HV100 (two tube pack sold for domestic usage) I.S.A.T. Series B Tropical weathering cycle (see p. 259)

continued

Trade source	Trade name or designation	Basic type	No. of compo-nents	Colour	Physical form, consistency or viscosity Ns/m^2	Working life	Method of application	Processing		Service temp. range °C	Main uses
								Curing cycle	Bonding pressure N/cm^2		
Araldite AY 105 + HY 953F—continued											
73	Araldite X83/348 + HY 953F	Epoxy polyamide	2	Amber	Liquid	2–3 h at 20°C	Spatula	1 h at 120°C	Contact to 140	−55 to 70	Bonding metals, ceramics, plastic rubber
73	Araldite AY 103 + HY 951	Epoxy resin cured with triethylene tetramine	2	Amber	Low viscosity liquid 1·6–2·0 at 21°C	3–4 h at 20°C	Brush Spatula	24 h at 20°C to 3 h at 60°C to 20 min at 100°C	Contact to 100	−60 to 65	Bonding metals, ceramics, plastic rubber
73	Araldite Strain gauge cement + HY 951	Epoxy resin cured with catalyst triethylene tetramine	2	Amber	Paste	1½ h at 20°C	Spatula	24 h at 20°C to 3 h at 60°C to 20 min at 100°C	Contact	65	Fixing and prot ing foil and wir type strain gaug

Adherends	Test spec.	Test conditions	Test temp. °C	Physical data	Key	Remarks
el r (mild, stainless) inium (alloys)	ASTM D696–44 ASTM D790–61 BS 2782: 102G	I.S.A.T. Series B After, 1 (1 wk) cycle 14 cycles After, 10 000 h at 130°C Cured, 1 h at 120°C	22 22 130	2 500 N/cm² 2 020 N/cm² 1 750 N/cm² 40–60 × 10⁻⁶ cm/cm/°C 0·30–0·65 (× 689476 N/cm) 55–70°C	b b b r E T	N.B. Data taken from an extensive programme of environmental testing. Further information available **from manufacturers**
rend materials listed for ldite AY 105 grade		Aluminium alloy BS 2L73 cured 1 h at 120°C, 6·9 N/cm² After, 1 d in water, 22°C 7 d in water, 22°C 3 mth in water, 22°C After, 4 d in water, 100°C 7 d in water, 100°C After, weathering, 3 mth I.S.A.T., Series B After, 1 (1 wk) cycle 14 cycles	−55 22 40 60 80 100 22 22 22 22 22 22 22 22	2 370 N/cm² 3 940 N/cm² 2 740 N/cm² 1 330 N/cm² 538 N/cm² 390 N/cm² 2 740 N/cm² 2 760 N/cm² 1 970 N/cm² 1 830 N/cm² 1 340 N/cm² 2 950 N/cm² 2 820 N/cm² 2 720 N/cm²	b b b b b b b b b b b b b b	Recent addition to the Araldite range. Has advantages over grades AY 105, AY 103, AV 121, with respect to ease of application and processing. More resistant to heat, hot water, and tropical conditions than AY 105 See N.B. above I.S.A.T. Series B Tropical **weathering Cycle (see p. 259)**
rend materials listed for ldite AY 105 grade	ASTM D696–44 ASTM D790–61 BS 2782: 102G	Aluminium alloy BS 2L73, cured 20 min At 100°C Cured At 120°C	 20 90 −60 20 20 100 150	Complies with DTD900/4365 D.C.I. AFS/774 1 930 N/cm² 207 N/cm² 1 450 N/cm² 5 500 N/cm² 8 500 N/cm² 1 720 N/cm² 275 N/cm² 90–95 × 10⁻⁶ cm/cm/°C 0·45–0·5 (× 689 476 N/cm²) 45–55°C	 b b b e e e e r E T	Cold or hot curing adhesive more flexible than AY 105 when cured. Alternative catalyst HY 956 available which lessens risk of dermatitis. Contains plasticiser. Similar mineral filled grade AV 121 has better gap filling properties for rough surfaces and joints with irregular gap widths, but shorter pot-life (1–1½ h at 20°C)
end materials listed for ldite AY 105 grade				Conforms to AFS/24		Optimum properties when post cured at 100°C. Alternative catalyst HY 956 lessens risk of dermatitis

Trade source	Trade name or designation	Basic type	No. of compo-nents	Colour	Physical form, consistency or viscosity Ns/m^2	Working life	Method of application	Processing		Service temp. range °C	Main uses
								Curing cycle	Bonding pressure N/cm^2		
73	Araldite + HZ 107	Epoxy polyamide	2	Amber	Liquid 4·3 at 21°C	8 h at 20°C	Spatula	24 h at 20°C to 3 h at 60°C to 20 min at 100°C	Contact	70	Resilient adhesi for bonding me glass, etc., to pe materials. Resil of cured materi controlled by catalyst: resin r
73	Araldite AW 127 + HW 127	Epoxy resin with catalyst (triethylene tetramine) and filler	2	Brown	Thick liquid 20 at 20°C	1 h at 20°C	Spatula	24 h at 20°C to 3 h at 60°C to 20 min at 100°C	Contact	40	Bonding low strength materi firm base mate Building applications
73	Araldite X33/1180	Epoxy resin with catalyst	2	Amber	Thick liquid 52 % w/w solids	1 wk at 20°C	Spatula Brush	Dried and heat cured, 1 h at 160°C + 1½ h at 180°C	Contact to 138	200	Pre-impregnati of glass and oth base materials laminating. Ad applications at elevated tem-peratures
73	Araldite X33/1189 + MY 750 + HT 976	Polyfunc-tional epoxy resin + liquid epoxy resin and catalyst (powder)	3	Brown	Thick liquid	2–3 h at 100°C	Spatula Roller	Applied to surfaces at 100°C. Cured 3 h at 180°C	Contact	240	Hot setting adh with high resist to heat. Metals ceramics and h resistant mater Strain gauge attachment

Adherends	Test spec.	Test conditions	Test temp. °C	Physical data	Key	Remarks
eel lass eather ubber (microcellular) astic foams uminium – wood (beech)	ASTM D696–44 ASTM D790–61 BS 2782: 102G	Cured 4 d at 20°C Cured 3 h at 70°C	20 20	1 170 N/cm² 1 380 N/cm² 65×10^{-6} cm/cm/°C 0·4 ($\times$ 689 476 N/cm²) 55–65°C	b b r E T	Unsuitable for use with certain plastics and rubbers which are attacked by solvent present in the adhesive. Preliminary trials recommended
astics (foam) bestos ood asterboard uminium (alloy) – aluminium	ASTM D696–44 ASTM D790–61 BS 2782 102G	Cured 24 h at 20°C Cured 20 min at 100°C	−60 20 −60 20	412 N/cm² 1 030 N/cm² 1 380 N/cm² 1 240 N/cm² 40×10^{-6} cm/cm/°C 0·13 ($\times$ 689 476 N/cm²) 30–40°C	b b b b r E T	Short post cure at 100°C is recommended where optimum properties are required. Formulation includes a tar component. Resistant to water and chemicals. Unsuitable for bonding chromium plates, glass, magnesium, nickel alloys, silver, tungsten carbide and fluorocarbons
ass cloth laminates el (stainless) – steel uminium – aluminium	BS 2782: 304B	Glass cloth normal to laminate parallel to laminate	20 150 200 220 20 100 200 20 20 20 20 20	57×10^3 N/cm² 46×10^3 N/cm² 41×10^3 N/cm² 37×10^3 N/cm² 11 200 N/cm²* 8 600 N/cm²* 4 150 N/cm²* 8 300 N/cm²* 14·4 kV/cm 19·5 kV/cm 5·6 (at 50 Hz) 0·012 (at 50 Hz) *Sira evaluation	f f f f e e e e m m n o	Epoxy novolac type. Intended primarily as a laminating adhesive, material is useful metal to metal bonding agent up to 200°C. Good retention of laminate bond strength after prolonged ageing. (55% at original strength retention after 6 mth ageing at 190°C)
herend materials listed for ldite ATI grade						

Trade source	Trade name or designation	Basic type	No. of compo-nents	Colour	Physical form, consistency or viscosity Ns/m^2	Working life	Method of application	Processing		Service temp. range °C	Main uses
								Curing cycle	Bonding pressure N/cm^2		
73	Araldite **ATI***	Epoxy resin with latent dicyan-diamide catalyst	1	Light Brown	Solid (m.pt. 80–100°C) in rod or powder form	Indefinite	Applied to hot surface at 100–120°C Berk flame-spray pistol	7 h at 140°C to 1 h at 180°C to 10 min at 240°C	Contact to 100	−50 to 120	Bonding metals, ceramics, plastics and heat resistant materials, for structural applications
73	Araldite X83/329	Epoxy resin with latent catalyst	1	Silver	Paste	Indefinite	Spatula	2 h at 100°C to 30 min at 150°C to 10 min at 180°C	Contact to 100	−55 to 120	Bonding metals, ceramics, plastics and heat resistant material for structural applications

Adherends	Test spec.	Test conditions	Test temp. °C	Physical data	Key	Remarks
...stos (rigid)				Complies with DTD 861A		Alternative grade AU1, con-taining aluminium filler, avail-
...mics (porcelain)		Aluminium alloy	−50	3 000 N/cm²	b	able. Grade AZ15+catalyst
...s fibre composites		BS 2L73 cured 1 h	0	3 125 N/cm²	b	HZ15 is equivalent to ATI in
...lyester)		at 180°C	50	3 275 N/cm²	b	solution. This alternative
...s fibre composites			100	1 890 N/cm²	b	system gives a less rigid
...oxy)			125	700 N/cm²	b	polymer when cured.
...on (+graphite)		After 4 mth in				Unsuitable for polyamides,
...etrafluorethylene (treated)		water 22°C	22	3 040 N/cm²	b	acrylics, polyacetals, poly-
...er (treated)		15 mth in water, 22°C	22	3 070 N/cm²	b	esters, polyolefines, poly-
...sten carbide		After weathering,				styrene, polyvinyl chloride,
...er and alloys (treated)		4 mth	22	3 160 N/cm²	b	wood and concrete.
		1 yr	22	3 270 N/cm²	b	ATI grade has greatest
...mium (treated)		4 mth	75	2 530 N/cm²	b	strength of Araldite series.
Tin and Solder		1 yr	75	2 260 N/cm²	b	Resistant to water and
...esium (alloys)		I.S.A.T. Series B				chemicals with excellent
...l		After, 14 (1 wk) cycles	22	2 870 N/cm²	b	electrical insulation
		40 (1 wk) cycles	22	2 540 N/cm²	b	properties.
...(mild, stainless)		64 (1 wk) cycles	22	2 580 N/cm²	b	I.S.A.T. Series B Tropical
...inium (alloys)		After, 10 d at 100°C	20	3 100 N/cm²	b	weathering cycle is described
		6 mth at 100°C	20	2 880 N/cm²	b	elsewhere (p. 259)
		Fatigue life	20	10⁶ cycles (250–670 Hz) for 480±480 N/cm²		N.B. Data taken from an extensive programme of environmental testing. Further information available
	ASTM D696–44			60×10^{-6} cm/cm/°C	r	from manufacturer
	ASTM D790–61				E	
...end materials listed for ...te ATI grade		Aluminium alloy	−55	2 550 N/cm²	b	Recent addition to the
		BS 2L73 cured.	22	3 300 N/cm²	b	Araldite range. Has
		30 min at 150°C	50	3 720 N/cm²	b	advantages over ATI grade for
			75	3 770 N/cm²	b	ease of application and
			100	2 880 N/cm²	b	processing. More economical
			125	690 N/cm²	b	curing schedule.
			150	240 N/cm²	b	Unsuitable for those adherend
		After, 4 mth in				materials which are incom-
		water 22°C	22	3 620 N/cm²	b	patible with ATI.
		15 mth in water 22°C	22	3 180 N/cm²	b	More resistant to water and
		After, weathering				weathering than ATI (tropical
		4 mth	22	3 470 N/cm²	b	exposure performance is
		1 yr	22	3 420 N/cm²	b	comparable)
		4 mth	75	3 490 N/cm²	b	
		1 yr	75	3 300 N/cm²	b	
		I.S.A.T. Series B				
		After 14 (1 wk) cycles	22	3 000 N/cm²	b	
		40 (1 wk) cycles	22	2 230 N/cm²	b	
		64 (1 wk) cycles	22	2 220 N/cm²	b	
...nium – aluminium			20	8 100 N/cm²*	e	
...(stainless) – steel			20	8 550 N/cm²*	e	
	ASTM D1002–64T		100	5 250 N/cm²*	e	
			150		e	
				*Sira evaluation		

Trade source	Trade name or designation	Basic type	No. of components	Colour	Physical form, consistency or viscosity Ns/m^2	Working life	Method of application	Processing		Service temp. range °C	Main uses
								Curing cycle	Bonding pressure N/cm^2		
73	Araldite* AY 18 + HZ 18	Epoxy resin with methylene diamine catalyst	2	Amber	Liquid 0·02 at 21°C	2 wk at 20°C	Spatula	Dried for 50 h at 15°C to 12 min at 120°C and cured for 2 h at 80°C to 30 min at 100°C to 15 min at 180°C	Contact to 100	165	Bonding metals, ceramics, plastic laminates and he resistant materia for structural applications. Impregnating varnish for electrical equipment
54	PR 905	Epoxy/ poly-sulphide	2	Dark Amber	2	25 min at 24°C and 50% relative humidity	Spatula Extrusion gun	45 min at 24°C for 450 g mass Heat cure up to 82°C	Contact	18 to 149	Non structural adhesive/encapsu lant for electrica components and printed circuit boards. Designed for applications requiring semi-flexible compoun with the propert of an epoxy resin
73	Araldite AY 111 + HY 111	Epoxy/ poly-sulphide	2	Dark Amber	Low viscosity liquid 7 at 21°C	20–30 min at 20°C	Spatula	24 h at 20°C to 3 h at 60°C to 20 min at 100°C	Contact		Cold setting adhesive especia suitable for bonding materia with differing expansion properties
133	Hermetite Double Bond No. 1	Epoxy	2	Trans-parent	Liquid	2–2·5 h at 20°C	Brush Spatula	24 h at 20°C to 30 min at 100°C	Contact		Bonding of meta rubber, plastics, glass, ceramics

Adherends	Test spec.	Test conditions	Test temp. °C	Physical data	Key	Remarks
erend materials listed for dite ATI grade, and acetals ester (film)		Aluminium alloy BS 2L73 cured ½ h at 100°C + 1 h at 180°C	20 150	Complies with D.C.I. AFS/864 2 400 N/cm² 1 550 N/cm²	b b	May be applied direct to the bonding surfaces or to paper or glass-cloth carriers. Outstanding resistance to water and chemicals
		Cured 2 h at 120°C	20 100 160	0·009 (at 10³ Hz) 0·004 (at 10³ Hz) 0·008 (at 10³ Hz)	o o o	
	ASTM D696–44 ASTM D790–61 BS 2782 : 102G			60–70 × 10⁻⁶ cm/ cm/°C 0·35–0·45 (× 689476 N/cm²)	r j	
		Deflection temperature, Cured, 2 h at 80°C		95–100°C 160–170°C	T T	
er y laminates olic laminate			24 49 121	5 × 10¹² Ω/cm 3 × 10¹¹ Ω/cm 1 × 10⁷ Ω/cm 136 kV/cm	p p p m	Addition of filler materials such as carbon black, silica, alumina, titanium oxide, up to 20% v/v, will alter physical properties. Cured resins have
			24 24	4·0 (at 10⁶ Hz) 0·016 (at 10⁶ Hz) 1·2 × 10⁻⁴	n o r	good resistance to non-oxidising acids, strong bases, chlorinated hydrocarbons, ketones, esters and
		After 24 h	24	0·35% water absorption 95J for 2·54 cm notch 20%	g k	petroleum. Volume shrinkage on cure is 3·5%.
inium – aluminium		After 30 min at 121°C		67 825 N/cm² 1 100 N/cm²	l d b	Alternative grade PR 905–1 has longer working life (1 h at 24°C)
tos (rigid) nics fibre composites n etrafluorethylene (treated) ster (film) tyrene (treated) er (treated) er (treated) sten carbide esium alloys inium – aluminium (stainless) – steel	ASTM D696–44 BS 2782 : 102G	Aluminium alloy BSL73 cured, 20 min at 100°C 24 h at 20°C	−60 20 −60 20	Complies with D.C.I. AFS/730 1 860 N/cm² 1 380 N/cm² 482 N/cm² 1 240 N/cm² 125–135 × 10⁻⁶ cm/ cm/°C 55–65°C	b b b b r T	Good adhesion to copper and brass. Cures to a flexible material. Resistant to water, petroleum, alkalies, and mild acids. Obnoxious odour produced during processing; adequate ventilation required for large assembly work
			20 20 100 150	2 900 N/cm² 5 000 N/cm² 1 380 N/cm² 70 N/cm²	e e e e	
inium – aluminium		After 30 min cure at 100°C	−60 −40 −20 0 20	723 N/cm² 757 N/cm² 826 N/cm² 826 N/cm² 1 584 N/cm²	b b b b b	Available as low viscosity liquid (Double Bond No 2) or paste (Double Bond No 3). All products are based on 1:1 weight ratio of resin:catalyst

Trade source	Trade name or designation	Basic type	No. of compo-nents	Colour	Physical form, consistency or viscosity Ns/m^2	Working life	Method of application	Processing		Service temp. range °C	Main uses
								Curing cycle	Bonding pressure N/cm^2		
149	Epoxy Preforms 615	Epoxy	1	Tan	Solid cylindrical pellets of mm dimensions 3·54 (O.D.) 2·27 (I.D.) 0·64 (height)	Indefinite	Inserted into joint part	30 min at 105°C (gel time) +16 h at 105°C to 2 min at 160°C (gel time) +15 min at 160°C	Contact		Bonding of meta tubes, glass-to-m seals, and variou dissimilar materi Assembly of electronic components and small artifacts. Printed circuit board applicatio Maximum servic temperature 155
105	Eccobond 276	Epoxy	2	Black	Thixotropic liquid 100% solids	24 h at 20°C	Spatula	8 h at 120°C to 2 h at 232°C	Contact		Bonding of meta glass, ceramics, e where service temperature approaches 250°
29	BS No. 8	Epoxy resin with Flexibiliser	3	Clear Yellow	Liquid	5 h at 20°C	Spatula	6 h at 60°C or 3 h at 78°C	Contact to 4	−50 to 80	Assembly of precision optica components
243	Sellobond Gum	Gum Arabic	1	Trans-parent	Liquid	Indefinite	Brush or Spreader cap on dispenser	Cold setting	Contact		Stationery uses
160	Samson No. 1	Gum Arabic	1	Pale Straw	Low viscosity liquid	Indefinite	Brush Machine	Cold setting	Contact		Paper sheet and strip gumming f remoistening Labelling Litho-plate preparation
107	Evo-stik **Thermaflo** 6820	Ethylene vinyl acetate co-polymer + resins	1	Yellow	Solid in block form	6 mth at 5–25°C	Special equip-ment available for appli-cation of molten adhesive. Roller applicator Extruder Hot spatula	Applied as a melt at 160–190°C which sets on cooling	Contact to 20		General purpos adhesive for use ambient tempera tures. Metals, plastics, rubber textiles

Adherends	Test spec.	Test conditions	Test temp. °C	Physical data	Key	Remarks
uminium ass eel ermoset stics ass ramics						One of a range of six products in various colours (red, tan, or black) and pellet sizes of mm dimensions. Good resistance to chemicals, oils and fuels
ass pper uminium eel ermoset plastics ass uminium—aluminium			 20 150	$1 \cdot 38$ J mm^{-2} s^{-1} $1°C^{-1}$ 11 700 10^{15} Ω/cm 15×10^{-6} 1 860 N/cm^2 1 446 N/cm^2	q f p r b b	Low shrinkage during cure. Applicable to vertical or overhead surfaces without sagging or flaw
ass ramics etals			20 20	7×10^{-5}/°C 2 755 N/cm^2	r	Refractive index $= 1 \cdot 586$. Conforms to F.V.R.D.E. specification No 2022 and S.R.D.E. specification DS 2144. Additional information available from Sira Institute
per, Cardboard						Fast drying glue
per						Sets to a medium flexibility material with moderate water resistance. Medium tack gum with slightly acid reaction
tton (duck)—cotton bber—rubber bber (gum stock)—resin rubber lyvinyl chloride (flexible)— PVC ather (chromic)—leather (tan) od—wood uminium—aluminium el (mild)—steel ypropylene (rigid)— polypropylene ythene—polythene rylic (sheet)—acrylic ass—glass			20 20 20 20 20 20 20 20 20 20 20 20	70–80 N/cm 35 N/cm 70 N/cm 19 N/cm 70 N/cm 155 N/cm^2 93 N/cm^2 95 N/cm^2 176 N/cm^2 100 N/cm^2* 110 N/cm^2* 110 N/cm^2* *Sira evaluation	a a a a a b b b b b b b	

Trade source	Trade name or designation	Basic type	No. of compo-nents	Colour	Physical form, consistency or viscosity Ns/m^2	Working life	Method of application	Processing		Service temp. range °C	Main uses
								Curing cycle	Bonding pressure N/cm^2		
45	Cascomelt HMR–2/2D	Synthetic thermo-plastic resins	1	Light brown	Strip material 100% solids 100 at 200°C	Indefinite	Machine	Applied as a melt at 200–220°C	Contact to 4		General purpose edge veneering and lipping using wood veneers, plastic laminates and thermoplastic edgings
281	Unilite R252	Thermo-plastic synthetic resins	1	White	Emulsion 16 to 17 solids content 52% w/w	Indefinite	Manual or machine	Joint assembly after 5–10 min at 20°C	14–69 cold or hot pressed		Designed for usage with high speed machinery in wood bonding, lamination etc.
88	Croid 660	Synthetic thermo-plastic resins	1	Yellow opaque	Solid 100% w/w solids 1·8 at 177°C		Machine	Applied as a melt at 140–177°C	Contact	−40 to 45	Hot melt adhesive for packaging applications. Carton and case sealing; plastic films; coated papers
107	Evo-stik Thermaflo 8428	Ethylene vinyl acetate copolymer + resins	1	Cream to yellow	Cylindri-cal solid sticks 100–120 at 195°C	6 mth at 5–25°C shelf life	Edge veneer machinery spatula	Applied as melt at 190–200°C	Contact to 20		Edge veneering applications using wood, melamine laminates or polyester veneers. Maximum service temperature 70°C
47	Bostik 68 GA 131	Polyester	2	Brown	Liquid 10 70% w/w solids	Pot life 1 d at 20°C	Nip roller	Up to 7 d at 20°C	Contact to 4	−30 to 150	Laminations of plastics materials for electrical insulation and other purposes
47	Thermogrip 9105	Polyester	1	Amber	Film 0·1 mm thick. Sp. gr. 1·2	Indefinite at 20°C	Placement between substrates and bonded within 30 s at tem-perature	Hot melt film bonded at 110–140°C under pressure. Sets on cooling	Contact to 10 with hot pressure plate		Bonding of fabrics and clothing materials, plastics, films, metals, rubbers, wood. Sealing seams of waterproof garments

Adherends	Test spec.	Test conditions	Test temp. °C	Physical data	Key	Remarks
ʼood ʼastics (decorative laminates)						Alternative product HMR–3/ZD available for rigid and low plasticiser content PVC edgings
ʼood—wood (beech) ʼastic laminate—wood ʼood—wood (beech) ʼastic laminate—wood (beech) ʼneer—wood ʼastic laminate—wood	BS 4071: 1966	After 5 min cold press After 5 min cold press After 15 min cold press After 15 min cold press After 1 min/100°C, hot press After 1 min/70°C, hot press Bonding pressure = 69 N/cm²	20 20 20 20 20 20	327 N/cm² 76 N/cm² 482 N/cm² 344 N/cm² 96 N/cm² 55 N/cm²	b b b b b b	Designed to give fast initial bond strengths with cold pressing and rapid bonding with hot press and low voltage heating units. Maximum strength in 1 h after assembly
ʼnipboard—metal foil laminate ʼax coated boards ʼlypropylene laminate ʼated papers ʼards (lacquered, varnished)						Non toxic, suitable for use under FDA packaging regulations for foodstuffs. May be applied as film for subsequent heat sealing
ʼelamine laminate—plywood			25 60	34 N/cm² 21 N/cm²	b b	Adheres well to hardwoods such as teak which have high natural oil content. Good melt stability with low increase in viscosity temperature rises. Resistant to water and detergent solutions
ʼlyethylene terephthalate ʼelinex) to, ʼlyimide (Kapton) ʼllophane ʼllulose acetate ʼlyethylene (treated) ʼlypropylene (oriented) ʼpper foil		For coatings, 30–35 g/m², bonded at 80°C	 20 20 20 20 20 20	300% 7·0 N/cm 5·2 N/cm 4·4 N/cm 8·7 N/cm 2·6 N/cm 8·7 N/cm	k a a a a a a	Satisfied requirements of **BS 2757 (1956) class B (130°C)** for electrical insulation laminates. Good resistance to Kerosene, perchloroethylene but not ketones. Swells in chlorinated hydrocarbons. Bostik Thinner 6316 available Coverage 2–12 g/m² or up to 35 g/m². Flash Pt. −17 to −7°C
ʼastomers: neoprene, natural, nitrate ʼastics: **Nylon, PVC, poly-styrene, poly-urethane (foam), polyester ABS** ʼbrics: Cotton, wool, polyester						Available in 150 m reels of width 10 mm–1 m. Flexibility retained below −30°C. Softening Pt. 125°C. Resistant to water at 90°C and perchloroethylene

Trade source	Trade name or designation	Basic type	No. of compo-nents	Colour	Physical form, consistency or viscosity Ns/m^2	Working life	Method of application	Processing		Service temp. range °C	Main uses
								Curing cycle	Bonding pressure N/cm^2		
47	Thermogrip 9410	—	1	Yellow	Solid as extruded strip roll 6–47 mm diameter	Indefinite at 20°C	Applied from extrusion applicator	Hot melt bonded at 180–250°C	Contact to 10		Rapid assembly of plastics mouldings, films, coated paper
47	Thermogrip 9440	—	1	Off-white	Solid. Extruded strip 6 mm wide	Indefinite at 20°C	As for Thermo-grip 9410	Hot melt bonded at 176°C	Contact to 10		Carton sealing. Bonding of plastics film and carton materials
107	Evo-stik* Thermaflo 6876	Ethylene vinyl acetate co-polymer + resins	1	Pale Yellow	Solid in film form	Indefinite 6 mth at 5–25°C	Heated nip rolls or hot iron	Film transfer at 70–80°C followed by bond-ing at 150–160°C	Contact to 20	60 or 1 h at 90	Metals, laminated plastics and textiles. Fabrication of leather goods. Specialised adhesive for acrylic assemblies and lamination work
107	Evo-stik Therma-flow 8428	Ethylene vinyl acetate co-polymer + resins	1	Cream to Yellow	Solid in cylindrical stick form 100–120 at 195°C	6 mth at 5–25°C	Edge veneering machin-ery, Hot spatula	Applied as a melt at 190–200°C	Contact to 20	70	Edge-veneering applications using wood, melamine laminates or polyester veneers
107	Evo-stik Therma-flow 8445	Ethylene vinyl acetate co-polymer + resins	1	White to Cream	Solid in cylindrical stick form 100 at 170°C	6 mth at 5–25°C	Edge veneering machin-ery, Hot spatula	Applied as a melt at 150°C	Contact to 20	70	Edge veneering with flexible and rigid polyvinyl chloride veneers

Adherends	Test spec.	Test conditions	Test temp. °C	Physical data	Key	Remarks
Metals Glass Polythene (film) Papers (waxed, glassines) PVC (film) Boards						Flexible joints with good resistance to moisture, acids, alkalies, chemical salt solution, oils, fats. Unstable towards sunlight and ultraviolet with polythene film. Poor resistance to hydraulic oils, hydrocarbons and chlorinated solvents. Thermogrip applicator available
Polythene (film) Polyester (film) Polypropylene Cellophanes						Comparatively low operating temperature for a hot melt adhesive. Unsuitable where bonds are under stress above 40°C. Poor resistance to organic solvents
Cotton (duck) – cotton Resin rubber – leather Melamine laminate – plywood Steel (mild) – steel Acrylic (sheet) – acrylic			25 60 25 20 20 20	8·7 N/cm 8·7 N/cm 35 N/cm 250–300 N/cm^2 144–179 N/cm^2	a a a b b b	Hot-melt adhesive film cast onto release paper. Thickness may be 0·065 mm or 0·115 mm. Infra-red lamp heating of film followed by bonding is feasible. Metallic substrates should be preheated to 50°C. Unsuitable for fluorinated polymers and acetal polymers. Good electrical insulation resistance with very high dielectric loss permits it to be used as high frequency welding adhesive. Withstands normal cleaning processes utilising water or detergent solutions
Cotton (duck) – plywood Melamine laminate–plywood			25 60 25 60	52 N/cm 19 N/cm 345 N/cm^2 21 N/cm^2	a a b b	Contains mineral fillers. Adheres well to hardwoods, such as teak, which have high natural oil content. Good melt stability, the rate of increase in viscosity with temperature rise being unusually low. Resistant to water and detergent solutions

Polyvinyl chloride (flexible) –
  Polyvinyl chloride (flexible)
Polyvinyl chloride (rigid) –
  plywood
Melamine laminate–plywood
			25
			25
			25
			60
				52–61 N/cm
				40–50 N/cm
				200–215 N/cm^2
				3–7 N/cm^2
					a
					a
					b
					b
						Resistant to effects of plasticiser migration when bonding polyvinyl chloride veneers. Non-staining. Temperature resistance related to substrate nature. (90°C for rigid pvc; 65°C for melamine or polyester laminates)

Trade source	Trade name or designation	Basic type	No. of compo-nents	Colour	Physical form, consistency or viscosity Ns/m^2	Working life	Method of application	Processing		Service temp. range °C	Main uses
								Curing cycle	Bonding pressure N/cm^2		
88	Croid 663		1	Light brown	Solid 100% w/w solids		Hot spatula Machine	Applied as a melt at 190–220°C	Contact	−10 to 100	Veneering and laminating of wood and plastics
103	Eastobond M3	Synthetic polymer blend	1	Light Yellow	Solid in granule form 100% w/w solids 3·6 at 149°C to 1·3 at 191°C	Open time 2–3 s at 177°C	Hot spatula Extruder	Applied as a melt at 177°C	Contact	71	Carton and paper bag sealing. Bonding polythene coated container. Packaging applications
234	Brimor U 527	Aluminium phosphate + silica filler	1	Green	Paste	20 min at 20°C	Spatula Brush	Dried $\frac{1}{2}$ h at 20°C, then $\frac{1}{2}$ h at 70°C + $\frac{1}{2}$ h at 100°C + 1 h at 200°C + 1 h at 250°C. Repeat for 2 over-coatings and finally cure 1 h at 350°C	Contact	750°C	Strain gauge attachment to heat resistant metals. Heater element bonding
234	Brimor U 529	Aluminium phosphate + silica filler	1	Green	Paste	20 min at 20°C	Spatula Brush	Dried $\frac{1}{2}$ h at 20°C then $\frac{1}{2}$ h at 70°C + $\frac{1}{2}$ h at 100°C + 1 h at 200°C + 1 h at 250°C. Repeat for 2 over-coatings and finally cure 1 h at 350°C	Contact	750	Strain gauges to heat resistant metals where the assembly is subject to stress, gas erosion, or temperatures exceeding 700°C, for prolonged period, e.g. gas turbine rotor components

Adherends	Test spec.	Test conditions	Test temp. °C	Physical data	Key	Remarks
Wood						Slight darkening in colour at 180°C with little change in viscosity. Flexible at low temperatures
Paper Cardboard Polythene (coated materials)						Viscosity change is less than 5% for +100 h exposure at 177°C. No odour at melt temperature Flash Pt. 277°C
Steels (low alloy) Iron Brass Titanium Copper Aluminium		Resistance of 0·1016–0·1270 mm layer on metal plate after curing	20–200 20–400 20–600 20–800 550 600 700 750 800	$9·2 \times 10^{-6}$ cm/cm/°C $12·8 \times 10^{-6}$ cm/cm/°C $18·4 \times 10^{-6}$ cm/cm/°C $15·3 \times 10^{-6}$ cm/cm/°C $1·0 \times 10^{6}$ Ω $0·3 \times 10^{6}$ Ω $0·03 \times 10^{6}$ Ω $0·02 \times 10^{6}$ Ω $0·01 \times 10^{6}$ Ω	r r r r	Particularly suited to heat resistant steels where surface oxidation of metal at high temperatures is less detrimental to adhesion. More suitable than the harder U529 grade for high frequency vibration conditions. Service temperature for copper-nickel alloy strain gauges is limited to 300°C
Steel Iron Brass Titanium Copper Aluminium		Resistance of 0·1016–0·1270 mm layer on metal plate after curing	20–200 20–400 20–600 20–800 550 600 700 750 800	$11·1 \times 10^{-6}$ cm/cm/°C $13·7 \times 10^{-6}$ cm/cm/°C $19·6 \times 10^{-6}$ cm/cm/°C $16·0 \times 10^{-6}$ cm/cm/°C $3·0 \times 10^{6}$ Ω $1·0 \times 10^{6}$ Ω $0·2 \times 10^{6}$ Ω $0·09 \times 10^{6}$ Ω $0·03 \times 10^{6}$ Ω	r r r r	Higher bond strength than U527 and more brittle when cured. Used up to 800°C when severe erosion conditions are present. Liable to craze under vibration

Trade source	Trade name or designation	Basic type	No. of compo-nents	Colour	Physical form, consistency or viscosity Ns/m^2	Working life	Method of application	Processing		Service temp. range °C	Main uses
								Curing cycle	Bonding pressure N/cm^2		
271	Denex-2 Ceramic cement	Inorganic	1		Paste	20 min at 20°C	Brush	Dried for $\frac{1}{2}$ h at 77°C and cured $\frac{1}{2}$ h at 100°C +1 h at 200°C+ 1 h at 250°C. Post cured, 1 h at 350°C	Contact	816	Strain gauges, temperature sensors for elevated temperature work
104	AS adhesive	Ceramic based	1		Paste	Up to 24 h at 20°C	Spatula	24 h at 20°C to 2 h at 400°C	Contact	400	Chrome-nickel alloy strain gauges for elevated temperatures
90	Onx* Silseal	Silicate	1	Off White	Fine grained paste	Indefinite	Spatula Trowel	Sets on heating	Contact	Up to 1650	Jointing of insulat-ing bricks. Protective surface coating for refractories
18	Fibrous Adhesive 81–27	Silicate	1	Off White	Very soft paste Bulk density is 1970 kg/m³	Indefinite	Brush	Sets in 8 h at 20°C on drying	Contact	10 to 430	Lagging asbestos cloth on high temperature insulation. Dense thermal insulations such as amosite, magnesia and asbestos block, to porous surfaces
114	Fortafix 'Ready mix'	Sodium silicate with filler	1	White	Paste	Indefinite	Dip coat Spatula Spray gun	Dried at 80°C before heat exposure	Contact	0– 1000	High temperature applications involv-ing jointing, sealing embedding, and repair of heat resistant and other materials
114	Fortafix 'Chemical Set' cements	Silicate blend with filler blend	2	White	Paste of variable viscosity	1 h when mixed	Brush Spatula Trowel	Chemical setting at 20°C. May be heat cured at 20–110°C	Contact	0– 1400	As for 'Ready mix' products with emphasis on large bulk applications
114	Chromix	Silicate with filler blend	1	Grey-black	Paste	Indefinite	Brush Spatula Trowel	Dried at 20–80°C before exposure to heat	Contact	0– 1400	

Adherends	Test spec.	Test conditions	Test temp. °C	Physical data	Key	Remarks
etals						Adhesive is usually given a final heat cure for 1 h at 38°C above intended service temperature. For platinum tungsten gauge final cure is at 760°C. Material may be thinned with distilled water
etals ncrete						Poor resistance to humidity
umina agnesia ickwork bestos			1400 1000	Resistant to high temperatures in boilers, furnaces, etc. 1 140 N/cm² 1 100 N/cm² Specific Heat 25 c.g.s. units	x x	Not subject to thermal expansion or contraction. Various other grades of cements and mortars available for service temperatures 500–1 800°C based on alumina or siliceous materials
bestos agnesia						Forms a hard, brittle in-combustible film which is water soluble (i.e. unsuitable adhesive where moisture encountered). Not recommended for glass or painted surfaces. Coverage is 1–1·7 m²/l
ass ramics bestos od rous materials—paper, board etals						Provide hard inflexible joints with good thermal and electrical insulation properties. Good resistance to organic solvents and acids, but unsatisfactory where high humidity or moisture is present
for 'Ready mix' products						Suitable for bonding non-porous materials since setting action is not dependent on loss of water. Similar joint performance to 'Ready mix' products but with slightly improved resistance to water
ass ramics etals						Available as two-part cement. Resistant to acids and solvents but not water. Unsuitable where high dielectric strength is required

Trade source	Trade name or designation	Basic type	No. of components	Colour	Physical form, consistency or viscosity Ns/m^2	Working life	Method of application	Processing		Service temp. range °C	Main uses
								Curing cycle	Bonding pressure N/cm^2		
286	M-Bond GA-100	Silica and aluminium phosphate	1	Grey	Paste	1 d at 24°C	Spatula	Precoat: 15 m at 175°C. Final: 1 h at 315°C. Heating rate of 95°C/h up to 135°C	None		For strain gauge assemblies intended to operate above 315°C
64	Autostic	Silicate with china clay filler	1	White	Paste	Indefinite	Brush Trowel Spray gun	Dried at 80°C before exposure to heat	Contact	− 180 to 1500	General purpose cement for bonding refractory materials and metals. Furnace repairs and gas tight jointing of pipework. Heat insulating materials
89	Pyramid No. 1	Sodium Silicate	1	White	Liquid 0·25–0·5 at 20°C 37·8 % w/w solids	Indefinite	Brush	Dried at 20–80°C before exposure to heat	Contact to 130	0 to 850	Fabrication of corrugated fibre-board. Wood bonding. Metal foil to paper lamination
89	Pyramid No. 3	Sodium Silicate	1	White	Liquid 0·2 at 20°C 33 % w/w solids	Indefinite	Brush	Dried at 20–80°C before exposure to heat	Contact	0 to 850	Cellulosic materials and silicate-type adherends for heat insulation
89	Pyramid No. 96	Sodium Silicate	1	White	Liquid 0·5 at 20°C 43·2 % w/w solids	Indefinite	Brush	Dried at 20–80°C before exposure to heat	Contact	0 to 850	Close grained cellulosic board bonding. Metal foil to paper laminating
89	Pyramid No. 84	Sodium Silicate	1	White	Viscous liquid 0·6–1·2 at 20°C 39·5 % w/w solids	Indefinite	Brush	Dried at 20–80°C before exposure to heat	Contact	0 to 850	Fabrication of rigid asbestos board. Insulation board bonding to furnace doors
89	Claysil No. 1	Sodium Silicate + china clay filler (25 % w/w)	1	White	High viscosity, thixotropic liquid 0·25–1·2 at 20°C	Indefinite	Brush	Dried at 20–80°C before exposure to heat	Contact	0 to 850	Insulation material to steel pipes and tanks

Adherends	Test spec.	Test conditions	Test temp. °C	Physical data	Key	Remarks
Metals Ceramics				Approximately 5% depending on linear expansion coefficient of substrate	k	Primarily used for free-filament strain gauges operating at elevated temperature. 0·03 mm base coat has leakage resistance of $> 10^6 \Omega$ at $+650°C$ and $0.3 \times 10^6 \Omega$ at $760°C$. Requires to be cured in air. Hygroscopic material. Installation requires skill (see Bulletin B–129)
Asbestos Ceramics Brickwork Glass Silver Aluminium Steel (mild) – steel			 20 20	For both d.c. and low a.c. frequencies is a good electrical insulant. Has inordinate loss factor at 20×10^6 Hz 172 N/cm² 310 N/cm²	 d b	Sets to a brittle, cement-like mass. Viscosity reduced by water dilution. Resistant to oil, petrol and weak acids
Aluminium (foil) Paper Wood (walnut) – wood Wood (ash) – wood Wood (beech) – wood Wood (persimon) – wood				 760 N/cm² 930 N/cm² 930 N/cm² 1 100 N/cm²	 b b b b	$SiO_2 : Na_2O$ ratio 3·3 : 1. Bonding of copper and tin sheets to walls reported. Suitable for glass to stone bonding. Coverage, 1 m² per 8–10 g
Vermiculite Glass (and glass wool) Paper						$SiO_2 : Na_2O$ ratio 3·65 : 1. Not prone to efflorescence. Fast setting at low temperatures
Aluminium (foil) Paper						$SiO_2 : Na_2O$ ratio 2·85 : 1 Gives water insoluble bond for metal–paper laminations
Asbestos (rigid) Metals (galvanised or otherwise)						$SiO_2 : Na_2O$ ratio 2·5 : 1. Coverage, 1 m² per 8–10 g
Glass fibre Steel						Based on Pyramid No. 1. Setting action retarded by dilution with 5 % water

Trade source	Trade name or designation	Basic type	No. of compo-nents	Colour	Physical form, consistency or viscosity Ns/m^2	Working life	Method of application	Processing		Service temp. range °C	Main uses
								Curing cycle	Bonding pressure N/cm^2		
39	Colset 185	Bitumen/latex emulsion	1	Brown black	Liquid	15–30 min open time at 20°C	Trowel Spreader	Dried in air to a tacky state	Contact to 7	0 to 66	Light weight thermal insulation boards, and preformed sections to porous and non-porous surfaces. Building applications
35	Bostik 4052	Bitumen/latex emulsion	1	Brown-black	Thin liquid		Brush Dipping	Dried in air to a tacky state	Contact to 7		Wood block flooring applications
269	Instantex	Latex	1	White	Low-medium viscosity liquid		Brush Machine	Air dried 15 min and contact bonded	Contact to 7	65	Bonding plastics to porous materials. Carpet bonding. Automobile trim work
47	Bostik 4051	Synthetic latex-resin in aqueous media	1	Yellow	Liquid 50% w/w solids	Indefinite	Trowel	Sets by drying 40–60 min drying period before bonding to non-porous base	Contact to 4	−15 to 75	Bonding tile materials for interior flooring applications
88	Croid 231	Natural rubber latex	1	White to clear when cured	Ammon-iacal aqueous solution of low vis-cosity. pH exceeds 10	Indefinite	Brush Roller Machine Stencil	Dried at room temp.	Contact to 4		For self-seal bonding of cartons and paper. Rapid assembly adhesive for carton stencilling machines in packaging industry. Envelopes and stationery applications
66	Chelsea B.100	Natural rubber-resin blend in solvent	1	Brown	Syrup 10 at 21°C	Open time 10 min on porous adherends	Brush	Contact bonding after drying	Contact to 4		Temporary and laminating procedures in footwear manufacture. Expanded polystyrene floortiles
47	Bostik C	Natural rubber-resin in mixed hydrocarbon solvent with filler additive	1	Black	Liquid. Solids content 65% w/w	Open time 20–30 min	Brush Roller Trowel	Dried to touch dry state and bonded. Porous materials are amenable to 'wet' assembly	Contact to 4	−50 to 70	General purpose adhesive for porous and non-porous materials

Adherends	Test spec.	Test conditions	Test temp. °C	Physical data	Key	Remarks
rk lystyrene (foam) lyvinyl chloride ncrete bestos						Dried film is tough and resistant to extension. Not recommended for constructions operating below 0°C. Colset bitumen primer 384 recommended for concrete and brick surfaces. Coverage, 1·85 m²/l
ood						Formulation gives strong flexible bond to resist lateral movement of flooring blocks. Coverage, 0·9 m²/l
lyvinyl chloride (rigid) trocellulose ywood rdboard brics						Must be stored above 0°C. Pre-coated components can be bonded after drying within 24 h Non-inflammable
oleum rk 'C ncrete nber						Good resistance to water, mild acids and alkalies. Bostik thinner 6020 available. Coverage 4 m²/l
per, board materials						Sets to a flexible, soft coating with self-stick properties. Dried coating is durable for 6 mth when stored adequately. Non-inflammable with good resistance to water. Should not be in contact with copper or brass containing materials.
ather (semichrome) brics (plastic coated) lystyrene (expanded) lyvinyl chloride						Compatible with cellular polystyrene materials. Flash Pt. below 0°C
od rk tals oleum t						Alternative Bostik 1255 available for spraying (45% w/w solids) or Bostik 1261 for heat resistance +100°C (53% w/w solids). Resistant to water but not oils, fuels or paraffins. Coverage 3 m²/l. Flash Pt. −17 to −7°C

Trade source	Trade name or designation	Basic type	No. of compo-nents	Colour	Physical form, consistency or viscosity Ns/m^2	Working life	Method of application	Processing		Service temp. range °C	Main uses
								Curing cycle	Bonding pressure N/cm^2		
47	Bostik 1279		1	Off white	Liquid. Solids content 60% w/w	Open time 15 min				−17 to 100	As for Bostik C where heat resist-ance is required
47	Bostik 1311		1	White	Liquid. Solids content 70% w/w		Dispensed from cartridge gun			−6 to 79	Rapid assembly of board and foam materials (except polystyrene). Interior application
57	BTR Cement 22462	Natural rubber in hydro-carbon solvent	1	Amber		15 min at 20°C	Brush	Air dried 15 min at 20°C and contact bonded	Contact		Bonding of fabrics and rubber to porous substrates
57	BTR Cement 41872/3	Natural rubber	2	Light Brown			Brush	Air dried 20 min at 20°C and heat cured 5 min at 140°C	Contact		Vulcanising cement for rubber bonding to textiles and rubbers
160	Samson 1602	Rubber latex-casein	1	White	Thick paste		Brush Roller machine	Air dried	Contact		General purpose adhesive for labelling and combining work with metals and plastics
98	Dunlop PT	Rubber latex-acry-lonitrile blend	1	Off White	Paste	5–8 min at 20°C	Spreader	Contact bonded when wet	Contact		Bonding vinyl plastics to floor surfaces
98	Dunlop A1020	Rubber latex	1	White	Liquid	Several hours at 20°C	Brush Spreader	Air drying within 15 min	Contact		Bonding textiles, papers, packaging materials. Carpet bonding
98	Dunlop SF	Styrene-butadiene rubber latices	1	Light Brown	Medium viscosity	15–25 min at 20°C	Brush Trowel Roller coater	Air drying	Contact		Bonding poly-styrene foams to porous surfaces
79	Copydex	Latex	1	White	Paste	Tacky after 10 min	Brush Spatula	Sets on air drying	Contact		Upholstery fabrics carpets, leather

Adherends	Test spec.	Test conditions	Test temp. °C	Physical data	Key	Remarks
s for Bostik C						Amenable to spray application when thinned with Bostik 6512 thinner. Coverage 2·8 m²/l. Flash Pt. −17 to −7°C
asterboard oftboard/hardboard astics foams						Gap filling up to 6 mm. Resistant to water but not oil. Flash Pt. below −2·8°C
luminium ubber (latex) anvas eather						May be thinned with benzene. Available in thicker grade 22463. Inflammable, Flash Pt. −16°C Coverage, 4 m²/l
ubber (styrene-butadiene) ubber (latex) luminium ardboard eather otton						Similar applications to grade 22462. Quick drying. May be thinned with toluene. Inflammable, Flash Pt. 0°C Coverage, 4 m²/l
ood extiles archment arnished surfaces						Corrodes copper and brass. Dries to a tough, flexible film. Slightly ammoniacal odour. Low tack adhesive
olyvinyl chloride ood oncrete phalt aster						Suitable for heated under floor application. Resistant to plasticiser migration. Coverage 5·5 m²/l
anvas aper brics aper llulosic materials						Bonds porous materials rapidly. Resistant to heat. Should be protected from frost, oils, and copper alloys
olystyrene (foam) ood ardboard bestos ickwork						Coverage, 2·2 m²/l
sorbent materials such as paper, fabrics and textiles (natural)				Stronger than materials joined		Semi-transparent in cured state. Resistant to washing, boiling in water

Trade source	Trade name or designation	Basic type	No. of compo-nents	Colour	Physical form, consistency or viscosity Ns/m²	Working life	Method of application	Processing		Service temp. range °C	Main uses
								Curing cycle	Bonding pressure N/cm²		
79	Chuk'ka		1	White	Semi-liquid	Indefinite	Spatula applica-tion of warm liquid at 90°C	Sets on air drying	Contact		China, ceramics, pottery, glass carpets, leather
210	Phillisol	Latex	1	White	Liquid	Tack-free in 15–45 min at 20°C	Brush	Contact bonded after drying at 20°C for 10 min	Contact to 20	40	Fabrics, paper. Plaster surfaces and elastomers. Porous materials
16	Holdtite RA 4126	Natural rubber in petroleum	1	Pale Cream	8–10 at 20°C	Tacky in 5–15 min	Brush Scraper	Contact bonding after drying at 20°C for 15 min	Contact to 20		Bonding plastic insulation tiles to plaster, masonry Natural and synthetic foams, metals
13	Holdtite RA 4110	Reclaim natural rubber in petroleum	1	Black	37·5–60 at 20°C	Tacky in 15–30 min	Stiff brush or scraper	Contact bonded after drying at 20°C for 20 min	Contact to 20		General purpose rubber cement
57	Silverlock B21	Reclaim rubber in petroleum spirit	1	Black	Medium viscosity liquid 50±2% solids	15 min at 20°C	Brush	Air dried 10 min at 20°C and contact bonded	Contact	104	Rubber to metal bonding. Automotive indust
98	Dunlop S 480	Reclaim rubber	1	Black	Medium viscosity liquid 50% w/w solids	15 min at 20°C	Brush Spreader	Air dried 10 min at 20°C and bonded	Contact		General purpose adhesive. Textiles, carpets and felts to metals, wood. Floor materials
57	Cement 21850 *	Cyclised rubber	1	Brown	Liquid	15 min at 20°C	Brush		Contact		Bonding rubber to metal when used i conjunction with rubber solutions
57	Cement 64092	Chlorinated rubber	1	Light tan	Liquid	15 min at 20°C	Brush		Contact		Priming solution f cold setting or hea curing in conjuncti with neoprene or nitrile cements

Adherends	Test spec.	Test conditions	Test temp. °C	Physical data	Key	Remarks
						Acts as a cement rather than an adhesive. Suitable for antique and ornament repair
Rubber – rubber Rubber – leather	ASTM D1876–61T		20 20	35 N/cm 44 N/cm	a a	Not resistant to oil. Also, available as an aqueous based adhesive
Rubber Concrete Wood Linoleum Polyurethane (foam) Leather Fabrics Chipboard				Conforms to M of D Spec. CS 2558		Inflammable, Flash Pt. 23°C Coverage, 3 m²/l
Plaster Aluminium Hardboard Latex (foam) Polyurethane (foam) Polystyrene (foam) Rubber Wood				Conforms to M of D Spec. 2558 and TS 412		Available in alternative grades, RA 4732 – shorter open assembly time; RA 4733 – longer open assembly time. These grades have better heat resistance properties. Inflammable, Flash Pt. 23°C Coverage, 3·3 m²/l
Steel Aluminium Glass Rubber (latex, sponge, foam) Polychloroprene Wood Leather						Non-sagging consistency. Resistant to water. Cured adhesive may be solvent reactivated with petroleum. Inflammable, Flash Pt. 43°C. Coverage, 14 m²/l
Rubber Canvas Leathercloth Wood Cork Linoleum Steel						Not recommended for poly-vinyl-chloride tiles. May be diluted with thinner T160. Resistant to weather, water and biodeterioration. Inflammable
Wood Rubber Metals						Can be vulcanised or air cured. Priming solution for metal or wood. Flash Pt. −32°C
Wood Metals Rubber						May be thinned with toluene. Flash Pt. 4°C

Trade source	Trade name or designation	Basic type	No. of compo-nents	Colour	Physical form, consistency or viscosity Ns/m^2	Working life	Method of application	Processing		Service temp. range °C	Main uses
								Curing cycle	Bonding pressure N/cm^2		
	Formula-tion 1	Acrylo-nitrile rubber + phenolic resin	1	Brown		1 h at 25°C	Brush	Dry for 1 h at 20°C or 20 min at 90°C then cured at 150°C for 25 min	90 to 140	150	General purpose adhesive
	Formula-tion 2	Acrylo-nitrile rubber + phenolic resin	1	Brown		1 h at 25°C	Brush	Dry for 1 h at 20°C or 20 min at 90°C then cured at 150°C for 20 min	90 to 140	150	Rigid adhesive for bonding metals
	Formula-tion 3	Acrylo-nitrile rubber + phenolic resin	1	Brown		4 h at 25°C	Brush	Dried for 4 h at 20°C and cured at 90°C for 20 min	35 for 3 min		Flexible adhesive for rubber, leather, and flexible plastics
193	Scotchweld AF 31 with EC 1459 primer	Acrylo-nitrile rubber + phenolic resin	1	Yel-low-Brown	Film supported on glass cloth (0·25 mm thick)		Brushed primer + press curing of film	Primer air dried 60 min at 20°C. Film cured 60 min at 175°C under pressure. Pressure released, on cool-ing, at 50°C	130	−40 to 120	Metal bonding for structural assemblie at elevated temperatures

Adherends	Test spec.	Test conditions	Test temp. °C	Physical data	Key	Remarks
eel (mild) – rubber ood – rubber ment – polyvinyl chloride (rigid) ood – leather eel – leather ood – cork eel – cork uminium alloy – phenolic laminate uminium alloy – polyester/ glass fibre eel (mild) – polyvinyl chloride pper (foil) – phenolic laminate eel (mild) – wood (oak)		Cured at, 120°C for 20 min 155°C for 20 min 100°C for 30 min 80°C for 2 min 100°C		 715 N/cm² 1 070 N/cm² 158 N/cm² 12·7 N/cm 1 300 N/cm²	 b b b a b	Formulated adhesives based on acrylonitrile rubber (Trade Source 50) and phenolic resin (Cellobond H833, Trade Source 55). Formulation / 1 / 2 / 3 : Breon 1001 100 100 100; Cellobond H833 100 140 40; Hexamine 10 14 4; Methyl ethyl ketone 210 254 144
uminium (alloy) – aluminium (alloy) agnesium – magnesium eel (mild) – steel ass – brass		Cured at 155°C for 20 min at bonding pressures up to 140 N/cm²		2 960 N/cm² 1 510 N/cm² 2 140 N/cm² 1 400 N/cm²	b b b b	Proportions by weight. Formulation 1 may be used without heat or pressure as a contact adhesive after air drying at 25°C
ather – leather		Cured at 90°C for 20 min		52·5 N/cm	a	
uminium (alloy) – aluminium o DTD 746	ASTM D1002– 53T	Etched in chromic/ sulphuric acid to DTD 915B Aged at 120°C, 1 000 h 10 000 h 30 000 h Aged at 130°C, 1 000 h 10 000 h 30 000 h 1 000 h 10 000 h 30 000 h Aged at 150°C, 1 000 h 3 000 h 1 000 h 3 000 h	20 125 200 225 120 120 120 −40 −40 −40 130 130 130 −40 −40 150 150	1 885 N/cm² 1 668 N/cm² 1 644 N/cm² 1 595 N/cm² 1 325 N/cm² 715 N/cm² 530 N/cm² 3 450 N/cm² 3 280 N/cm² 3 200 N/cm² 1 910 N/cm² 1 630 N/cm² 1 770 N/cm² 3 030 N/cm² 2 970 N/cm² 1 160 N/cm² 1 265 N/cm²	b b b b b b b b b b b b b b b b b	Bond strength retention for short periods up to 150°C. Bond strength deterioration for 1 000 h. Immersion in fluids at 130°C is less than that for thermal ageing in air, e.g. Kerosene (Spec. DERD 2494), Hydraulic oil (Spec. DTD 585), Chlorinated silicone oil (Spec. DTD 900/4725), Synthetic ester lubricant (Spec. DERD 2487) Data provided by B.A.C. (Operating) Ltd., Filton, Bristol

Trade source	Trade name or designation	Basic type	No. of compo-nents	Colour	Physical form, consistency or viscosity Ns/m²	Working life	Method of application	Processing Curing cycle	Processing Bonding pressure N/cm²	Service temp. range °C	Main uses
193	Scotchweld AF 30 with EC 1459 primer (10% w/w solids)*	Acrylo-nitrile rubber + phenolic resin	1	Light Tan	Unsup-ported film (0·28 mm thick)		Brushed primer + press curing of film	Primer air dried 60 min at 20°C. Film cured 60 min at 175°C under pressure. Pressure released, on cool-ing at 50°C	130	−40 to 130	Metal bonding for structural applications at elevated temperatures
287	Metlbond 402 Type 1 −402 Type 2	Acrylo-nitrile rubber + phenolic resin	1	Yellow	Unsup-ported film (0·25 mm thick)		Brushed primer + press curing of film	Primer air dried 60 min at 20°C+ 15 min at 120°C. Film cured 60 min at 177°C under pressure. Pressure released, on cool-ing, at 50°C	69	−40 to 130	Metal bonding for structural applications at elevated temperatures
47	Boscoprene 2402*	Acrylo-nitrile rubber in hydro-carbon ketone solvent	2	Brown to Purple	Medium viscosity liquid	8 h at 20°C	Brush Roller	Dried at 20°C in 20 min and contact bonded for 72 h or 15 h at 70 C	Contact to 20	−40 to 100	Bonding of rubber, plastics, metals and textiles. Heat resistant adhesive for plastic laminates. Rubber overshoes to propellor blades in aircraft industry

Adherends	Test spec.	Test conditions	Test temp. °C	Physical data	Key	Remarks
Aluminium (alloy) – aluminium to DTD 746	ASTM D1002–53T	Etched in chromic/ sulphuric acid to DTD 915B	20	2 925 N/cm²	b	Bond strength deterioration for 1 000 h immersion in fluids at 130°C is comparable with that for thermal ageing in air, e.g. Hydraulic oil (Spec. DTD 585), Kerosene (Spec. DERD 2494), Synthetic ester lubricant (Spec. DERD 2487), Chlorinated silicone oil (Spec. DTD 900/4725). Subject to creep at 100°C, 150°C for sustained loading (40% ultimate tensile shear strength produces up to 0·3 mm deformation, with standard lap joint, in 500 h) Data provided by B.A.C. (Operating) Ltd., Filton, Bristol
			125	1 160 N/cm²	b	
			200	652 N/cm²	b	
			225	435 N/cm²	b	
		Aged at 120°C, 1 000 h	−40	3 830 N/cm²	b	
		3 000 h	−40	2 760 N/cm²	b	
		30 000 h	−40	2 410 N/cm²		
		1 000 h	120	1 350 N/cm²		
		3 000 h	120	1 520 N/cm²		
		30 000 h	120	1 030 N/cm²	b	
		Aged at 130°C, 1 000 h	−40	2 760 N/cm²	b	
		3 000 h	−40	2 730 N/cm²		
		20 000 h	−40	3 090 N/cm²	b	
		1 000 h	130	1 270 N/cm²		
		3 000 h	130	1 260 N/cm²	b	
		20 000 h	130	1 220 N/cm²	b	
		Aged at 150°C, 1 000 h	−40	3 180 N/cm²		
		3 000 h	−40	3 120 N/cm²	b	
		1 000 h	150	1 380 N/cm²		
		3 000 h	150	1 310 N/cm²	b	
		Sustained loading at test temperature	100	Withstands 536 N/cm² (+ 500 h)		
			150	Ruptured by 513 N/cm² (200 h)		
Aluminium (alloy) – aluminium to DTD 746	ASTM D1002–53T	Etched in chromic/ sulphuric acid to DTD 915B	20	2 658 N/cm²	b	Subject to creep at 100°C, 150°C for sustained loading (40% ultimate tensile shear strength produces up to 0·3 mm deformation with standard lap joint, in 500 h) Data provided by B.A.C. (Operating) Ltd., Filton, Bristol
			125	1 075 N/cm²	b	
		Aged at 130°C, 1 000 h	−40	3 060 N/cm²	b	
		10 000 h	−40	3 540 N/cm²	b	
		30 000 h	−40	3 090 N/cm²	b	
		1 000 h	130	1 570 N/cm²	b	
		10 000 h	130	1 545 N/cm²		
		30 000 h	130	2 040 N/cm²	b	
		Aged at 150°C, 1 000 h	−40	2 290 N/cm²		
		3 000 h	−40	2 200 N/cm²		
		1 000 h	150	1 100 N/cm²	b	
		3 000 h	150	1 210 N/cm²		
		Sustained loading at test temperature	150	Withstands 560 N/cm² (+ 500 h)		
Steel Aluminium Natural rubber Polychloroprene Acrylonitrile rubber Butyl rubber Polyurethane rubber Leather Paper Fabric				Complies with DTD 900/4679 when used with primer 9247, 9250 and 9252		Resistant to water, petrol, oil, kerosene. Not resistant to ester, ketones, aromatic and chlorinated hydrocarbons. Withstands 5N sulphuric acid and 5N caustic soda solutions. Cured film is resistant to 100% R.H. at 38°C for 14 d. Inflammable, Flash Pt. is 0 to −7°C. Coverage, 4 m²/l

Trade source	Trade name or designation	Basic type	No. of compo-nents	Colour	Physical form, consistency or viscosity Ns/m^2	Working life	Method of application	Processing		Service temp. range °C	Main uses
								Curing cycle	Bonding pressure N/cm^2		
47	Bostik Clear adhesive	Acrylo-nitrile rubber in ketone solvent + synthetic resin	1	Trans-parent	Medium viscosity liquid 31±2% solids	Tack free time is 20 min at 20°C	Brush Roller Spray gun	Dried at 20°C and contact bonded. May heat reactivate dried film at 100°C and bond under pressure	Contact to 20	−5 to 85	General purpose adhesive for plastics, metals, and porous adherends. Toy manufacture, furniture, leather goods, upholstery
35	Boscoprene 2762	Acrylo-nitrile rubber base in ketone solvent + catalyst	2	Trans-parent blue or pink	Liquid 29% solids	Tack free time is 15–20 min Pot-life 8 h at 20°C	Stiff brush Roller Spray gun	48 h at 20°C to 16 h at 50°C. May heat reactivate dried film at 100°C and bond under pressure	Contact to 20	−40 to 100	General purpose adhesive for metals, plastics, rubber and textiles where tough resilient bond is required
47	Bostik 1762 and 1763	Nitrile rubber resin in ketonic solvent	1	Col-ourless		Open time 5–10 min at 20°C	Roller coat (1762) Knife blade (1763)	Applied to one substrate and as-sembled while wet. Sets by drying	Maxi-mum contact pressure		Bonding of poly-urethane and PVC foams. Lamination of plastics sheets to foams. Automotive applications
47	Bostik 2762	Nitrile rubber-resin in ester/ketone solvents	2	Col-ourless	Medium viscosity solution 10 29% w/w solids	Pot life 12 h at 20°C	Brush Spray Roller coat	Cures in 48 h to 7 d (optimum) at 20°C or 16 h at 50°C	Contact to 4	−40 to 100	Applications where tough resilient bond is required with good resistance to temperature and petroleum fuels. Nitrile rubber bag-tank manufacture

Adherends	Test spec.	Test conditions	Test temp. °C	Physical data	Key	Remarks
Aluminium Wood Leather Canvas Polyvinyl chloride				Adhesion to metal improved by use of Primer 9240		Non-staining adhesive. Resistant to water, petrol, oil, alcohols provided immersion is not prolonged. Good resistance to most acids and alkalies. Dried film may be solvent reactivated with acetone. Inflammable, Flash Pt. below 3°C. Coverage, 2·9 m²/l
Wood Aluminium Polyamide Polyvinyl chloride (flexible, rigid) Leather Canvas Acrylonitrile rubber				Complies with DTD 900/4666 for bonding nitrile rubber to itself. Metal adhesion improved by primer 9240 application		Non-staining adhesive with good resistance to petrol, oil, kerosene, Avtur, Avtag, Augas fuels, mild acids and alkalies. Swells in chlorinated hydrocarbons. Inflammable, Flash Pt. 0 to −7°C Coverage, 4 m²/l
Polyurethane Polyvinyl chloride						Unsuitable for non-porous surfaces such as metal or painted metals. Both products suitable for H.F. welding, e.g. PVC and substrate. Resistant to oils, fuels, water, alcohols in short term but not ketones and esters. Flash Pt. below −2·8°C
Metals Wood Nitrile rubber Nylon PVC (plasticised and rigid) Porous materials						Approved to DTD 900/4666A. Amenable to solvent at heat activation (at 100°C) application technique within 8 h of preparation. Poor resistance to esters, ketones, and chlorinated hydrocarbons. Flash Pt. −17 to −7°C. Coverage 2·8 m²/l

Trade source	Trade name or designation	Basic type	No. of compo- nents	Colour	Physical form, consistency or viscosity Ns/m^2	Working life	Method of application	Processing		Service temp. range °C	Main uses
								Curing cycle	Bonding pressure N/cm^2		
193	EC776	Acrylo-nitrile rubber and phenolic resin in ketone solvent	1	Amber	Low viscosity liquid 23% w/w solids	10–20 min at 20°C	Brush Flow gun	Air dried 10 min at 20°C and contact bonded. Heat reactiva-tion at 120–150°C	Contact to 20	−75 to 121	Synthetic rubbers to rigid materials where good hydrocarbon fuel resistance is required. Fuel tank sealant
12	X303	Acrylo-nitrile rubber and synthetic resin in solvent	1	Green-tan	Liquid	20 min at 20°C	Brush Trowel	Air dried 10–20 min at 20°C and contact bonded	Contact to 7		Rubber to metal bonding. Vinyl plastics and control fabrics
123	Pliobond 40	Acrylo-nitrile in methyl-ethyl ketone	1	Tan	Liquid 40% w/w solids		Brush Spray Roller coater	Dried at 20°C or in oven at 93°C for 5 min. Cured under pressure at 93–150°C for 15 min	68–210	121 to 16°	Bonding a variety of materials such as metals, woods, ceramics, fabrics and plastics. Attaching metal name plates to machinery. Assemblage of electrical components requiring high insulation
16	Holdtite RC 4220	Acrylo-nitrile in ketone solvent	1	Pale brown	Liquid 0·2–0·35 at 20°C	Open time 10 min at 20°C	Brush Scraper	Contact bonding after 10 min at 20°C or heat activation of dried film at 90–100°C for 2 min			High bond strength adhesive with good electrical insulation properties. Packaging, electrical furniture and engineering industries

Adherends	Test spec.	Test conditions	Test temp. °C	Physical data	Key	Remarks
Steel Glass Rubbers (Buna-N) Plastic laminates Cotton (duck) – aluminium			20	Conforming to DTD 900/4452 26 N/cm	a	Retains flexible properties at −75°C. Resistant to oil, petrol, hydrocarbon fuels and water (24 h at least) Inflammable. Flash Pt. 4°C. Coverage, 6 m²/l
Rubber (Buna-N) Aluminium Steel Polyvinyl chloride				275 N/cm²	b	Resistant to heat and oil. May be thinned with acetone. Inflammable. Flash Pt. −7°C. Coverage, 3·7 m²/l
Canvas – aluminium Canvas – steel (stainless) Canvas – wood (oak) Canvas – canvas Neoprene – neoprene Leather – leather Aluminium – aluminium Steel (stainless) – steel Wood (maple) – wood Wood (maple) – aluminium Wood (maple) – steel Wood (birch) – wood Plywood (fir) – plywood Phenolic (laminate) – phenolic		Shear strength data for specimens cured at 150°C for 10 min		27·5 N/cm² 34·3 N/cm² 68·9 N/cm² Cohesive failure of canvas 19·3 N/cm² 34·3 N/cm² 1 650 N/cm² 1 240 N/cm² 930 N/cm²* 930 N/cm²* 895 N/cm²* 825 N/cm²* 345 N/cm²* 618 N/cm² *cohesive failure of wood	b b b b b b b b b b b b b b	Cures to a permanently flexible bond which does not embrittle on ageing or exposure to low temperature. Resistant to water, weak acids and alkalies, salt solutions lubricating oils and most chlorinated solvents. May be thinned with acetone or cyclohexanone. Bonds subject to creep under sustained loading; 140 N/cm² in shear causes failure of aluminium joints in 200 h at 25°C. Not recommended for heavily loaded structural assemblies. Shear strengths decrease to 25% of 20°C value at 65°C, due to thermoplastic nature of adhesive
Concrete Porcelain Polyester (film) Glass Fabrics Brass Aluminium Steel Wood			15 15 15 15	$2·2 \times 10^9$ Ω/cm 194 kV/cm 9·3 (at 10^3 Hz) 0·028 (at 10^3 Hz) Conforms to A.F.S. Spec. 585	p m n o	Flexible film when cured. Resistance to moisture and many corrosive solutions. Non-staining. Available in alternative grades, RC 4230—higher viscosity RC 4240 – for elevated temperature use. Inflammable, Flash Pt. −17°C. Coverage, 3·3 m²/l

Trade source	Trade name or designation	Basic type	No. of compo-nents	Colour	Physical form, consistency or viscosity Ns/m^2	Working life	Method of application	Processing		Service temp. range °C	Main uses
								Curing cycle	Bonding pressure N/cm^2		
280	Phenoxy resin PAHJ	Phenoxy resin	1	White	Cylindrical pellets, $\frac{1}{8}$ in dia	Indefinite as pellets	Dissolve pellets in any common organic solvent, then brush on, or rollers, or spatula. A 'film' form can be obtained on request	Varied; 10 s at 340°, 3 min at 260° or 30 min at 190°	Enough to maintain alignment only	−70 to 120	As a fast bonding material for metals, wood and paper. Also for cloths, foils etc. For pipes and automobiles to
193	EC2214	Epoxy–nylon	1	Silvery	100% solids, a paste	6 mth at 25°C	Spatula, spreader or very stiff brush	40 min at 121°C	Light pressure	−60 to 170	As a flexible epoxy combining strength with good peel properties
91	FM34	Polyimide	1	Olive green	As a filled resin on a glass cloth i.e. a leathery consistency contains 14% volatiles	6 mth if kept below −20°C or 2 wk at R.T.	Cut out the amount required and sandwich in joint	30 min to heat up to 280°C, then 90 min at 280°C	280	−40 to 300, but for short spells up to 350	As a high tempera-ture structural adhesive; in aircraft Bonds metals, alloy and many other materials
91	Blooming-dale HT 424	Epoxy/phenolic resins + aluminium powder filler	1	Grey	Film supported on glass cloth (0·38 mm thick)	12 d at 20°C	Press curing of film	52 N/cm² applied at 20°C; heated to 165°C in 40 min under pressure. Pressure released, on cooling, at 50°C	52	−40 to 150	Metal bonding for structural applications at elevated temperatures

Adherends	Test spec.	Test conditions	Test temp. °C	Physical data	Key	Remarks
uminium – aluminium	ASTM D1002–53T			2 200 N/cm²	b	Good creep resistance. Used as a lacquer for wood or as a general coating for metals. Bonds metals like gold, zinc and nickel, platinum, silver and copper very well. Poor adhesion to most plastics. Chemical resistance generally good – withstands ketones, alcohols, water, salt spray but not jet fuel or hydraulic oils. Can be easily mixed with dyes to give a desired colour
				2 400 N/cm²		
				2 000 N/cm²		
				1 100 N/cm²		
rch – birch	ASTM D906	7 d at 25°C in 50% R.H.		450 N/cm²	b	
	ASTM D906	3 h in water at 25°C		350 N/cm²	b	
uminium – aluminium	ASTM D1002–53T		−40	2 200 N/cm²	b	Useful single-component adhesive retaining good strength at low temperatures. Chemical resistance as for epoxies
			25	3 500 N/cm²	b	
			120	1 500 N/çm²	b	
			160	500 N/cm²	b	
el – steel			25	50 N/cm	a	
el – steel	MMM–A-132 Type IV	10 min at 260°C Aged at 260°C for 3 y	25	2 400 N/cm²	d	Tested at Sira, and found to be an excellent high temperature adhesive. Correct storage is essential *for optimum strength.* A primer is available to increase strength. Contains volatiles that come off during cure. Chemical resistance is excellent
			260	1 460 N/cm²	d	
			260	1 400 N/cm²	d	
			25	14 N/cm	a	
			25	8 000 N/cm²	e	
			300	3 500 N/cm²	e	
uminium (alloy) – aluminium .o DTD 746	ASTM D1002–53T	Etched in chromic sulphuric acid to DTD 915B	20	2 562 N/cm²	b	Good bond strength retention, for short periods, up to 100°C. Bond strength deterioration for 1 000 h immersion in fluids at 130°C is less than that for thermal ageing in air, e.g. Kerosene (Spec. DERD 2494), Hydraulic oil (Spec. DTD.585) Chlorinated silicone oil (Spec. DTD.900/4725) Synthetic ester lubricant (Spec. DERD.2487). Not subject to creep at 100°C, 150°C, for sustained loads (40% ultimate tensile shear strength.) Fatigue strength exceeds 25% ± 10%. Ultimate shear strength at 10⁷ cycles (50 Hz) at 150°C (*See below)
			125	1 644 N/cm²	b	
			200	1 450 N/cm²	b	
			225	1 402 N/cm²	b	
		Aged at 120°C 1 000 h	−40	2 340 N/cm²	b	
		10 000 h	−40	910 N/cm²	b	
		30 000 h	−40	440 N/cm²	b	
		1 000 h	120	1 990 N/cm²	b	
		10 000 h	120	965 N/cm²	b	
		30 000 h	120	3 860 N/cm²	b	
		Aged at 130°C 1 000 h	−40	2 180 N/cm²	b	
		10 000 h	−40	855 N/cm²	b	
		20 000 h	−40	551 N/cm²	b	
		1 000 h	130	2 040 N/cm²	b	
		10 000 h	130	635 N/cm²	b	
		20 000 h	130	386 N/cm²		
		Aged at 150°C 1 000 h	−40	2 280 N/cm²		
		1 000 h	150	1 760 N/cm²		
		Sustained loading at test temperature	150	Withstands 1 070 N/cm² (+ 500 h)	b	

Trade source	Trade name or designation	Basic type	No. of compo-nents	Colour	Physical form, consistency or viscosity Ns/m^2	Working life	Method of application	Processing		Service temp. range °C	Main uses
								Curing cycle	Bonding pressure N/cm^2		
73	Hidux 1197A	Epoxy/ phenolic resins	1	Olive	Film supported on glass cloth (0·3 mm thick)		Press curing of film	Cured for 5 min at 100°C and 7 N/cm^2 pressure +30 min at 150°C and 70 N/cm^2 pressure. Pressure released, on cooling, at 50°C	69	−40 to 150	Metal bonding for structural applica-tions at elevated temperatures
191	Tego glue film	Phenolic formalde-hyde	1	Yellow	Paper based films (various) 60–80% resin con-tent of thickness, 0·06–0·1 mm	Indefinite	Films sand-wiched between adherends	Film thickness dependent 3 min at 157°C to 9 min at 130°C under pressure	Up to 1 600	150	Furniture manu-facture. Veneering and waterproof plywood fabrication. Bonding of decorative plastic laminates to wood. Aircraft and boat building applications
191	Meno resin film Grade 5	Melamine/ urea for-maldehyde blend	1	Yellow	Paper based films (various) 75% w/w resin (0·09 mm thickness)	Indefinite	Film sand-wiched between adherends	2 min at 95°C to ½ min at 130°C	69 to 450	130	Bonding and surfacing veneers, plywood and chipboard. Plastic decorative laminate bonding
45	Cascophen RS-216-M*	Resorcinol formalde-hyde + catalyst RXS-8	2	Brown	Liquid	8 h at 16°C to 20 min at 38°C	Brush Spatula Mechan-ical spreader	Cured at 16°C to 80°C under pressure	Contact to 70 5 min at 82°C or 4 h at 21°C or 10 h at 16°C		Constructional laminates for man craft. Building and Timber application Aluminium — ply-wood bonding. Laminated plastic

Adherends	Test spec.	Test conditions	Test temp. °C	Physical data	Key	Remarks
...minium (alloy) – aluminium ... DTD 746	ASTM D1002–53T	Etched in chromic – sulphuric acid to DTD 915B	20 125	2 949 N/cm² 1 619 N/cm²	b b	Good bond strength retention, for long periods, up to 150°C
		Aged at 120°C 1 000 h	−40	1 670 N/cm²	b	*Data provided by
		3 000 h	−40	860 N/cm²	b	B.A.C. (Operating) Ltd.,
		30 000 h	−40	620 N/cm²	b	Filton, Bristol
		1 000 h	120	1 340 N/cm²		
		3 000 h	120	725 N/cm²	b	
		30 000 h	120	483 N/cm²	b	
		Aged at 130°C 1 000 h	−40	1 725 N/cm²		
		3 000 h	−40	1 100 N/cm²	b	
		30 000 h	−40	585 N/cm²	b	
		1 000 h	130	1 340 N/cm²	b	
		3 000 h	130	895 N/cm²	b	
		30 000 h	130	518 N/cm²	b	
		Aged at 150°C 1 000 h	−40	1 270 N/cm²	b	
		3 000 h	−40	990 N/cm²	b	
		1 000 h	150	1 170 N/cm²	b	
		3 000 h	150	825 N/cm²	b	
...vinyl chloride (rigid) ...tic laminates ...board ...wood ...d (birch) – wood	BS 1455: 1963 BS 1203: 1963	Knife test Dry strength Wet strength Mycological test		Complies with BS Spec. for WBP Plywood 8–10 Dry 6–8 Wet 172 N/cm² 138 N/cm² 138 N/cm²	b b b	Available in rolls up to 220 cm wide. Resistant to boiling water and biodeterioration. Greater flexibility than liquid-resins for bonding plywoods. Steam-heated steel platen press required for laminating work with this adhesive
...wood ...board ...dboard ...tic laminates				Complies with BS 1203 : 1963 Types MR and BR		Available in rolls up to 220 cm wide and as alternative Grade 4. Resistant to cold water, fungi, but not steam or boiling water. Suitable for bonding assemblies requiring lower bonding pressures than phenolic resin film – Tego
...d (including Teak) ...estos ...minium ...nolic laminate ...styrene (foam) ...vinyl chloride ...amide (rigid)				Complies with BS 1204, Pts. I, II, types WBP/CC and WBP/GF. Also, BS 1203/WBP		Should be cured above 16°C. Open assembly time, 25 min at 16°C. Recommended for severe outdoor exposure conditions. Combines high strength with durability. Gap-filling. Resistant to cold or boiling water, moist or dry heat and biodeterioration. Withstands tropical environments without strength deterioration. Casco Primer DR-4 (casein latex) recommended for aluminium–wood jointing

Trade source	Trade name or designation	Basic type	No. of compo- nents	Colour	Physical form, consistency or viscosity Ns/m^2	Working life	Method of application	Processing		Service temp. range °C	Main uses
								Curing cycle	Bonding pressure N/cm^2		
45	Cascamite 'One Shot'	Urea formalde- hyde	1	White	Powder	Curing tempera- ture dependent	Brush Roller Spreader	9 h at 10°C to 1 h at 21°C after mixing powder with water (22%)	4–40 for 18 h at 10°C to 3 h at 21°C		Wood gluing and bonding of plastic laminates to wood Plywood, chipboard manufacture. Boat building and timber engineering
73	Melolam 295	Melamine formalde- hyde (plasticised)	1	White	Powder, added to solvent to give 0·03– 0·05 solution		Dipper Spreader	For laminates 135– 145°C under pressure	For lami- nates 690 to 1 030 for 18–20 min	180	Manufacture of post-forming laminates
73	Aerophen 1201 + Hardener HP1 or Hardener HP2	Phenol formalde- hyde	2	Red- Brown	0·25–0·42 43–45 % solids	4 h at 20°C	Roller coater	5 min at 110°C to 2 min at 140°C 4 min at 110°C to $1\frac{1}{2}$ min at 140°C	Timber density depen- dent 96 to 180		Manufacture of plywood and similar materials
73	Aerodux 185 + Hardener HRP150 or Hardener HRP 155	Resorcinol formalde- hyde	2	Red- Brown	Liquid	7–8 h at 10°C to 1 h at 30°C	Brush Spreader	1 h at 40°C to $1\frac{1}{2}$ min at 100°C	12 h at 10°C to $1\frac{1}{2}$ h at 30°C for pres- sures of 34–70		Wood and porous cellulosic material plastics. Bonding synthetic and natural rubbers to wood. Bonding Asbestos-based boards to wood veneer and plastic
73	Aerolite H–F resin + Hardener HF	Urea formalde- hyde	2		Low vis- cosity liquid	18–20 h at 21°C	Spreader	8 min at 71°C to 1 min at 110°C			Wood laminates

Adherends	Test spec.	Test conditions	Test temp. °C	Physical data	Key	Remarks
od nolic laminate						Complies with BS 1204/MR. May be extended with 20% w/w wood flour for veneering applications. Excess glue may be removed with soapy water before setting has taken place. Avoid copper, brass, or ferrous metal vessels for mixing the adhesive. Coverage, 7 m²/kg
er, od fibre boards pboard						Prepared as a solution in water–alcohol mixture (10–20% alcohol) up to 40–60% w/w resin, for impregnation of paper for subsequent laminate manufacture. Laminates may be prepared with clamping pressures down to 206 N/cm². Laminates are resistant to light, boiling water (+2 h) and alkali staining (10 min in hot 10% caustic soda)
od (veneers of moisture ontent 5–7%). % moisture content not ecommended				Complies with plywood specs. BS 1455, BS 1088, DIN 68 705, CS 250, IS 710		Hardener HP1 suitable for absorbent timbers (beech). Hardener HP2 is preferable for 'difficult' timbers (Canadian birch). Mixture remains liquid for 24 h (up to 40°C) and can then be mixed in with fresh resin/hardener
estos crete amics and porcelain nglazed) od (inc. teak) yvinyl chloride ystyrene (expanded) yurethane (expanded) nite (expanded) ther yamide stic laminates				Complies with BS 1204. Parts 1, 2, (Type WBP) also, BS 1203, Part 1		Cold setting, gap filling material which should be applied 5 min after mixing. Pressure application should be within 2½ h at 10°C to 15 min at 30°C. Hardener HRP 155 is recommended where maximum assembly time is required for large bonding areas. Not suitable for silicone rubber bonding. Coverage, 1 m² per 225 g
						Fast curing adhesive when used with glue-line heating under radio-frequency conditions. Alternative catalysts, L38, L48, L50, may be used

Trade source	Trade name or designation	Basic type	No. of compo-nents	Colour	Physical form, consistency or viscosity Ns/m^2	Working life	Method of application	Processing		Service temp. range °C	Main uses
								Curing cycle	Bonding pressure N/cm^2		
38	Beetle Cement W2	Urea formalde-hyde + catalyst	2		Liquid 2–3 at 25°C	Pot-life varies from 10 min at 10°C to 10 h at 32°C for catalyst used	Brush Spreader Roller	Cured under pressure 10–82°C	Up to 70 Press times vary from $1\frac{1}{2}$ min at 82°C to 15 h at 10°C for catalyst used		General purpose wood adhesive
38	Beetle Cement W37	Urea formalde-hyde + catalyst	2		Heavy fluid paste	Pot-life varies from 12 min at 32°C to 18 h at 10°C for catalyst used	Brush Spreader Roller	Cured under pressure 10°C to 99°C	Up to 70 Press times vary from $1\frac{1}{4}$ min at 99°C to 60 h at 10°C for catalyst used		Bonding fabric paper-board laminates to wo Not intended for non-absorbent materials such a glass fabrics or thermoplastic laminates. Gap filling resins. Boat building
38	Beetle Cement W54	Urea formalde-hyde + catalyst	2		Low vis-cosity liquid $68\pm2\%$ solids content	Pot-life varies from $\frac{3}{4}$ h at 32°C to +48 h at 10°C for catalyst used	Brush Spreader	Cured under pressure 70°C to 150°C	Up to 70 Press times vary from 15 s at 149°C to 90 s at 71°C for catalyst used		Wood working where long worl life with short pressing times are required
73	Aerolite 300 + Hardener GU.X	Urea formalde-hyde	2	Green Blue Brown for vari-ous cata-lysts	Liquid	10–20 min	Brush Spreader	Cured under pressure from 15–30°C	Press times vary from $5\frac{1}{2}$ h at 15°C to $2\frac{1}{4}$ h at 30°C		Bonding lamina plastics to wood by the separate application method
	Hardener GB.Q X					5 min			6 h at 10°C to 1 h at 32°C		
	Hardener GB.P X					10 min			2 h at 21°C to 1 h at 32°C		
	Hardener GB.M X					20 min			$3\frac{1}{2}$ h at 21°C to 2 h at 32°C		

Adherends	Test spec.	Test conditions	Test temp. °C	Physical data	Key	Remarks
ulosic materials dboard board voods and veneers per of moisture content tween 5 and 15% w/w	BS 1203 MR BS 1204 MR/GF BS 1203 BR	3 h in water at 67°C 3 h in water at 67°C with joint gap 1·25 mm 3 h in water at 100°C		In general, selection of suitable catalyst ensures that adhesives comply with these specifications. Melamine hardener is available to satisfy BS 1203 BR		W2 is available in powder form (W36) for reconstitution with water. Other grades are available for use with powder catalysts. Liquid catalysts enable a 'separate application' method to be used, i.e. hardener applied to one surface and resin to the other. W52 grade has been developed for high frequency curing methods. Mixing vessels of mild steel, enamelware or polythene are suitable; copper and brass should not be used. Alkaline glues such as casein and sodium silicate degrade these adhesives. Good water resistance and excellent resistance to mould growth. Coverage, 1 m² per 150–250 g
d c laminates corative and non-corative)				Complies with BS 1204 Part 1 and Part 2 (Type MR)		Gap filling and moisture resistant. Resistant to biodeterioration. Hardeners can be supplied in various colours to assist application. Hardeners GU (green), GBP (green) GPQ (brown), GBM (blue). Contamination with iron should be avoided, since wood staining will occur with high tannin content woods (e.g. oak, ash). Hardener GU.X permits separate application of resin and catalyst to each adherend before bonding

Trade source	Trade name or designation	Basic type	No. of compo-nents	Colour	Physical form, consistency or viscosity Ns/m²	Working life	Method of application	Processing		Service temp. range °C	Main uses
								Curing cycle	Bonding pressure N/cm²		
57	Aerolite 306 + Hardener GU.X	Urea formalde-hyde	2	as for Aero-lite 300	Powder	10–20 min	Brush Spreader	Cured under pressure from 15–30°C	Press times vary from 5½ h at 15°C to 2¼ h at 30°C		Bonding laminate plastics to wood by the separate application method
287	Narmco Imidite 1850	Polybenz-imidiazole	1	Green	As a filled resin on a glass cloth	3 mth at RT longer if refriger-ated	Cut out the amount required and sandwich in joint	Varied, depend-ing on area being bonded, e.g. Heat to 370°C apply pressure and leave for 3 h cool slowly. Post-cure recom-mended	Max. of 1 400. Min of 400	−40 to 550, but for short spells up to 650	As a high temper-ture structural adhesive; in aircraft, missiles, etc.
66	Chelsea 391	Poly-chloro-prene	1		Syrup $9·5 \times 10^{-3}$ m²/s at 18°C	Open time is 35 min at 20°C	Brush	Contact bonding within 35 min of drying in air. Heat re-activate 2 min at 90°C	15 s at 45		Footwear manu-facture, leathers, rubbers and plastic coated fabrics
47	Bostik 1 GA 186	Polychloro-prene-resin in hydro-carbon solvent	1	Buff	Medium viscosity liquid	Tack life for one-way wet bonding 1–5 min; two-way wet bonding 5–10 min; two-way dry bonding with solvent activation 15–40 min	Brush	Sets by drying at room tempera-ture. Bonding by wet, semi-wet or solvent activation of dried coatings depend-ing on substrate	Contact to 4		Adhesives for use automobile indu Car-trim applica for porous and non-porous adherends
47	Bostik 1 GA 516		1	Light Brown	Liquid 7 Solids content, 36% w/w		Brush Sticker Roller coater				1 GA 186—gene purpose adhesive 1 GA 516 } severe 1 GA 585 { temps up to 150°C short perio
47	Bostik 1 GA 585	Polychloro-prene-resin in hydro-carbon/ ester solvent	1	Light brown	Liquid 3 30% w/w solids		Brush Scraper Pressure equip-ment				

Adherends	Test spec.	Test conditions	Test temp. °C	Physical data	Key	Remarks
r Aerolite 300				Complies with BS 1204. Parts 1, 2, (Type MR)		Powder version of Aerolite 300 with similar properties. Powder adhesive is mixed with water and applied to the surface. May be used with Hardeners GBP.X, GBQ.X GBM.X
—steel		Aged 1 mth Aged 1 mth Aged 1 mth	25 150 350	2 700 N/cm² 2 600 N/cm² 2 000 N/cm²	b b b	An excellent high temperature adhesive but rather awkward processing condition are required for optimum performance i.e. post curing in nitrogen atmosphere. Available as a resin. Expensive material. Contact with supplier recommended for additional data. Also, recent papers in BODNAR, M. J., (Ed.), *Structural Adhesives Bonding, Polymer Symposia, No. 3,* Interscience (1966)
her (chrome)—leather her (chrome)—leather her (chrome)—resin rubber her (chrome)—latex blend her (chrome)—rubber BR)		After 48 h in water at 40°C SATRA Tensiometer 180°C Peel test	20 20 60 20 20	56 N/cm 31 N/cm 33 N/cm 82 N/cm 70 N/cm	a a a a a	Adhesive may be solvent reactivated with acetone. Low temperature storage should be avoided. Inflammable, Flash Pt. 16°C
(painted) ral rubber prene rubber vas d						Tack life considerably reduced in humid environment. Consistency reduced by use of Bostik Thinner 6001. Flash Pt. −3°C
						Suitable for painted surfaces based on cellulose or synthetic enamel. Good resistance to sunlight, water and humidity. Fair resistance to oils and hydrocarbon fluids. May be thinned with Bostik 6020. Coverage 4·5 m²/1. Flash Pt. −17 to −7°C. As for 1 GA 516. Alternative product (36% w/w solids) available as 1 GA 581

Trade source	Trade name or designation	Basic type	No. of compo-nents	Colour	Physical form, consistency or viscosity Ns/m^2	Working life	Method of application	Processing		Service temp. range °C	Main uses
								Curing cycle	Bonding pressure N/cm^2		
47	Bostik 1440		1	Red	Liquid 27% w/w solids	Open time 15–20 min at 20°C	Brush	Sets by drying at room tempera-ture		−40 to 60	Specially formula for bonding Bost foam pads to a variety of surface
47	Bostik 1444	Polychloro-prene-resin in hydro-carbon solvent	1	Red	Liquid 20% w/w solids	Open time normally 20 min at 20°C	Spray (cold or hot)			−40 to 120	Sprayable contac adhesive for bon laminated plastic boards, and sandwich panels
47	Bostik 3403		1	White		Open time 30 min at 20°C	Brush Sticker			−20 to 70	Polyurethane (es or ether) foam adhesive for assembly to fabri and other porou materials
47	Bostik 2402	Polychloro-prene	2	Brown	Liquid 3 28% w/w solids	Pot life 8 h at 20°C. Tack life 20–30 min at 20°C	Brush Roller coater	Cures in 72 h at 20°C to 15 h at 70°C. Optimum strength after 7 d		−40 to 80	Wide range of industrial applications for joining rubber, plastics, and oth materials. Heat resistant adhesive for laminated plastic Aircraft applicat involving rubber overshoes to propeller blades. Inflatable rubber products
66	Chelsea 381	Polychloro-prene-resin blend in solvent	1	Light beige	Syrup 6 at 20°C	Open time 1 h at 20°C	Brush	Contact bonded within 1 h after drying. Heat activation 20 s at 90°C	55 (15 s)		Footwear manu-facture, leathers, rubbers and plas fabrics (coated) wood laminatior
107	Evo-stik Impact 528	Polychloro-prene + Synthetic resins in hydro-carbon solvents	1	Off-white to amber	Liquid 3–4 at 25°C	Tacky 15 min after applica-tion to porous substrate or 30 min on non-porous material	Brush Spreader	Contact bonding of touch-dry adhesive or re-activate dried film at 85–95°C with heat lamp	Contact to 20		Bonding of laminated plastic and rigid plastic wood, metal, and boards. Fibre and cane f boards. Plastic covering

Adherends	Test spec.	Test conditions	Test temp. °C	Physical data	Key	Remarks
ter ·d ·kwork ·tic laminates ·als		After 24 h	20 20 20 20	27 N/cm² 27 N/cm² 41 N/cm² 41 N/cm²	b d b d	Bostik Thinner M 501 available. Good resistance to oils, fuels, alcohols, mild acids and alkalies. Coverage 2 m²/1. Flash Pt. less than −3°C
·tic laminate ·rd materials ·d						As for Bostik 1440. Coverage 7–10 m²/1. Flash Pt. −17 to −7°C
·vas ·ric materials ·urethane foams						Soft, resilient bonds. Thinner 6020 available to reduce viscosity. Coverage 7 m²/1. Flash Pt. −17 to −7°C
·prene ·ile ·l ·alon ·urethane ·: (rigid and plasticised) ·d						Tough, resilient bonds. Approved to DTD 900/4679A when used with Bostik 9252 primer. Thinners M501 or 6104 available to reduce viscosity. Resistant to oils, fuels, water but not esters and ketones. Unaffected by caustic soda (5N). Coverage 2–4 m²/1. Flash Pt. −17 to −7°C
·her (chrome)—leather ·her (chrome)—rubber/resin ·her (chrome)—latex blend ·her (chrome)—SBR rubber ·her (chrome)—leather		After 48 h in water (40°C)	20 60 20 20 20	31 N/cm 33 N/cm 82 N/cm 70 N/cm 56 N/cm	a a a a a	Peel data refer to SATRA (p. 332) tensiometer 180° peel test. Adhesive may be solvent reactivated prior to bonding. Low temperature storage should be avoided. Flash Pt. below 0°C
·on (duck)—cotton			20 20 20	61–78 N/cm 43–61 N/cm 241–275 N/cm² 206–413 N/cm²	a a b b	Unsuitable for bituminous substrates or expanded polystyrene. About 50% reduction in bond strength at 80°C. Resistant to 120°C if 7·5% w/w accelerator DF added (pot life is then 6 h at 20°C). Resistant to water, dilute acids alkalies and many aliphatic hydrocarbon solvents. Coverage 5·5 m²/1 on non-porous surface. Approved to DTD 900/4564

Trade source	Trade name or designation	Basic type	No. of compo-nents	Colour	Physical form, consistency or viscosity Ns/m^2	Working life	Method of application	Processing		Service temp. range °C	Main uses
								Curing cycle	Bonding pressure N/cm^2		
276	Tuf Stick 37	Neoprene rubber in solvent blend	1	Gold/green	Medium viscosity syrup	Indefinite	Stiff brush or scraper	Dried for 20 min at 20°C and joint made by contact bonding	Contact to 4		General purpose impact adhesive
18	Contact Bond Cement 3A4	Polychloro-prene	1	Buff	Liquid solids content 21% w/w	Open time 10 min at 20°C	Brush Spray	Sets at room temp. after 0–15 min drying time	Contact to 4	−20 to 90	Contact adhesive for porous and impermeable materials
98	Dunlop RF	Poly-chloro-prene	1	Light Brown	Heavy syrup	10–60 min at 20°C	Brush Spreader	Air dried 10–20 min at 20°C	Contact		Bonding all types of rubber flooring to metals, woods and masonry
177	Clam* No. 3	Poly-chloro-prene in spirit base	1	Off White	Syrup	Tacky 15 min after appli-cation	Brush Spreader	Contact bonding or Pres-sure on dried adhesive at 110°C for 2 min	15	Softens at 70°C	General purpose adhesive
259	Spectra-bond	Poly-chloro-prene in hydro-carbon solvents	1		Syrupy liquid	Tacky 15 min after appli-cation	Brush	Contact bonding of tack-free adhesive	Contact to 20		General purpose contact adhesive for metals, plastic laminates, insulation boards fabrics and rubbe Cement products and wood bondin
113	Instabond	Poly-chloro-prene in hydro carbon solvents	1		Syrup	Up to 6 h at 20°C	Brush Spreader	Contact bonding of tack-free adhesive	Contact to 20	4 to 50	Plastic laminates to metal sheets, plywood.

Adherends	Test spec.	Test conditions	Test temp. °C	Physical data	Key	Remarks
ous materials ther ber od—wood l—steel	BS 4071: 1966	After 3 d at 20°C	20 20	270 N/cm² 110 N/cm²	b b	Highly inflammable solvent base (Flash Pt. −12°C). Maximum service temperature of 45°C for short periods
estos ster crete als						Rapid drying adhesive based on flammable solvents (Flash Pt. below 10°C). Fire resistant, when dry, to BS 467 Part 7 (Class 1)
ber l od crete						Good heat resistance; suitable for underfloor heating applications. Stored at 4–21°C. Inflammable, Flash Pt. 40°C. Coverage, 2·7 m²/l
amine sheet to wood, metal. rdboard to plaster. ther, glass and non-bsorbent materials				Generally stronger than any porous materials bonded		Adhesive coated surfaces may be heat or solvent reactivated. Not suitable for unbacked polyvinyl chloride. Store in sealed containers. Short tack time. Coverage, 2·5 m²/l. Resistant to moisture and ageing. Not resistant to plasticisers
estos rdboard ss fibre ther t od—wood od—plastic laminate l (stainless)—steel		 After 1 d at 20°C After 2 d at 20°C After 3 d at 20°C After 5 d at 20°C After 3 d at 20°C After 3 d at 20°C	 20 20 20 20 20 20	 175 N/cm² 186 N/cm² 220 N/cm² 268 N/cm² 200 N/cm² 110 N/cm²	 b b b b b b	Good retention of adhesion after 10 cycles of following, 24 h at 49°C +24 h in water at 32°C. +24 h in dry ice chamber. Inflammable. Coverage, 6 m²/l
od—wood l—steel		After 24 h at 20°C After 48 h at 20°C After 72 h at 20°C After 5 d at 20°C After 24 h at 20°C After +5 d at 49°C After 72 h at 20°C After, 72 h at 20°C+ After, 2 h at 121°C	20 20 20 20 20 20 20	176 N/cm² 186 N/cm² 220 N/cm² 270 N/cm² 110 N/cm² 110 N/cm² 21 N/cm²	b b b b b b b	Not damaged by freezing but will set at temperatures below 4°C. Bond strength for plastic to steel retained after 10 cycles. of, 24 h at 50°C+24 h in water +24 h in dry ice. Inflammable. Coverage, 6 m²/l

Trade source	Trade name or designation	Basic type	No. of compo-nents	Colour	Physical form, consistency or viscosity Ns/m^2	Working life	Method of application	Processing		Service temp. range °C	Main uses
								Curing cycle	Bonding pressure N/cm^2		
16	Holdtite RB 4380	Poly-chloro-prene in hydro carbon mixture	1	Pale fawn	Liquid 1·5–2 at 20°C	15 min at 20°C	Brush	Contact bonding of tack-free adhesive, or heat reactiva-tion at 95–105°C for 2 min under pressure	Contact to 20		Decorative laminates, upholstery materials and cellulosic substrates. Metals
38	Bexol 1528	Chlorinated rubber in solvents	1	Light brown	9 Solids content 29·0% w/w	Open-time is 30 min	Brush Spreader	Dried at 20°C for 10 min	Contact to 4		A general purpos contact adhesive joining plastics t non-metallic substrates
172	UHU All-purpose	Polyvinyl ester in mixed ester solvent	1	Col-ourless	Liquid 3 35% w/w solids	Unspeci-fied	Brush	Sets by solvent loss and optimum strength after 24 h	Contact to 4		Domestic applica tions and genera purposes. Service temperat 70°C max
79	Invisibond	Polyester resin + catalyst	2		Paste	30 min after mixing	Spatula	Cold curing in 30 min	Contact to 7		Metal frames an edges to stone flooring
50	Bakelite SR 17449	Polyester resin + hardener 17448 + catalyst 17447	3		Liquid	2 h at 25°C	Spatula	60°C for 1 h and then slowly increased to 140°C and held for 1 h. Post cured for 1 h at 10°C above service temp	Contact	170	Strain-gauges ba on copper-nicke alloys
104	P2 Adhesive	Polyester resin + catalyst	2		Liquid	20 min at 20°C	Spatula	2–3 h at 20°C to 2 h at 100°C	Contact	180	Strain gauges ba on copper-nicke alloys with poly or cellulosic car material

Adherends	Test spec.	Test conditions	Test temp. °C	Physical data	Key	Remarks
‧od ‧yvinyl chloride (rigid, ‧exible) ‧thercloth ‧reboard ‧minium ‧rics				Sira tested as non-fabric tendering		Available in alternative grades, RB 4330 – high viscosity, 2·5–3·5 Ns/m² RB 4383 – spray version, 0·45–0·65 Ns/m² RB 4744 – non-flammable Inflammable, Flash Pt. − 5°C. Coverage, 3 m²/l
‧yvinyl chloride ‧ystyrene ‧ plastics ‧od ‧ber (natural)						Bonds given are tough with excellent durability. Water proof joints which are also resistant to oils and fuels. Based on inflammable petroleum solvents. Shelf life, 6 mth below 24°C
‧or UHU Hard adhesive		At 0·1 mm thickness film At 0·2 mm thickness film At 0·2 mm thickness film	20 20 20 20	$10^{14}\Omega$/cm 440 kV/cm $2 \times 10^{11}\Omega$ (surface res.) 490 N/cm²	p m d	Resistant to oils, fuels, and mild acids or alkalies, but not alcohol and esters. Flash Pt. − 24°C
‧minium ‧el ‧ncrete ‧ramics						Good gap filling properties
‧tals						Three constituents are liquids which may be dispersed with syringes. Shrinkage on curing is greater than is usual with epoxy resins. Unsuitable for use above 170°C
‧tals ‧yesters ‧od						Good resistance to humidity. For high temperature use, a post cure at 180°C for 2 h is recommended. Inflammable

Trade source	Trade name or designation	Basic type	No. of compo-nents	Colour	Physical form, consistency or viscosity Ns/m^2	Working life	Method of application	Processing		Service temp. range °C	Main uses
								Curing cycle	Bonding pressure N/cm^2		
104	PS Adhesive	Polyester resin + catalyst	2		Liquid	20 min at 20°C	Spatula	2–3 h at 20°C	Contact	180	Strain gauge attachment to concrete
47	Boscoprene 2115–5	Poly-sulphide rubber in ketone solvent and catalyst	2	Buff	Thixotropic paste 88% solids	2–3 h at 25°C	Extruder gun Brush Spatula	3 d at 25°C		−50 to 130 With-stands higher temps for short periods	Sealant for fuel ta and pressurised cabins in aircraft where good weat proof and water-proof properties required
94	Flexane 95*	Poly-urethane	2	Black	Liquid 31 000 cps	47 min	Spatula Trowel	24 h at 25°C	Contact	120	Repair of rubber rolls, hoses, belti casting and caul Metal bonding
94	Flexane 60*	Poly-urethane	2	Black	Liquid 2 400 cps	30 min	Spatula Trowel	24 h at 25°C	Contact	94	As for Flexane 9
94	Flexane 30*	Poly-urethane	2	Black	Liquid 3 000 cps	25 min	Spatula Trowel	24 h at 25°C	Contact	66	As for Flexane 9

Adherends	Test spec.	Test conditions	Test temp. °C	Physical data	Key	Remarks
crete						Good resistance to moisture. Adhesive is used to pre-coat concrete prior to strain gauge attachment. Alkaline surfaces hamper curing
als		After 7 d at 25°C	25 25	510 53 (share A) 275 N/cm²	k l d	Optimum properties obtained after 7 d cure. Resistant to petrol, oil, hydraulic fluids, ester lubricants. Moderate resistance to acids and alkalies; swells in chlorinated hydrocarbons and benzene. Good adhesion to metal. Cured film is softened in ethylene dichloride. Inflammable. Flash Pt. below −3°C
l ninium od ber ss vas	ASTM D1004– 59T ASTM D638– 58T ASTM D395– 55	After 48 h in water After 48 h in water		90 kV/cm 85 kV/cm 1·53 × 10¹³ Ω/cm 2·78 × 10¹² Ω/cm 12·2 × 10⁻² Jm/m²s/°C 103 N/cm² 700 N/cm 265% 95 14%	m m p p q d h k l v	Cures to a hard semi-rigid solid. Primer available for improving metal adhesion. Good water, oil, fuel resistance but not generally recommended for use with acetone and similar solvents. Linear shrinkage is 0·0003 cm/cm at 24°C
or Flexane 95	ASTM D1004– 59T ASTM D638– 58T ASTM D395–55			15·5 × 10⁻² Jm/m²/s/°C 33 N/cm² 87 N/cm 150% 60 14%	q d h k l v	Cures to a more flexible, semi-rubber-like solid. Good water, oil and fuel resistance but not suitable for use with acetone. Linear shrinkage is 0·0001 cm/cm at 24°C. Data on electrical characteristics not available
or Flexane 95				92 kV/cm 1·61 × 10¹¹ Ω/cm 13·8 × 10⁻² Jm/m²/s/°C 27·5 N/cm² 35 N/cm 130% 30 15%	m p q d h k l v	Flexible, rubbery, solid when cured. Good resistance to oil, water and fuels. Linear shrinkage is 0·0001 cm/cm at 24°C

Trade source	Trade name or designation	Basic type	No. of components	Colour	Physical form, consistency or viscosity Ns/m^2	Working life	Method of application	Processing		Service temp. range °C	Main uses
								Curing cycle	Bonding pressure N/cm^2		
47	Bostik 2064	Polyure-thane in hydro-carbon solvent	2	Brown	Liquid 0·33 23% w/w solids	Pot life 8–16 h	Brush Roller coater	3–7 d at 20°C	Contact	Up to 100	Bonding plastics materials and fabr when high heat resistance is requi
66	Chelsea c/22	Polyure-thane in solvent	1	Off white trans-lucent	Liquid 6 at 20°C	Open time 5 min after applica-tion	Brush	Heat activa-tion of dried film at 80°C	55 (15 s)	60 max	Bonding of PVC (plasticise to itsel or leather, plastic rubber (treated). Shoe trade materials, convey belting, textiles
47	Bostik 3206	Polyure-thane in ketone hydro-carbon solvent	1	Light amber	Liquid 4 at 25°C 24% w/w solids	Open time 10–20 min	Brush Trowel Spray	Dried to tacky state and bonded at room tempera-ture	Contact to 4	Up to 100	Decorative laminated plastic porous materials and coated fabric foam materials
54	PR 1527	Poly-urethane + catalyst	2	Black	25	5 h at 24°C	Spatula Extrusion gun	5 d at 24°C to 6 h at 82°C	Contact to 20	−62 to 149 For short per-iods, 204	Applications requiring flexible, cold flow resistan rubber with good electrical propert Metals, rubbers a plastics
250	Gelva 264	Polyvinyl acetate maleate copolymers in ethanol	1	White	7–11 at 25°C 55% w/w solids	Indefinite	Brush				Lamination of plastic films to metal foils

Adherends	Test spec.	Test conditions	Test temp. °C	Physical data	Key	Remarks
(fabrics) rethane foam						Bostik Thinner M576 available. Good resistance to water and most solvents, when fully cured. Flash Pt. -17 to $-7°C$
er (chrome)—PVC ded)			20	107 N/cm	a	SATRA 180° Peel data. Excellent resistance to plasticisers.
er (chrome)—PVC cised)			20	117 N/cm	a	Rubber failure where substrates treated by halogenation
rubber (halogenated)—PVC			20	49 N/cm (rubber fails)	a	primer (details SATRA p. 332)
rethane—polyurethane			20	49 N/cm	a	Flash Pt. below 0°C
rethane—aluminium			20	68 N/cm	a	
s arbonate s (neoprene, PVC, rethane) -melamine formaldehyde			20	275 N/cm²	b	Excellent adhesion to rigid and plasticised PVC. Resistant to oils, fuels, and water. Thinned with Bostik 6011 for spraying. Flash Pt. -17 to $-7°C$
		Thickness 3·15 mm	24	113 kV/cm	m	May be used as an encapsulant or potting compound. Primers are available to improve adhesion to neoprene (PR 1523) or polyvinyl chloride (PR 1534). Good resistance to moisture and biodeterioration
		Thickness 0·63 mm	24	165 kV/cm	m	
			24	6·6–6·9 (at 10^3–10^6 Hz)	n	
			24	0·03–0·09 (at 10^3–10^6 Hz)	o	
			24	$1·5 \times 10^{12}$ Ω/cm	p	
				450%	k	
				393 N/cm	h	
				24%	v	
			148	$1·3 \times 10^9$ Ω/cm	p	
27 (cured)—aluminium		Metal primed with PR1531	24	87 N/cm	a	
27 (cured)—cadmium		Metal primed with PR1531	24	87 N/cm	a	
27 (cured)—Neoprene		Abraded	24	43 N/cm	a	
27 (cured)—polyvinyl ride		tackified with acetone	24	52 N/cm	a	
rethane (foam) yrene ster (film) ose acetate inyl chloride (foam) nium (foil) nium (foil)—aluminium						Non tacky, flexible adhesive. Dried film is resistant to light and water. One of a range of multipolymer solutions based on polyvinyl acetate. Tech. Bulletin 6081 lists potential uses of materials in formulations for hot melt and pressure sensitive adhesives

Trade source	Trade name or designation	Basic type	No. of compo-nents	Colour	Physical form, consistency or viscosity Ns/m^2	Working life	Method of application	Processing		Service temp. range °C	Main uses
								Curing cycle	Bonding pressure N/cm^2		
276	Tuf Prime bond	Polyvinyl acetate emulsion with plasticiser	1	White	Liquid with 50% solids content	Indefinite	Spatula Trowel	24 h at 20°C	Contact to 4		General building industry requirements
47	Bostik PVA Adhesive	Polyvinyl acetate	1	White	Emulsion 51% w/w solids	Open time 5 min at 20°C	Brush Spreader	Sets by drying. Optimum strength 12–24 h at 20°C	Contact to 4	−20 to 70	General purpose adhesive for gluing variety of substr in building and other trades
88	Croid 80A 62	Polyvinyl acetate	1	White to clear when cured	Aqueous emulsion of high viscosity pH = 4–5		Brush Roller	Wet assembly and pressure until set within few hours at room temp.	20–35		General woodw assembly and veneering with wood and plasti laminates
45	Cascorez VN-33 + catalyst VX-3	Polyvinyl acetate	2	Pale pink	Liquid 8–10 at 25°C. Solids content 57±1%	Shelf life is 3 mth at 20°C. Usable life is 24 h at 20°C	Brush Spreader	1 h at 20°C to 30 min at 30°C or hot pressed for 1·5 min at 95°C to 0·5 min at 115°C	34–120		Wood gluing of cabinets, doors, furniture
200	Duro-lok 150*	Polyvinyl acetate + catalyst	1	White	Liquid 4 at 20°C 48% w/w solids	24 h at 20°C	Brush Spreader	Cold press 20–40 min at 20°C or Hot press 1–10 min at 65–150°C	10–20 or 70 for hot cure	100	Wood gluing. Decorative lamin bonding

Adherends	Test spec.	Test conditions	Test temp. °C	Physical data	Key	Remarks
us materials, rete d—wood (beech)	BS 4071: 1966	After 7 d cure period	20	860 N/cm²	b	Cures in 24 h with maximum strength properties after 7 d. Unsuitable for use below 2°C and above 37°C
d crete tics als stos						May be diluted up to 50% v/v with water. Cured adhesive is resistant to water, oils, acids, and alkalies. Freeze-thaw stable. Approved to BS 4071: 1966. Coverage 2 m²/1 to 5–10 m²/1 (diluted)
d board materials amine laminate d—wood	BS 745		20	630 N/cm²	d	Especially suitable for cold press veneering. Unsuitable for unsupported joints or where high temperatures are likely. Low resistance to water. Conforms to BS 4071
d (moisture content tween 5 and 15%)						Cross linked PVA which gives joints resistant to boiling water (for several hours). Amenable to radiofrequency heating. Alternative product VN–31 available for production assembly in flow line presses
ber (with water content less an 12% w/w) d (birch, pine, fir) ood						Thermosetting emulsion adhesive with water resistant properties comparable to phenolic wood glues. Resistant to boiling water and weathering. Suitable for R.F. processing. Not gap-filling. Unsuitable for permanently stressed joints or laminated beams. Fair freeze-thaw stability. Coverage, 7 m²/kg

Trade source	Trade name or designation	Basic type	No. of compo-nents	Colour	Physical form, consistency or viscosity Ns/m^2	Working life	Method of application	Processing		Service temp. range °C	Main uses
								Curing cycle	Bonding pressure N/cm^2		
200	Kor-lok 3000	Polyvinyl acetate with vinyl cross link catalyst	1	White	Liquid 3 at 20°C 45% w/w solids	24 h at 20°C	Machine spreader	Sets in air above 24°C or Heat set at 65–150°C under pressure in ½–10 min	20 to +150		Production of curved plywood, edge bonding with hot strip heaters. Laminating timber
200	Wood-lok 140–0212	Polyvinyl acetate	1	White	Liquid 3 at 20°C 51% w/w solids	Indefinite	Brush Roller Trowel Pressure gun	Air setting in 20 min			Assembly gluing of wood
107	Evo-stik 8148	Poly-urethane in solvent	1	Trans-parent	Medium viscosity liquid		Brush Machine	Dried for 30 min at 20°C and heat re-activated prior to bonding	Contact		Vinyl plastics to porous materials. Footwear manu-facture
66	Chelsea C/14	Poly-urethane in solvent +2% catalyst	1	Trans-parent	Liquid	10 d at 20°C	Brush	Dried in air 1 h at 20°C then heat activated 40 s at 85°C	55	60	Vinyl plastics to porous materials footwear manu-facture
84	Cow PA 72	Poly-urethane in solvent +catalyst	2	White	Liquid	45 min at 20°C	Brush	Dried in air 1 h at 20°C and cured 24 h at 20°C to 1 h at 80°C	Contact	70	Bonding to metal rubbers and syn-thetic textiles. Protective coating lacquer
57	Silverlock A1187-B	Poly-urethane in ethyl acetate	1	Trans-parent	Liquid		Brush Spray (preferred)	30 min at 164°C to 5 min at 205°C	Contact	100	Bonding of flock materials to rubber

Adherends	Test spec.	Test conditions	Test temp. °C	Physical data	Key	Remarks
od (birch, fir, pine) wood eers						Thermosetting emulsion adhesive. Suitable for R.F. processing. May be thinned with water although it is not recommended. Coverage, 8·5 m²/l
od (above 7°C)				Complies with BS 4071		Non-gap filling and suitable for interior use with close contact joints. Bonds may be light machined after 20 min. Resistant to humid, tropical environments. Freeze-thaw stable. Withstands biodeterioration. Coverage, 5 m²/kg
yvinyl chloride (flexible) ther vas yvinyl chloride (flexible)— VC a evaluation				78 N/cm	a	Maximum pre-cemented life after drying should not exceed 7 d. Resistant to plasticisers. Catalyst available for accelerated curing
rics ber ather (chrome)—polyvinyl hloride a evaluation		SATRA Tensiometer 180° Peel test	18 60	112 N/cm 37 N/cm	a a	Accelerator available as separate component. Excellent resistance to plasticiser migration
minium yamide textiles (nylon) yurethane coated fabrics						Available in other colours. Lacquer with good adhesive properties for porous substrates. Inflammable, Flash Pt. below 23°C
on amide (nylon) on ber (latex)						Cures to thermoset bond. Abrasion properties exceed those of the flocking material. Resistant to ozone. Good adhesion to uncured rubber when cured with adhesive simultaneously. Inflammable

Trade source	Trade name or designation	Basic type	No. of compo-nents	Colour	Physical form, consistency or viscosity Ns/m^2	Working life	Method of application	Processing		Service temp. range °C	Main uses
								Curing cycle	Bonding pressure N/cm^2		
177	Clam* No. 7	Polyvinyl acetate	1	White	Emulsion	Open time is 4 min	Brush	Sets on drying in 15 min	Contact		Multi-purpose adhesive for absorbent material and some plastics
248	Sellobond Clear	Polyvinyl acetate	1	Clear	Liquid		Dispenser spreader cap	Dries in 1 h at 25°C	Contact to 3 for 30 min		Household repairs and handicrafts
144	Dufix	Polyvinyl acetate	1	Trans-parent	Emulsion with high solids content		Dispenser container Brush Roller Spray gun	Dries in 1 h at 25°C	Contact to 3 up to 24 h	−20 to 65	Porous adherends. Bonding of plaster and concrete repair. Gap filling adhesive for wood work and tiles
278	Cerafix	Polyvinyl acetate + filler	1	Off White	Thick paste	8 h to 5 d Sub-strate dependent	Notched trowel	Dries in 8–16 h at 20°C (with porous materials) to 5 d at 20°C (non-porous)	Contact to 3	−20 to 65	Interior bonding of ceramic tiles and mosaics to glazed painted surfaces
278	Unibond	Polyvinyl acetate	1	Trans-parent	Emulsion	30 min to 3 h. Substrate dependent	Brush	Dries in 30 min to 3 h. Allow 24 h for setting	Contact to 3 up to 24 h	−20 to 65	General purpose adhesive for building and civil engineering applications. Woodwork and masonry rendering

Adherends	Test spec.	Test conditions	Test temp. °C	Physical data	Key	Remarks
ement, concrete, plaster, wood, cork, leather, paper backed PVC to polystyrene (expanded), paper ardwood – hardwood			 20 20	 1 520 N/cm² 470 N/cm²	 b g	Variations in temperatures have no great effect on viscosity or handling of the adhesive. Coverage 10 m²/l. Resistant to petrol, paraffin, white spirit, oil, water. Unaffected by sunlight
⸱brics ⸱ather ⸱ass ⸱per ⸱lyvinyl chloride ⸱ood – wood	BS 1204			265 N/cm²	b	Good tack properties. Resistant to water and hydrocarbons. Non-porous adherends require 12 h period before handling
⸱per and cardboard ⸱ttery ceramics ⸱lystyrene (foam) ⸱xtiles ⸱nvas ⸱rk ⸱bestos ⸱ncrete ⸱ood – wood	BS 4071: 1966			690–1 370 N/cm²	d	Freeze-thaw resistant but advisable to protect from frost. Prolonged exposure to moisture should be avoided. May be water diluted up to 1 : 5 ratio. Unsuitable for bonding two impervious adherends. Resistant to acids and alkalies. Coverage, 32–93 m²/l
⸱umina ⸱ickwork ⸱ncrete ⸱asonry ⸱w plaster ⸱ramic tile – tile		At 25°C and 95% R.H. after, 1 wk 4 wk After 10 cycles at 25° to −10°C at 95% R.H. Conditioned at 70°C, 1 wk 4 wk	 25 25 25 25 70 70	Complies with BSS CP212, Pt. 1 (1963) 117 N/cm² 74 N/cm² 88 N/cm² 58 N/cm² 151 N/cm² 100 N/cm²	 d d d d d d	Cures only under frost free conditions. More resistant to ageing than traditional cement mortar. Odourless and non-staining. Not intended for outdoor use or immersion in water Coverage, 1·6–1·8 m²/l
⸱ipboard, hardboard ⸱umen surfaces ⸱rk ⸱lystyrene (foam) ⸱ass ⸱ad ⸱c ⸱estos ⸱od – wood ⸱od – concrete ⸱ncrete – concrete ⸱ncrete – ceramic tiles ⸱n (galvanised) – iron ⸱oleum – concrete				 651 N/cm² 158 N/cm² 124 N/cm² 31 N/cm² 65 N/cm² 85 N/cm²	 b b b b b b	Maximum bond strength in few days. Unsuitable for rubber polythene or polyvinyl chloride bonding. Resistant to water, petrol, hydraulic oils, weak acids and alkalies, and bio-deterioration. Can be diluted with water in ratio 1 : 2 water for non-load bearing general adhesion purposes

Trade source	Trade name or designation	Basic type	No. of compo-nents	Colour	Physical form, consistency or viscosity Ns/m^2	Working life	Method of application	Processing		Service temp. range °C	Main uses
								Curing cycle	Bonding pressure N/cm^2		
58	Bal-Tad	Polyvinyl acetate	1	Trans-parent	Liquid	Open-time of 20 min	Brush	Air drying	Contact to 7	82	Ceramic wall tiles to interior surfaces. Building industry
88	Croid 843	Polyvinyl acetate	1	Trans-parent	High viscosity emulsion pH 6–7	Indefinite	Brush Roller	Air drying	20–35		Veneering and joinery. Furniture construction
88	Polystik	Polyvinyl acetate	1	Trans-parent	High viscosity emulsion pH −6·5	Indefinite	Brush	Air drying	20–35		Joinery, cabinet work. Floor materials and other building application
2	Tufskin	Polyvinyl acetate	1	White	Medium viscosity emulsion	Indefinite	Brush Machine	Cold setting 30 min to 2h under pressure	Contact to 100		Box making. Back sizing of carpets. Porous materials
54	PR 340	Poly-sulphide polymer + catalyst	2	Alum-inium	Liquid 1·3	2 h at 24°C and 50% relative humidity	Spatula Extrusion gun	72 h at 24°C or Heat cure up to 48°C	Contact	−54 to 107	Applications requiring tough rubber adhesive/ sealant with good weathering proper-ties up to 3 yr
193	EC 1120 PC	Poly-sulphide polymer + catalyst EC-1031	2	Tan	Medium viscosity liquid 93% w/w solids	90 min at 25°C	Spatula Extrusion gun	24 h at 25°C	Contact	−54 to 82	Bonding and sealin of low voltage electrical equipmen against moisture and corrosive agen Aircraft industry

Adherends	Test spec.	Test conditions	Test temp. °C	Physical data	Key	Remarks
Cement Aluminium Plywood Asbestos Brickwork Ceramic tile – brickwork				Complies with BS CP 212 for wall-tile adhesives 210 N/cm²		Unsuitable for damp surfaces. Freeze-thaw stability. Applicable to hot surfaces up to 80°C. Non-staining. Coverage, 0·9–1·26 m²/l
Wood Plywood Chipboard Plastic laminates – wood Wood – wood	BS 745			Conforms to requirements of FIRA, 'Note 25 for chair assembly' Complies with BS 4071 620 N/cm²	b	Cures to slightly extendible hard film. Suitable for R.F. heating processes. Durable but not recommended for water exposure. Applicable under winter workshop conditions. Non-inflammable
Polystyrene (foam) Brickwork Concrete Wood (walnut, sapele, mahogany) Wood (beech) – wood	BS 745			690 N/cm²	b	Resistant to damp, petrol, oil, grease and mild acids and alkalies. May be diluted with water for cement renderings. Non-inflammable. Coverage, 167 g/m²
Wood (beech) – wood		Cured for 48 h at 20°C, Undiluted Diluted, 25% water 50% water 100% water	20 20 20 20	830 N/cm² 523 N/cm² 515 N/cm² 413 N/cm²	b b b b	Fast setting with high initial bond strength. May be diluted with water. Resistant to biodeterioration
Aluminium Cadmium Acrylics PR 340 (cured) – aluminium Aluminium – aluminium		After 7 d in water at 24°C	24	100% 50 138 N/cm² 17 N/cm 55 N/cm²	k l d a b	No evidence of corrosion on aluminium or cadmium after 14 d at 80°C and 95% relative humidity. Does not craze acrylic plastics. Retention of adhesion when immersed in water, hydraulic fluids and fuels. Resists biodeterioration
EC1120PC (cured) – aluminium	ASTM D1304–60		26 82 26 82 26 26 25 20	78 kV/cm 7·6–7·5 (at 60–10⁴ Hz) 7·3–7·0 (at 10³–10⁴ Hz) 0·05–0·002 (at 60–10⁴ Hz) 0·013–0·02 (at 10³–10⁴ Hz) 10^{12} Ω/cm at 85 V $3·7 \times 10^{11}$ Ω/cm at 500 V 20 (shore A) +21 N/cm	m n n o o p p L a	Cures to highly resilient material which withstands vibration and thermal shock. Resistant to water, hydrocarbon fuels and oils. May be stored at −29°C for 24–36 h before application

Trade source	Trade name or designation	Basic type	No. of components	Colour	Physical form, consistency or viscosity Ns/m²	Working life	Method of application	Processing		Service temp. range °C	Main uses
								Curing cycle	Bonding pressure N/cm²		
97	Silastoseal	Silicone rubber	1	White	Paste	Tack-free time is less than 1 h at 20°C	Spatula Dispenser pack Mastic gun	Exposure to air. 0·6 mm films cure in 90 min 3 mm films cure in 24 h	Contact	−75 to 250	Flexible heat stable bonds between metals, ceramics, plastics, wood, etc. Repairing silicone rubber moulds. Electrical and electronic equipment. Furnace and refrigerator doors. Plastic pipe bonding
144	Silcoset 150	Silicone rubber	1	Red	Paste 250	Tack-free time, 15–30 min	Cartridge pack, Dispenser, Dipping Spraying Spatula	Exposure to air. 3 mm films cure in 24 h. 0·3 mm films may be heat cured at 120°C	Contact	−60 to 250	Flexible heat stable elastomeric adhesive and sealing agent. Electronic component bonding and caulking. Coating cloth for anti-stick surface, conveyor belting. Repair of wire and cable insulation. Hot air ducting assembly
97	Dow Corning 282	Silicone resin in xylene	1	Translucent	70–150	Indefinite	Brush	5 min at 150°C (catalysed) or solvent loss at 94°C to tacky state	Contact	−73 to 290	Used primarily for manufacture of pressure sensitive tapes with wide variety of backing materials. High temperature electrical insulation moisture and corrosion resistant wrappings

Adherends	Test spec.	Test conditions	Test temp. °C	Physical data	Key	Remarks
Metals (tinned and galvanised)	BS 903, C4 1957			156 kV/cm	m	Use of primer MS 2402 is recommended for some materials. Fair bonds to polyester and phenolic resins (50–80% cohesive failure of adhesive). Inconsistent bonds with teak and other oily woods. Does not liberate corrosive acetic acid during cure. Resistant to moisture, low pressure steam, oxidation and weathering. Physiologically inert. Grade ESP 2470 is suitable for similar applications where non sag/flow character of Silastoseal B presents process difficulties (e.g. for silicone foams)
Polychloroprene	BS 903, C3 1956			3·0 (at 10^6 Hz)	n	
PVC (primed)	BS 903, C3 1956			0·001 – 0·001 (at $10^3 – 10^6$ Hz)	o	
Acrylics	BS 903, C2 1956			6×10^{15} Ω/cm	p	
Ceramics	BS 903, A2 1956			300%	k	
Glass	BS 903, A7 1957			30 BS°	l	
Epoxide resins	BS 903, A2 1956			240 N/cm²	d	
Polystyrene						
Wood (oak)						
Aluminium – aluminium			20	207 N/cm²	b	
		After 7 d at 200°C	200	103 N/cm²	b	
		7 d at 250°C	250	124 N/cm²	b	
steel (mild) – steel			20	172 N/cm²	b	
		After 7 d at 200°C	200	103 N/cm²	b	
		7 d at 250°C	250	103 N/cm²	b	
Copper – copper			20	207 N/cm²	b	
		After 7 d at 200°C	200	117 N/cm²	b	
steel*		Cured 24 h at 20°C	20	207 kV/cm	m	Available in white colour as Silcoset 151. Primer OP is recommended for pretreatment of some materials; Chemlok 607 for others (Trade source No. 79). Acetic acid liberated during cure (possible corrosion hazard) which later disappears. May be diluted with cyclo-hexane, toluene, trichloro-ethylene and other dry solvents. Resistant to ageing, weathering, ozone, corona discharge and moisture. Data on chemical resistance available
Aluminium†		Cured 24 h at 20°C + 4 h at 250°C	20	200 kV/cm	m	
Copper, Brass*†		After 24 h in water	20	180 kV/cm	m	
Tinplate*†						
Glass				3·1 – 2·6 (at 50 – 10^7 Hz)	n	
Wood*				0·2 – 0·003 (at 50 – 10^7 Hz)	o	
Phenolic resins*		Cured 24 h at 20°C	20	1×10^{15} Ω/cm	p	
Polyester resins*		After 24 h in water	20	2×10^{14} Ω/cm	p	
Polyamide*		following cold cure		$1·88 \times 10^{-5}$ Jm/m²/s/°C	q	
Nitrite rubber*				200%	k	
Polychloroprene*				45–50 BS°	l	
silicone rubber				3·5 N/cm	h	
Concrete*		24 h at 20°C	20	255 N/cm²	d	
		10 d at 250°C	20	179 N/cm²	d	
OP primed						
Chemlok 607 primed						
Glass cloth (either plain or silicone varnished)		96 h at 23°C and 50% RH		582 kV/cm	m	Aggressive tack. High electric strength and arc resistance, low loss factor. Good resistance to moisture, weather, sunlight, and ozone. Withstands contact with many oils and chemicals including acids and alkalies. Catalysts such as benzoyl peroxide may be used to accelerate curing rates or to allow lower curing temperatures
silicone rubbers						
Polyethylene terephthalate	ASTM D150					
PTFE				2·93 – 2·87 (at $10^2 – 10^5$ Hz)	n	
Polyimide				0·0051 – 0·0036 (at $10^2 – 10^5$ Hz)	o	
Aluminium foil				$7·2 \times 10^{15}$ Ω/cm	p	
	ASTM D257	96 h at 23°C and 90% RH		565 kV/cm	m	
				3·19 – 3·08 (at $10^2 – 10^5$ Hz)	n	
				0·0047 – 0·0038 (at $10^2 – 10^5$ Hz)	o	

continued

Trade source	Trade name or designation	Basic type	No. of compo-nents	Colour	Physical form, consistency or viscosity Ns/m²	Working life	Method of application	Processing		Service temp. range °C	Main uses
								Curing cycle	Bonding pressure N/cm²		
Dow Corning 282—continued											
97	Silastic 735 RTV	Silicone rubber	1	Black	Uncured rubber. No sag or flow	<½ h Tack-free time	Spatula Trowel	24 h at 20°C (20% R.H.) Full cure in 5 d	Contact	−65 to 260	General purpose bonding and sealing application. Adheres to metals, glass, ceramics, painted surfaces. Adhesive/sealant for situations where material is expected to support con-siderable suspended weight. High pres-sure exposure conditions
97	Silastic 140 RTV	Silicone rubber	1	Trans-lucent	Non-sag Uncured rubber	<1 h Tack-free time	Spatula, Extrusion from capsule	72 h at 20°C for 20% R.H.	Contact	−66 to 230	Adhesive/sealant for metals, glass, natural and synthetic fibres. Sealant for oven doors; bonding gaskets in heating and refrigerating units, insulation

Adherends	Test spec.	Test conditions	Test temp. °C	Physical data	Key	Remarks
ressure sensitive tape with aluminium foil backing	ASTM D1000			1.9×10^{15} Ω/cm	p	
			−70	>11 N/cm	a	
			−20	>11 N/cm	a	
			100	8.8 N/cm	a	
			200	Cohesive failure	a	
			250	Cohesive failure	a	
		No ageing		9.3 N/cm	a	
		After 7 d at 150°C	25	10 N/cm	a	
		After 7 d at 250°C	250	Cohesive failure	a	
uminium				2.6 Jm/m²s/°C	q	Cures without liberating cor-
tanium				220 N/cm	h	rosive acetic acid to a tough
eel (stainless)	ASTM D412	After 5 d at 25°C		850%	k	rubbery solid. Silicone 1200 primer available for improve-
lass						ment of adhesion. Not suitable
ork, phenolic bonded	ASTM D412			535 N/cm²	d	for plastics which exude plasticisers.
licone rubber	ASTM D412	After 5 d at 260°C		100%	k	Resistant to weathering and moisture.
	ASTM D412			262 N/cm²	d	Non-slumping.
ured rubber–aluminium				14.8 N/cm	a	Requires evaluation before use as an electrical insulator in
ured rubber–titanium		Data obtained on		12.2 N/cm	a	view of its conductive
ured rubber–steel (stainless)		surfaces primed with		10.5 N/cm	a	properties
		DC 1200 primer		12.2 N/cm	a	
luminium–aluminium (2024 Alclad)				310 N/cm²	b	
ork–cork (phenolic bonded)				Cork failed cohesively	b	
	ASTM D412			148 N/cm²	d	Produces acetic acid during cure which may corrode some
	ASTM D412			350%	k	metals.
	ASTM D816			280 N/cm²	b	Uncured material represents an eye hazard.
licone rubber – silicone rubber	MIL-5-7502			35 N/cm	a	Optimum bonding where primer DC A-4094 is used.
licone rubber – fluorosilicone rubber				17.5 N/cm	a	Good resistance to heat ageing and hot oil or water.
licone rubber – neoprene				5.25 N/cm	a	Withstands cycles of freezing and thawing, oxygen, ozone
licone rubber – glass cloth				14 N/cm		and ultra-violet rays.
licone rubber – steel (stainless)				26.2 N/cm	a	Resistant to biodeterioration
licone rubber – aluminium				35 N/cm	a	
licone rubber – copper				17.5 N/cm	a	
licone rubber – glass				35 N/cm	a	
licone rubber – PTFE				17.5 N/cm	a	
licone rubber – aluminium		Not aged		35 N/cm	a	
		After 7 d at 200°C	20	35 N/cm	a	
		After 1 d at 250°C	20	8.7 N/cm	a	
		After 3 d in water at 100°C		26.2 N/cm	a	
		After 3 d in ASTM No. 1 oil at 150°C	20	17.5 N/cm	a	

Trade source	Trade name or designation	Basic type	No. of components	Colour	Physical form, consistency or viscosity Ns/m^2	Working life	Method of application	Processing		Service temp. range °C	Main uses
								Curing cycle	Bonding pressure N/cm^2		
153	RTV-118	Silicone rubber	1	Trans-lucent	Self-levelling rubber 35	Tack-free in 30 min at 20°C	Brush Spray Extruder gun	Cures on exposure to air at 20°C in 24 h	Contact	−65 to 150	Thin film rubber coating of electrical and electronic components
200	National 311–0330	Starch	1	White	Paste 9 at 20°C 29% w/w solids		Brush Machine	Dries at 90–150°C		100	Aluminium foil laminating Packaging industry
219	Nylaweld	Unspeci-fied	1	Trans-parent	Liquid	Open time is 2–5 min	Brush Spray Dipping	Dried in air for 2–5 min at 20°C and cured for 1 h at 65°C or $\frac{1}{2}$ h at 150°C	200 to 410		Bonding nylon components together
107	Evo-stik 863	Synthetic rubber latices + synthetic resin with inorganic fillers	1	Off White	Stiff cream 54–58% w/w solids in water	30 min at 20°C	Spreader Trowel	Contact bonding within 15–30 min	Contact	7 to 60	Covering and panelling of expanded poly-styrene parts to vertical and hori-zontal absorbent surfaces
107	Evo-stik 873	Synthetic resin/rubber latex with inorganic fillers	1	Off White	25–40 at 20–30°C 40% w/w solids	15–25 min at 20°C	Spreader Trowel	Contact bonding within 15–30 min	Contact	7 to 50	Bonding flooring materials

Adherends	Test spec.	Test conditions	Test temp. °C	Physical data	Key	Remarks
	ASTM D149			210 kV/cm	m	Higher temperatures and humidity accelerate the cure; optimum cure conditions are 33°C and 90% relative humidity.
	ASTM D150			2·70 − 2·60 (at 60 − 10^6 Hz)	n	
	ASTM D150			0·0004 − 0·0018 (at 60 − 10^6 Hz)	o	
	ASTM D257			2×10^{15} Ω/cm	p	Acetic acid liberated during cure which may cause corrosion of metals.
	ASTM D412	After 1 wk at 25°C		357 N/cm²	d	Withstands 260°C for short periods.
		After 1 d at 149°C		310 N/cm²	d	Primer SS 4004 recommended for nickel and stainless steel surfaces to improve adhesion and inhibit corrosion.
		After 1 d at 177°C		302 N/cm²	d	
		After 1 d at 204°C		275 N/cm²	d	
Steel (20 mesh screen) − copper		70 h cure at 25°C		117 N/cm²	b	
Steel (20 mesh screen) − aluminium		70 h cure at 25°C		117 N/cm²	b	Linear shrinkage of 0·3% after 7 d cure
Steel (20 mesh screen) − glass		70 h cure at 25°C		Cohesive failure of glass	b	
Steel (20 mesh screen) − acrylic (rigid)		70 h cure at 25°C		Cohesive failure of acrylic	b	
Steel (20 mesh screen) − steel (stainless)		70 h cure at 25°C		96 N/cm²	b	
Steel (20 mesh screen) − aluminium		7 d cure at 25°C		35 N/cm	a	
Aluminium (foil)						High initial tack. Resistant to heat. May be diluted up to 25% w/w with water
Paper						
Cardboard						
Polyamide − polyamide (Nylons)		Cured at 65°C	20	3 770 N/cm²	d	Curing above 100°C in inert atmosphere recommended. Bond strengths approach the ultimate material strength where heat curing is employed. Joints are flexible and resistant to boiling water or strong alkalies
		Cured at 93°C	20	5 900 N/cm²	d	
		Cured at 100°C	20	6 530 N/cm²	d	
		Cured at 150°C	20	6 800 N/cm²	d	
Cement						Dilution of adhesive not recommended. Bond strength deteriorates above 60°C. Resistant to damp, steamy environments. Adhesive has freeze-thaw stability. Coverage 1·8 m²/l
Hardboard						
Plaster						
Polystyrene (foam) − wood			20	55 N/cm²	d	
Polystyrene (foam) − concrete			20	37 N/cm²	d	
Linoleum						Suitable for under-floor heating surfaces. Resistant to detergent floor cleaners and moisture. Adhesive has freeze-thaw stability. Coverage 4−5·5 m²/l
Asphalt						
Cement						
Hardboard						
PVC flooring-wood/concrete				17−26 N/cm	a	
PVC flooring-wood/concrete				27−41 N/cm²	b	

Trade source	Trade name or designation	Basic type	No. of compo-nents	Colour	Physical form, consistency or viscosity Ns/m^2	Working life	Method of application	Processing		Service temp. range °C	Main uses
								Curing cycle	Bonding pressure N/cm^2		
5	Adsol 120	Synthetic rubber resins blend in hydro-carbon solvents	1	Light Brown to red	Liquid	Open time 5–10 min at 20°C	Brush Roller Spray gun	Contact bonding when tacky or heat re-activation of dried film at 80°C	Contact to 7	100	Plastics combinations, metals, ceramics
13	Surestik 1470	Synthetic rubber/ resins in hydro-carbon solvents	1		Liquid	1 h at 20°C	Brush Trowel	Air dried and contact bonded within 1 h	Contact to 7		Bonding of vinyl plastics to rigid materials
13	Surestik 875	Synthetic resin blend in solvents	1		Liquid	15–20 min at 20°C	Brush Trowel	Air dried for 10 min and contact bonded within 20 min	Contact	−45 to 70	Plastic foams to rigid materials
47	Bostik C Adhesive	Rubber/ resin in petroleum	1	Black	Thick liquid	Tack free time is 30 min	Brush Spray gun Extruder Trowel	Contact bonding after air drying	Contact to 4	−47 to 71	General purpose adhesive for porous and non-porous materials
47	Bostik 1261	Rubber/ resin in hydro-carbon solvent	1	Dark Brown	Medium viscosity liquid	Tack free time is 15 min	Brush Trowel Spray gun	Contact bonding after air drying	Contact to 4	−18 to 100	General purpose bonding — as for Bostik C
47	Boscoprene 2763	Synthetic rubber/ resin in ketone solvent + catalyst	1	Trans-parent	Medium viscosity liquid 31±2% solids	8 h at 20°C	Brush	Dried 10 min at 20°C and cured 3 d at 20°C or 16 h at 70°C	Contact to 20	−5 to 100	Used in vacuum covering process of wide range of articles such as cases, cabinets, automotive components
47	Bostik 1762	Synthetic rubber/ resin in ketone solvent	1	Trans-parent		Tack free time of dry film is 20 min	Brush Extruder Roller coater	Dried 10 min at 20°C and contact bonded	Contact to 7	−5 to 32	Bonding plastic foams, laminate sheets, and porous materials. Motor car headlining materials

Adherends	Test spec.	Test conditions	Test temp. °C	Physical data	Key	Remarks
Glass Enamel Wood Rubber Melamine laminate Aluminium – aluminium			20 20	61 N/cm 1 380 N/cm²	a b	May be solvent reactivated. Resistant to water, oil and ageing. Inflammable, Flash Pt. below 23°C. Non-inflammable grade available. Coverage, 4–7 m²/l
Aluminium Polyvinyl chloride (rigid) Asbestos Cardboard Wood						Optimum bond strength in 24 h. Resistant to oil, petrol and moisture. Non-staining with plastic materials. Inflammable
Polystyrene (foam) Wood Aluminium Concrete						Solvent does not deform foam material. Not suitable for copper and alloys. Resistant to water but not oil or petroleum
Rubber Cork Linoleum Felt Wood Aluminium						Resistant to water but not oil. Softens at 49°C. Inflammable, Flash Pt. below −3°C. Coverage, 3 m²/l
s for Bostik C						Better heat resistance than Bostik C grade. Poor resistance to oil. Softens at 77°C. Inflammable. Coverage, 3 m²/l
Wood Canvas Leather Acrylonitrile rubber Polyvinyl chloride (rigid, flexible) Polyamide Glass ABS plastics Polystyrene (rigid)				Boscolite primer 9240 improves adhesion to metals		Resilient bond. Resistant to petrol, oil, water, acids and alkalies, but not esters and ketones. Adhesive yellows on exposure to sunlight. Inflammable, Flash Pt. −3°C. Coverage, 1·4 m²/l
Polyurethane (foam) Polyvinyl chloride (foam)						Resistant to petrol, oil, alcohols but not ketones and esters. Not suitable for non-porous adherends. Amenable to H.F. welding processes. Inflammable, Flash Pt. −3°C. Coverage 10 m²/l

Trade source	Trade name or designation	Basic type	No. of compo-nents	Colour	Physical form, consistency or viscosity Ns/m^2	Working life	Method of application	Processing		Service temp. range °C	Main uses
								Curing cycle	Bonding pressure N/cm^2		
107	Evo-stik 5007	Acryloni-trile rubber/ resin in ketone solvents	1	Pale Yellow	Liquid 15% w/w solids	Open time is 7 min	Brush Spray gun	Contact bonding within 2–7 min	Contact to 7	−30 to 70	Bonding cellular plastics to themselves and other substrates
98	S854	Synthetic rubber	1	Trans-parent	Medium viscosity liquid		Brush	Air dried 15 min at 20°C	Contact		Pressure sensitive adhesive for labelling. Auto-motive trim work. Laminating of packaging films
98	Dunlop LP	Synthetic rubber	1	Golden Brown	Heavy syrup	2–40 min at 20°C	Brush	Air dried up to 40 min at 20°C	Contact		Bonding laminated plastics to metals and rigid materials. Decorative plastic
213	PA127/2	Synthetic polymer	1	Yellow to brown	Liquid 25% w/w solids	Open time 3 min at 20°C	Brush	Air dried 1–3 min at 20°C and contact bonded	Contact	155	Bonding foam insulation to metal and painted surfaces for elevated tem-perature use
213	Plusbond	Synthetic polymer	1		Medium viscosity liquid 40% w/w solids	Open time 3 min at 20°C	Brush Spatula Roller coater Extruder	Air dried 2–3 min 20°C and contact bonded	Contact to 20		Polyurethane foam bonding to itself and other substrates. Automotive industry
18	Polystyrene Adhesive	Unspecified	1	Off-white	Liquid. Solids content 39% w/w	Open time 10 min at 20°C	Brush Spray Trowel	Sets at room temp. 0–15 min drying time	Contact to 4	−28 to 82	Insulation material including poly-styrene. Galvanised steel air conditioning duct
38	Bexol 101 Polystyrene Cement	Polystyrene in solvents	1	Clear	10 Solids content 33·5% w/w	Open time indefinite (minutes at 20°C)	Brush	Wet bonded and allowed to set at room temp.	Contact to 4		Bonding polystyrene to itself and porous substrates. Insulating lacquer and coating for electrical components such capacitors, coils, cables

Adherends	Test spec.	Test conditions	Test temp. °C	Physical data	Key	Remarks
olyvinyl chloride (foam and rigid) olyester-urethane (foam) olyether-urethane (foam) ood uminium						Available in alternative grades of viscosity from 25–30% w/w solids. Resistant to water, oil, dilute acids/alkalies and detergent solutions. Non-staining on exposure to Xenon lamp radiation (200 h). Inflammable, Flash Pt. below 10°C Coverage 5–8 m^2/l
olythene (film) per etals						Aggressive tack. Inflammable, Flash Pt. 40°C. Coverage, 5·5 m^2/l
olyester (laminates) ood etal (painted, unpainted) athercloth oncrete						Dried film may be reactivated with solvent T559. Inflammable, Flash Pt. 40°C. Coverage, 3·7 m^2/l
olyurethane (foam) — metal		3 d in water at 20°C 3 h in steam at 105°C 16 h at 140°C	20 20 20	Satisfactory adhesion — cohesive failure of foam		Non-inflammable. Coverage, 4 m^2/l
olyurethane (foam, rigid) ood brics ssian						Aggressive tack. Non-staining. Not recommended for expanded polystyrene for which alternative spray grade 145/S is available. Inflammable
sulation materials based on res, plastics foams or films. bestos						Based on flammable solvent (Flash Pt. 18°C). Fire resistant when set
olystyrene (film)		1 mm thick polystyrene film cast from 101. Tested at 1·6 MHz	20 20 20	Permittivity 2·62 Power factor 0·0004 0·0011	o	Electrical properties compact well with normal values for polystyrene. Properties vary little with temperature, humidity, and exposure to alcohols, alkalies and mineral acids. Contains highly flammable petroleum mixture. Shelf life, 9 mth below 24°C

Trade source	Trade name or designation	Basic type	No. of components	Colour	Physical form, consistency or viscosity Ns/m²	Working life	Method of application	Processing		Service temp. range °C	Main uses
								Curing cycle	Bonding pressure N/cm²		
47	Bostik 1160	Synthetic resin in alcohol	1	Amber	Liquid. Sp. gr. 0·9	Indefinite	Brush Dip coat	Dried up to 27°C and heat cured for 20 min at 141°C	Contact to 4	−40 to 300	Fabrication of ho... air ductings in aircraft condition... units. Bonding of butyral sheeting
47	Bostik 4141	Synthetic resin in aqueous solvent	1	Transparent when cured	Liquid 55% w/w solids	Indefinite	Brush Roller coat Knife blade Spray	Dried film cured at 80 to 110°C	Contact to 4	−10 to 70	Bonding PVC to itself and other substrates
160	Samson 35 Heatfix	Synthetic resin	1	Transparent	Low viscosity liquid		Brush Roller coater	Air dried and heat sealed at 82°C	Contact to 7		Heat sealing adhesive for pape... and board
163	Klingerflon	Synthetic polymer	1	White	Fiim	Indefinite	Hand	Contact bonded	Contact	4 to 38	Pressure sensitive adhesive tape wit... fluorocarbon film backing. Electrica... insulation, metal masking
46	Arabol Lagging adhesive 60–89–05	Synthetic resin with plasticiser	1	White	Cream 14±2 at 21°C 57% solids	4–6 h at 20°C	Brush	Air dries in 4–6 h	Contact	93	Lagging of pipes, storage tanks an... boilers with ther... insulating materi... Civil engineering and shipping industries
213	Laminating adhesives PA/78/13	Synthetic resin lacquers + catalyst	2		Low viscosity liquid		Brush Roller coater	Heat reactivation of dried film at 40–50°C	Contact to Pressure nip rollers		Plastic film laminating to foi... and other substr... for packaging applications

Adherends	Test spec.	Test conditions	Test temp. °C	Physical data	Key	Remarks
lass cloth for ducting anufacture. etals olyvinyl butyral (sheet)						Good resistance to water, oil and fuels. Bostik Thinner 63 1b available and 6104 for metal priming. Flash Pt. 5–15°C. Coverage 3·6 m²/l for both surfaces
etals ood VC abrics oard materials						Alternative Bostik 4142 (45% w/w solids) available. **Bostik Thinner M754 or 6104** for dry adhesive or surface preparation. Non flammable. Resistance to oils but not organic solvents. Coverage 8·2 m²/l
ellulosic materials						Sets to a colourless film rapidly. Resistant to water. Wet sealing adhesive where water based products are precluded. May be diluted with methanol. Inflammable, Flash Pt. above 23°C
etals lass lastic laminates				5 cm × 2·54 cm sustains 100 g load without creep. 4 mm creep for 6·25 cm² shear loaded, with 500 g, in 24 h		May be exposed to 177°C for short periods. Resistant to water, brine, detergents. Stable up to 2 yr to light exposure
lass cloth sbestos etals anvas (duck) – canvas		After, 8 wk at 40°C	20	56 N/cm², canvas ruptured	b	Dries to a semi-white gloss film. Resistant to steam, water and biodeterioration, but not frost. Non-inflammable. Coverage, 2–2·7 m²/kg
		8 wk at 85°C	20	2·1 N/cm², canvas ruptured	b	
		12 h in steam	20	35 N/cm²	b	
		1 wk in steam	20	23 N/cm²	b	
		12 h in water at 20°C	20	24 N/cm²	b	
		24 h in water at 20°C	20	21 N/cm²	b	
ellulose (film) ellulose acetate (film) olythene (film) olypropylene (oriented) olyvinyl chloride (flexible) luminium (foil) aper						Maximum bond strength in 5 d. Resistant to heat, water vapour, water, solvents. May be thinned with acetone or ethyl acetate free from water or alcohols. Available in alternative low viscosity grade PA/76/13. Coverage, 1 m²/1·5–2·5 g

Trade source	Trade name or designation	Basic type	No. of compo-nents	Colour	Physical form, consistency or viscosity Ns/m²	Working life	Method of application	Processing		Service temp. range °C	Main uses
								Curing cycle	Bonding pressure N/cm²		
213	Plustex 276	Synthetic resins in aqueous media	1		High viscosity liquid 55% w/w solids		Spreader Roller coater	Contact bonding of wet adhesive	Pres-sure nip rollers	100	Continuous lamina-tion of fibre and wood boards to polyvinyl chloride and other flexible plastics
47	Bostik 1777	Synthetic rubber/ resin in ester solvent	1	Off White	Medium viscosity liquid	Indefinite	Brush Spray gun	Dried 1 h at 50–70°C and cured for 4 h at 100°C to 5 min at 250°C under pressure	Up to 110	−50 to 250	Bonding brake linings to metal shoes in automotive industry
47	Bostik 1160	Synthetic rubber/ resin in alcohol	1	Amber	Thin liquid	Indefinite	Brush	Dried 1 h at 27°C and cured 20 min at 141°C	Contact to 20	−40 to 300	Fabrication of hot air ductings for aircraft conditioning units. Metals, plastic sheets
23	Bexol 1563	Synthetic resins in hydro-carbon solvents	1	Grey	Thixotropic liquid 21% w/w solids	Indefinite	Brush	Contact bonded whilst wet. Sets ½–24 h at 20°C	Contact	−20 to 60	Bonding of vinyl plastic pipe fittings
23	Bexol 1528	Chlorinated rubber in hydro-carbon solvents	1	Light Brown	Liquid 9 at 23°C 29% w/w solids	30 min at 20°C	Brush	Air dried 10 min at 20°C and contact bonded	Contact to 7	−20 to 60	General purpose contact adhesive
241	SBD Certite 19–19	Polyester resins + catalyst + filler	3	Grey	Slurry to paste according to filler content	15 min at 20°C	Trowel Spreader	Cold setting in 25 min at 4–27°C according to grade	Contact		Jointing applica-tions in building and civil engineering construction. Bonding pre-cast concrete and setting in dowel bars and bolts

Adherends	Test spec.	Test conditions	Test temp. °C	Physical data	Key	Remarks
Millboard Hardboard Chipboard Asbestos (rigid) Plywood Polyvinyl chloride						Laminates require 6 h storage period before handling. No drying required before bonding. Resistant to water. Alternative lower viscosity grade available. Non-inflammable. Flame spread is Class 1
Asbestos (rigid) Steel						May be thinned with Bostik 6846 for spray gun application. Improvement in bond strength on heating. Special equipment required for brake lining processes. Resistant to oil, petrol and water. Inflammable, Flash Pt. below −3°C. Coverage, 3 m²/l
Steel Aluminium Polyvinyl butyral (sheet) Glass cloth						May be thinned with Bostik 6919. Resistant to oil, paraffin, petrol, water. Inflammable, Flash Pt. below −3°C. Coverage, 3·7 m²/l
Acrylonitrile butadiene styrene Polyvinyl chloride (chlorinated) Polyvinyl chloride (rigid) − pvc		Shear loads on 2·5 cm diameter pipe after curing, 5 min 24 h 30 d	 20 20 20	 274 N/cm² 2 100 N/cm² 3 100 N/cm²	 b b b	Available in other colours. Resistant to water, oils, petroleum. Inflammable Coverage, 5 m²/l
Polyvinyl chloride Acrylonitrile butadiene styrene Polystyrene Rubber Wood						Additive Bexol 1529 available for metal substrates. Resistant to ageing, water, oils, petroleum. Immediate green strength is good. Inflammable. Coverage, 5 m²/l
Concrete Brickwork Timber Steel Asbestos		After 1½ h cure After 2 h cure	20 20	1 375 N/cm² 6 170 N/cm² 7 550 N/cm² 0·9 × 10⁶	d c c j	Other grades available for winter–tropical environments. Good resistance to water and frost. Withstands corrosive soils and industrial atmospheres. Unsuitable for rubber, bitumen impregnated materials and plastics such as polystyrene or polyurethane

Trade source	Trade name or designation	Basic type	No. of compo-nents	Colour	Physical form, consistency or viscosity Ns/m^2	Working life	Method of application	Processing		Service temp. range °C	Main uses
								Curing cycle	Bonding pressure N/cm^2		
107	Evo-stik Resin 'W'	Synthetic resin	1		Liquid	Open assembly time, 10–25 min	Brush Spatula Spreader	60 min at 20°C to 5 min at 80°C	Contact to 20	80	Cabinet making and joinery. Hard and soft wood bonding
193	EC 711	Synthetic rubber in toluene	1	Light tan	Medium syrup 32% w/w solids.	20 min at 20°C	Brush Flow gun	Air dried 5–20 min at 20°C and contact bonded. Heat curing 30 min at 135°C	Contact to 20	−30 to 80	Rubber to metal bonding (gaskets, shock pads) Automotive and electrical industry.
193	EC 750	Synthetic rubber in ketone solvent	1	Red	Heavy syrup 54% w/w solids	5 min at 20°C	Flow gun	Air dried 5 min at 20°C. Full cure in 1–3 d	Contact	−53 to 93 or 150 for short periods	Weathertight sealing and caulking of closed vessels under internal pressure. Aircraft industry
193	EC 847	Synthetic rubber in acetone	1	Dark Brown	Medium syrup 40% w/w solids	3–10 min at 20°C	Brush Spray gun	Air dried 10 min at 20°C and contact bonded. Heat curing 10 min at 120–150°C	Contact	−40 to 95	Vinyl plastics, nitrile rubbers to metals. Metal-metal bond-ing in automotive industry. Zinc foil to enamelled metal. Vinyl plastic lamination to metals and porous materials
193	EC 1828	Synthetic rubber in solvent blend	1	Red	Thin syrup 22% solids		Brush Spray gun	Air dried and heat cured under pressure for 2 min at 177°C	100	−34 to 121 For short periods 200°	High strength adhesive for bond-ing honeycomb sandwiches, building panels and similar applications

Adherends	Test spec.	Test conditions	Test temp. °C	Physical data	Key	Remarks
Wood – various Wood (beech) – wood			20	Up to 1 300 N/cm² 690 N/cm²	b b	Optimum bond strength developed in 20 h for 20°C cure. Bonds hardwoods by application to one surface only. Coverage, 6·4–9·6 m²/l
Polychloroprene Steel Aluminium Canvas Hardboard Cork Cellulose (sponge)				Conforms to DTD/900/4522		Resistant to water, petrol, oil. Inflammable, Flash Pt. 5°C. Coverage, 5–6 m²/l
Aluminium (anodized or primed with zinc chromate)				Conforms to DTD/900/4451		Alternative grade EC802 for brush application available. Flexible joint seal at 53°C. Resistant to oil, water, petrol. Primer EC 776 improves metal adhesion. Inflammable, Flash Pt. 10°C. Coverage 1·30 metres × 6 mm strip per litre.
Rubber (Buna-N) Polyvinyl chloride (rigid) Leather Wood Metal (enamelled) Zinc (foil) Steel Aluminium – aluminium		Cured at 150°C for 60 min	25 50 80 120	1 200 N/cm² 620 N/cm² 345 N/cm² 83 N/cm²	b b b b	High resistance to plasticisers present in vinyls and nitrile rubber. Withstands oil, fuel, aliphatic hydrocarbons. Dried adhesive may be solvent reactivated 6 mth after coating. Inflammable, Flash Pt. −18°C. Coverage, 6 m²/l
Polystyrene Asbestos Wood Plastic laminate Steel Aluminium – aluminium			−34 20 82 107 20 121	940 N/cm² 650 N/cm² 410 N/cm² 276 N/cm² 555 N/cm² 790 N for 64 min before rupture	b b b b b	Displays good heat resistance and is suitable for applications bearing sustained loads at high temperatures. Resistant to low temperatures and water immersion. Inflammable, Flash Pt. below −18°C. Coverage, 7·4–9·8 m²/l. Edgewise compression test carried out on sample, 76 cm × 20 cm × 2·54 cm end blocked with steel plates to prevent peel type failure. Sandwich flexure test performed on sample, 76 cm × 20 cm × 2·54 cm. Core shear test performed on sample, 76 cm × 45·6 cm × 2·54 cm.
Paper honeycomb (18% phenolic impregnated), 1·27 cm × 2·54 cm cell size with steel skins 0·81 cm thick	MIL–S–401A	After 7 d in water Sustained loading. Edgewise compression of sample. After 7 d at 20°C After 7 d at 20°C After 7 d at 20°C After 6 FPL cycles After 6 FPL cycles Sandwich flexure of sample	−34 20 107 20 107	32·6 kN 21·3 kN 6 300 N 25·75 kN 15·8 kN		

continued

Trade source	Trade name or designation	Basic type	No. of compo- nents	Colour	Physical form, consistency or viscosity Ns/m^2	Working life	Method of application	Processing		Service temp. range °C	Main uses
								Curing cycle	Bonding pressure N/cm^2		
EC 1828—continued											
193	EC 1917	Synthetic rubber in petroleum naphtha	1	Amber	Medium syrup 30% w/w solids	Open time 90 s at 20°C	Brush Spray gun	Air dried and contact bonded within $1\frac{1}{2}$ min	Contact	−23 to 80	Bonding insulation materials, and vinyl plastics to painted steel. Automotive trim work
73	Redux 775	Phenolic-polyvinyl formal	2	White pow- der + trans- parent liquid	Liquid with powder	30 min drying period. Prepared substrates can be stored for 3 d before assembly	Brush or roller to apply liquid. Wet substrate is then dipped into powder	Cured for 30 min at $150 \pm 5°C$ Pressure removed when joints cooled to 70°C	17 to 70 depends on sub- strate	Up to 120	Bonding metals, alloys, wood, thermosetting plastics, friction materials, rubbers in aircraft and industrial applications

Adherends	Test spec.	Test conditions	Test temp. °C	Physical data	Key	Remarks
		After 7 d at 20°C	−34	2 270 N		Forest Products Laboratory (FPL) weathering test (see p. 259)
		After 7 d at 20°C	20	1 530 N		
		After 7 d at 20°C	107	740 N		
		After 6 FPL cycles	20	1 680 N		
		After 6 FPL cycles	107	975 N		
		Panel shear of sample				
		After 7 d at 20°C	−34	6 950 N		
		After 7 d at 20°C	20	6 900 N		
		After 7 d at 20°C	107	2 370 N		
		After 6 FPL cycles	20	7 200 N		
		After 6 FPL cycles	107	4 250 N		
olystyrene (foam) etals (painted) olyvinyl chloride − steel (painted)		0·12 mm wet coat on painted steel with open time,		Complies with D.C.I. Spec AFS 658		Fast tack adhesive with non-staining properties
		30 s		8·2 N/cm²	b	
		60 s		13·8 N/cm²	b	
		3 min		9 N/cm²	b	
		5 min		7 N/cm²	b	
oth − steel (painted)		open time				
		30 s		17·3 N/cm²	b	
		60 s		17·3 N/cm²	b	Inflammable, Flash Pt.
		3 min		10·4 N/cm²	b	below −18°C.
		5 min		7 N/cm²	b	Coverage, 4 m²/l
uminium and alloys eel ood agnesium and alloys atural rubbers olychloroprene (neoprene)				Complies with U.S. Federal Spec. MMM-A-132 Type 1, Class 2 adhesives		Alternative Redux systems are available as liquid/powder or film systems. These adhesives have been the subject of considerable investigation during the past decade. The
eel (mild)−steel (mild)			24	3 460 N/cm²	b	user is strongly recommended
eel (stainless)−steel (U.S. Type 301)			24	3 900 N/cm²	b	to consult the manufacturer for the relevant data sheets
agnesium alloy−magnesium alloy (HK31A-H24)			24	2 225 N/cm²	b	pertaining to the application envisaged. e.g. Redux Liquid
tanium−titanium (K1, 130)			24	2 825 N/cm²	b	K6 + Powder C is excellent
024-T3) Alclad−Alclad	DTD 775B MIL-A-5090B		24	3 200 N/cm²	b	for vulcanised rubbers or thermosetting plastics. Redux
024-T3) Alclad−Alclad } 024-T3) Alclad−Alclad			24	3 470 N/cm²	b	64 is suitable for bonding friction linings to brake shoes
			82	1 590 N/cm²	b	and clutch discs. Redux joints have good
		Fatigue strength; cycles to failure at 413 N/cm²	24	No failure at 10⁷ cycles		resistance to fuels, anti-icing fluids, hydraulic oil, salt spray, tap water and 100% relative
		Creep rupture; deformation after 192 h at 1 105 N/cm²	24	Less than 0·0254 cm max deformation		humidity at 50°C. A Redux accelerator is available to promote rapid curing, or
		T-peel	24	53 N/cm	a	reduce the curing temperatures
			24	0·33 × 10⁶ N/cm²	j	to 110°C for K6/C or 120
				2 884 Jm/m²/s/°C	q	systems

In addition to the manufactures adhesives listed above in this section, there are unlimited possibilities for compounding 'home-made' adhesives. The following references will be found to contain recipes which have been used successfully over a number of years for the laboratory, industrial and domestic purposes.

HURD, J., *Adhesive Guide*, Research Report, M39, Sira (1959).

STANDAGE, H. C., *Cements, Pastes, Glues and Gums*, Technical Press Ltd., London, (1931).

Handbook of Chemistry and Physics, (Hodgman, Ed.), 35th edition, Chemical Rubber Publishing Co., Cleveland, Ohio, (1953).

STRUCTURAL ADHESIVES

Important metal to metal structural adhesives which have been extensively used for aircraft and space applications are summarised below. These adhesive products qualify under the U.S. Federal military specification MMM-A-132 Types I to IV (Adhesives, Heat Resistant, Metal to Metal) briefly described below. Most of these products are manufactured abroad but can generally be ordered from the suppliers listed.

$MMM-A-132$	Minimum average tensile shear strengths (N/cm^2)*				
Test temperature and exposure time	TYPE 1 Class 1	TYPE 1 Class 2	TYPE II	TYPE III	TYPE IV
260°C for 10 min	3 100	1 720	1 550	1 550	1 550
260°C for 192 h	1 720	860	—	—	—
Room temperature, 23·8°C	—	—	1 380	1 380	1 380
82·2°C for 10 min	—	—	1 380	1 380	1 380
194·4°C for 102 min	—	—	—	950	1 380
194·4°C for 192 h	—	—	—	—	690

*measured as lb/in²

Trade Source (see p. 322)	Trade Name or Designation	Basic Type	Physical form	Qualification to $MMM-A-132$	
				Type	Class
91	FM 1000	Epoxy-nylon	Film (unsupported)	1	1
287	Metlbond 408	Epoxy-nylon	Film (unsupported)	1	1
193	AF 32/EC 1660	Phenolic-nitrile	Film (unsupported)+liquid primer	1	1
193	AF 6	Phenolic-nitrile	Film (unsupported)	1	2
193	AF 30/EC 1660	Phenolic-nitrile	Film (unsupported)+liquid primer	1	2
193	AF 41/EC 1956	Epoxy-nylon	Film (unsupported)+liquid primer	1	1
193	AF 40/EC 1956	Epoxy-nylon	Film (unsupported)+liquid primer	1	1
193	AF 300/EC 2254		Film+liquid primer	1	1
91	FM 238/BR 238			1	2
193	AF 30/EC 1660	Phenolic-nitrile	Film (unsupported)+liquid primer	1	2
193	AF 120/EC 2320	Epoxy-nitrile	Film (unsupported)+liquid primer	1	2
193	AF 126			1	2
73	Redux Film 775	Phenolic-polyvinyl formal	Film (unsupported)	1	2
73	Redux 775	Phenolic-polyvinyl formal	Liquid+powder	1	
122	Plastilock 620/A626B	Phenolic-nitrile	Film (unsupported)+liquid primer	1	2
122	Plastilock 639			1	2
287	Metlbond 406	Epoxy-nylon	Film (unsupported)	1	2
287	Metlbond 400			1	3
91	FM 47	Phenolic-Polyvinyl formal		1	3
91	FM 48			1	3
91	FM 54				
91	FM 61/BR 2227		Composite film (supported)+ liquid primer	1	3
91	FM 86/BR 86			1	3
91	FM 96	Epoxy-nitrile	Film (supported)	1	3
91	FM 97			1	3
91	BR 92			1	3

Trade Source (see p.322)	Trade Name or Designation	Basic Type	Physical form	Qualification to MMM−A−132	
				Type	Class
91	FM 97–1070			1	3
91	FM 245			1	3
91	FM 250			1	3
193	AF–40T	Epoxy-nylon	Film (unsupported) + liquid primer	1	3
193	AF–110	Epoxy-nitrile	Film (supported)	1	3
193	AF–110/EC 1682	Epoxy-nitrile	Film (supported) + liquid primer	1	3
193	AF-111	Epoxy-nitrile	Film (supported)	1	3
193	AF-114			1	3
193	AF 204/EC 1660			1	3
193	Scotchweld EC 1469			1	3
193	Scotchweld EC 1471	Phenolic-polyvinyl formal	Solvent solution	1	3
193	EC 2186	Epoxy-nitrile	Paste (100% solids)	1	3
193	AS 5640			1	3
193	AS 8233			1	3
91	FM 132			1	3
287	Narmco			1	3
287	Metlbond 328	Epoxy-nitrile	Film (supported)	1	3
287	Metlbond 4021	Phenolic-nitrile	Film (unsupported) + liquid primer	1	3
287	Metlbond 329			2	
193	Scotchweld AF 31/EC 1459			2	
193	AF 31	Phenolic-nitrile		2	
193	Scotchweld AS 9795			2	
193	Scotchweld EC 1469			2	
193	Scotchweld EC 1595	Epoxy	Paste (100% solids)	2	
122	Plastilock 650	Phenolic-nitrile	Film (unsupported) + liquid primer	2	
91	HT 424	Phenolic-epoxy	Film (supported)	2	
91	HT 424	Phenolic-epoxy	Film (supported) + liquid primer	3	
193	AF 7431			3	
193	AF 31/EC 2174	Phenolic-nitrile	Film + liquid primer	4	

KEY TO ADHESIVE PRODUCT CHARTS

The following alphabetical symbols have been used in the tabulated description of properties and performance data of adhesive products.

Symbol	Property	Symbol	Property
a	Peel strength	o	Dissipation factor
b	Shear strength	p	Volume resistivity
c	Compressive strength	q	Thermal conductivity
d	Tensile strength	r	Thermal expansion
e	Torsion shear	s	Viscosity (dynamic)
f	Flexural strength	t	Viscosity (kinematic)
g	Impact strength	T	Heat Deflection Temperature
h	Tear strength	u	Capacity
j	Modulus of elasticity	v	Compression set ASTM D395–55
k	Elongation	w	Fatigue strength after load cycling
l	Hardness (Shore D or L)	x	Modulus of rupture
m	Dielectric strength	y	Bulk density
n	Dielectric constant	z	Refractive Index

SI AND RELATED UNITS OF MEASUREMENT: READY REFERENCE CONVERSION FACTORS

The following table is selective and concerned with those units in use throughout the handbook. It is intended to facilitate rapid reference to conversions frequently needed in everyday work.

A number of publications have been issued by the Ministry of Technology in collaboration with the British Standards Institution and the National Physical Laboratory. The following list will assist those requiring further information on metric units in general.

1. The Metric System in the United Kingdom: The Use of SI Units. British Standards Institution. PD 5686. January 1969.

2. Changing to the Metric System. Conversion Factors, Symbols and Definitions, P. Anderton and P. H. Bigg.
Ministry of Technology, National Physical Laboratory H.M.S.O. (1966).

3. The International System (SI) Units B.S. 3763: 1964.

4. Conversion Factors and Tables.
B.S. 350: Part 1: 1959, Bases of Tables, Conversion factors
B.S. 350: Part 2: 1962, Detailed conversion tables
B.S. 350: Supplement No. 1: 1967, Additional tables for SI conversions.

Property	Non SI unit	Factor for conversion to SI unit
MECHANICAL		
Modulus of elasticity	$E(\times 10^6 \text{ lb/in}^2)$	$E(\times 689476 \text{ N/cm}^2)$
Peel strength	1 lb/in	1·75127 N/cm
	1 kg/cm	9·80665 N/cm
Shear strength	1 lb/in^2	0·689476 N/cm^2 or 6·89476 kN/m^2
	1 kgf/cm^2	9·80665 N/cm^2 or 98·0665 kN/m^2
Compressive	1 lb/in^2	0·689476 N/cm^2
Tensile strength	1 lb/in^2	0·689476 N/cm^2
Flexural strength	1 lb/in^2	0·689476 N/cm^2
Impact strength	ft. lb/inch notch	1·355 J for 25·4 mm notch
Tear strength	1 lb/in	1·75127 N/cm
Fatigue strength after load cycling	1 lb/in^2 at X Hz	0·689476 N/cm^2 at X Hz
	1 kgf/cm^2 at X Hz	9·8066 N/cm^2 at X Hz
ELECTRICAL		
Dielectric strength	1 volt/mil	0·39 kV/cm
Dielectric constant	1 at 10^n c.p.s.	1 at 10^n Hz
Dissipation factor	1 at 10^n c.p.s.	1 at 10^n Hz
Volume resistivity	1 Ω/cm	1 Ω/cm

Property	Non SI unit	Factor for conversion to SI unit
HEAT		
Thermal conductivity	1 Btu in ft^{-2}h^{-1} °F^{-1}	0·144228 J mm^{-2} s^{-1} 1°C^{-1}
	1 Cal cm^{-1} sec^{-1} 1°C^{-1}	418·68 J mm^{-2} s^{-1} 1°C^{-1}
Thermal expansion	1°F^{-1}	1·8 C^{-1}
OTHER PROPERTIES		
Viscosity (dynamic)	1 p	0·1 Ns/m^2
Viscosity (kinematic)	1 St	10^{-4} m^2/s
Capacity	1 lb UK gal^{-1}	0·09978 kg/l
	1 lb US gal^{-1}	0·0831 kg/l
Bulk density	lb/ft^3	16·0185 kg/m^3
Length	1 in	2·54 cm
	1 ft	30·48 cm
Area	1 in^2	6·45 cm^2
	1 ft^2	929 cm^2 or 0·0929 m^2
Weight	1 oz	28·35 g
	1 lb	453·6 g
	1 g	10^{-3} kg
Volume	1 pt	0·569 l
	1 gal	4·546 l

| # Surface Preparation

6.1. INTRODUCTION

The synopsis of methods for the surface preparation of adherends presented here is divided into sections concerned with pretreatments for Metals, Plastics, Rubbers, Fibre products, Inorganic materials, Wood and Painted Surfaces. Each section includes a short summary of those aspects which are particularly relevant to the class of material under discussion. Summaries are followed by detailed descriptions of pretreatment methods as the basis for their subsequent evaluation in connection with specific adherend-adhesive materials. Before referring to a specific section, the reader should first consider the introductory remarks on Surface treatments. It is important to adhere to the conditions of recommended pretreatments for optimum results unless the manufacturer indicates otherwise for a particular adhesive.

6.2. SAFETY PRECAUTIONS

Safety precautions should be strictly observed where chemical solutions and solvents are employed for a pretreatment procedure. Possible hazards include:

Skin and eye burns—where acid reagents are used —always wear rubber gloves and safety goggles. Inhalation of acid fumes should be avoided. In preparing aqueous acid reagents, *always* add the acid slowly to the water to avoid the risk of explosion.

Fire or explosion—dry parts should be cleaned with inflammable organic solvents in the open-air or in a fume cupboard and *never* in an oven which is likely to ignite vapours. Inhalation of solvent vapours should be avoided.

6.3. SURFACE TREATMENT

The strength of a joint depends not only on the cohesive strength of the adhesive (or adherends), but also on the degree of adhesion to the bonding surfaces. Adhesion at the adhesive-adherend interface occurs within a layer of molecular dimensions and the bond strength can be reduced to nothing by surface contaminants which are themselves weakly adherent and which prevent contact between adhesive and substrate. For optimum adhesion, the surfaces to which any adhesive is applied must be cleaned or converted to a suitable condition before bonding and this is the main purpose of all surface pretreatments. There is no overall theory covering the wide range of pretreatments recommended in the literature and the choice of treatment becomes a matter of evaluation of processes practicable for a given adherend-adhesive system. It follows that any change in one component of the system calls for a reassessment of a particular treatment specified for the original system. The choice of treatment largely depends on such factors as the adherend material and its condition, the adhesive type selected, joint loading requirements, service environments and service life, processing costs and resources available. The degree of surface preparation required may also be indicated by the effects of previous storage and processing or earlier service and repair of an assembly.

The composition of the surface layer of an adherend is determined by the previous history of the material and is usually an unknown quantity. Metals are invariably annealed or heat-treated, acid-pickled, anodised, coated with lubricating oils, oxides or rust, treated with rust inhibitors, or protected with polymer or lacquer coatings at some stage of their processing. Plastics surfaces are contaminated with release agents, plasticisers or possess weak surface layers. Similar conditions apply to other materials and one can be certain that the composition of the substrate is quite unlike that of the bulk material. The simplest treatments rely on the cleaning action of solvents or abrasives to remove surface contamination which otherwise impedes wetting of the base material by the adhesive. Other surface treatments are more involved and tend to increase adhesion by promoting the process by which adhesion occurs, e.g. chemical treatments modify the surface physically and chemically to increase its specific adhesion properties, whereas mechanical roughening creates a surface which interlocks with the adhesive to provide better mechanical adhesion. Prepared surfaces can often be preserved for later use by coating with a strippable protective primer. Otherwise it is advisable to coat the treated substrate with adhesive immediately or within a few hours.

Where high bond strengths are required, the surface preparation must be thorough. With structural adhesives it is important to note that the surface preparation employed may be critical for the fabrication of permanent bonds and, therefore, a careful evaluation is required. For example, solvent-cleaned aluminium joints may show a rapid decrease in strength on weathering whereas acid-etched joints retain high strengths for many years in similar environments. Similar considerations apply to structural adhesives intended for high temperature and other extreme service conditions. Non-structural adhesives, such as the rubbers and pressure-sensitive materials, usually perform satisfactorily if the surface is given a solvent cleaning treatment. It should be noted that where the load bearing requirements are less exacting it may be possible to dispense with pretreatment processes. For instance, certain plastisol or other oil-resistant adhesives are applied directly to oiled steel in the bonding of automobile body parts. In general, surface preparations recommended by adhesives manufacturers should not be ignored where the adhesive is for a special purpose.

Mechanical treatments are useful when the number of assemblies to be bonded is small or where the state of the bonding components forbids the use of chemical agents (e.g. corrosion of precision-made parts or paint-masked areas adjacent to bonding areas). Typical techniques include abrasion with sand or emery paper, grit or vapour blasting, and steel wool abrasion. Chemical treatments are preferable where the production quantity justifies the installation of necessary equipment (etching tanks, rinsing baths and drying ovens, etc.). Chemical procedures vary widely with the type of material and normally involve the immersion of degreased surfaces into one or more baths containing solutions of suitable reagents, which may be hot, after which the surfaces are washed in water and dried. When aqueous solutions are used in pretreatments and rinse baths, for best results they should be prepared from distilled water (de-ionised water is satisfactory provided non-ionic organic impurities are absent) in order to avoid deposition of residues of soluble salts on drying the substrate. For spray rinsing, water should have a specific conductance below 10 micromhos. Tank rinse water can have higher conductance values up to 30 micromhos. Clean gloves or tools should be used to handle treated parts and the bonding areas should not be touched. The cleaning area should be well vented where solvent vapour and acid fumes are present and it may become necessary to control the humidity and temperature where treated components are likely to be exposed to the air for long periods.

Mechanical and chemical treatments have recently been supplemented by physical methods which include flame treatments, ionic bombardment in a vacuum (corona discharge) and electric discharge. These techniques have proved successful with inert plastics like polyethylene, polypropylene and polytetrafluoroethylene which are difficult to bond effectively without some kind of pretreatment. The nature of the surface changes brought about by these processes is not fully understood but there is evidence that electrical methods change the molecular structure of the substrate and effect oxidation of the surface in the case of polyethylene.

Satisfactory pretreatment results can be obtained for small objects with simple laboratory equipment although these techniques are more suited to mass production processes which require the continuous preparation of surfaces (e.g. on-line fabrication of plastics laminated films). Thus, an oxidising bunsen flame or the high voltage spark discharge from an induction coil (e.g. Tesla leak-tester) render the surface of polyethylene receptive to adhesives. The extent to which water will wet the surface often provides a convenient method for determining the efficacy of a treatment.

Recent evaluation work at Sira on these physical methods with selected plastics has shown that reproducible adhesion results are not easily obtained unless strict attention is paid to the standardisation of treatment times, the interval before adhesive application to the treated surface, and humidity. Activated surfaces are prone to aerial contamination and should be bonded within a short time.

6.4.   METALS

Surface preparation is of major significance to the integrity and durability of metallic joints where these are subjected to high stresses and severe service conditions simultaneously. For various surface treatments the initial bond strengths of joints may be similar but this cannot be taken as an indication of the performance of these joints in service. The pretreatment appropriate to the service requirements of a bonded assembly is a matter for critical evaluation where many structural applications are concerned. After chemical treatment, metal surfaces are highly active and can easily attract aerial contaminants such as dust and moisture. It is expedient to bond treated surfaces as soon as possible after surface preparation.

Degreasing
Metal surfaces are invariably contaminated with oils

and greases and an essential pretreatment is de-greasing either with steam or by wiping the surface with solvent wetted cloths (until fresh cloths show no signs of soiling) or, more effectively, by vapour degreasing. The degreasing must be thorough and contaminants must not be redeposited on the surface as the solvent evaporates; a final surface rinse with fresh solvent is always advisable. Commonly used organic solvents, such as trichloroethylene or perchloroethylene, should be checked periodically for the formation of corrosive acid. Solvent de-greasing is a preliminary treatment for any more extensive pretreatment processes which follow, though it is normally sufficient for joints which will not be subject to abnormal service conditions (e.g. high stresses or severe service environments). The effectiveness of degreasing treatments for metals can be checked by applying water to the treated surface. In the absence of oil or grease, a uniform water film is produced which flows off evenly from the tilted surface without forming droplets or discontinuities. Droplet formation indicates the presence of con-tamination. Alkaline cleaners and/or detergent solutions are usually employed, after solvent treat-ments, to remove dirt and inorganic soils. Un-inhibited (etching-type) strong alkaline solutions are used for ferrous metals, titanium and certain copper alloys, but aluminium generally requires inhibited solutions if etching is to be avoided. The alkaline cleaning process is usually followed by a chemical or mechanical treatment because it leaves surfaces incompatible with most adhesives. The surface must be dried with hot air before bonding.

Mechanical methods
Mechanical means of abrasion are employed where the metal surface is badly oxidised or scaled. Oxide formation is rapid with base metals such as alumi-nium but is not usually a problem with inert metals like titanium or stainless steel. The methods depend on the abrasive action of wire-brushes, sand- and emery-papers, or shot blasting techniques to remove unwanted surface layers. Generally, these methods are more difficult to control than chemical methods and, except with certain metals and alloys, joint strengths tend to be lower than those obtained by chemical processing. Abrasion produces surface roughening and as a result the joints are more susceptible to water and solvent penetration. For critically machined components, it may not be possible to use abrasive methods.

With dry abrasive blasting the best results are obtained with sharp, jagged grits, with good cutting edges, based on hard materials such as alumina, silica quartz or carborundum, whereas glass or metal beads of round shape tend to peen the metal surface and are unsuitable. Size and nature of the abrasive grit are important aspects of shot-blasting; for each metal and alloy, there is a preferred range of grit sizes. It is important to degrease the surface before and after abrasion, and to ensure that the grit particles are free from contamination likely to soil the surface. A final solvent or water rinse serves to remove abrasive dust which would otherwise weaken adhesion at the adhesive-adherend interface.

Similar considerations apply to wet abrasive blasting treatments. This method is preferable to dry-blasting for surface preparation since the range of abrasive materials which may be employed is greater. The use of water limits its applicability to non-corrosive metals unless anti-corrosion additives are included in the water. Control of water purity is also important.

Chemical methods
Chemical and electrochemical treatments are pre-ferable to mechanical methods where economic processing and easier control of reproducibility of surface (and joint) quality are important. In addition to cleaning action, these treatments may also produce a chemically resistant surface layer which improves bond strength retention in service. The treatments involve immersion in reagents (which range from dilute to concentrated acid or alkaline solutions) at room or elevated temperatures. The acids and bases attack surface oxides more rapidly than the base metal and in the case of acids, inhibitors may be added to enhance this effect in the removal of heavy oxide scales where abrasive blasting is unsuitable.

After the thicker oxide films are removed by pickling, the metal is usually treated under controlled conditions with milder solutions, such as acid-dichromates, to produce thin oxide layers of con-trolled structure and thickness.

Earlier remarks concerning the purity of water used in chemical treatments are particularly relevant to metals. The final rinse should be done in high purity water although the degree of purity needed depends on the application and adhesive system employed. Epoxy adhesives have a lower tolerance for acids than some other adhesive types and are more tolerant towards alkalinity and chlorides than phenolic-nitrile adhesives. Rinse water temperatures can be critical and should not exceed 70°C for some metals. For example, aluminium and its alloys have been extensively investigated in connection with aircraft industry requirements. A sulphuric acid-sodium dichromate etch treatment for aluminium produces a strong adherent surface layer of hydrated oxide, β-$Al_2O_3.3H_2O$ provided the temperature is below 60°C. Above this temperature (e.g. for reagent or subsequent rinse water) there is a change of oxide structure to α-$Al_2O_3.H_2O$; bond strength per-formances for epoxy-adhesives on this substrate layer are inferior to those based on the trihydrate oxide form. Thus, a seemingly unimportant process variation can reduce the quality or structural integrity of a bond. Similar phenomena can be anticipated for other base metals.

Table 6.1. SURFACE PREPARATION FOR METALS

Adherend	Degreasing solvent	Method of treatment	Remarks	R (p.2
Aluminium and Aluminium alloys	Trichloro-ethylene	1. Abrasion. Grit or vapour blast, or 100 grit emery cloth, followed by solvent degreasing	Suitable for general purpose bonding. Steel shot abrasion not recommended	1
		2. Vapour degrease for 10 min and immerse in 2–3% solution of alkaline detergent at 65°C. After washing, etch for 30 min at 60°C in, Parts by wt. Chromium trioxide 5 Sulphuric acid (93% sp.gr. 1·84) 15 Water 80 Rinse in water and distilled water, and air dry at 70°C.	Min. Tech. Specification DTD 915B employed for preparation of aluminium alloys in the aircraft industry	

		3. After treatment by Method 1, anodise in following aqueous solution at 40°C,
		
		            Grams per litre
		Total Chromium trioxide 100
		  Sulphuric acid 0·2
		  Sodium chloride 0·2
		Raise voltage from 0–40V in first 10 min, maintain at 40V for 20 min and then increase to 50V in next 5 min. Maintain at 50V for 5 min. Rinse in water and distilled water. Dry in air at 70°C.
			Min. Tech. Specification DTD 910C. Ion concentration must not be greater than limits specified. Free chromium trioxide should not exceed 30–35 g/litre	

		4. Etch for 30 min at 60–65°C in, Parts by wt. Sulphuric acid (93%, sp.gr. 1·84) 27·0 Sodium dichromate (crystalline) 7·5 Water (distilled) 2·5 Rinse in distilled water after washing in tap water and dry in air	For maximum bond strength, temperatures must not exceed 65°C when washing and drying	1, 2
		5. Immerse for 5 min at 65°C in, Parts by wt. Sodium metasilicate 1·0 Detergent (non-ionic) 0·1 Water (distilled) 40·0 Rinse in distilled water after washing in tap water and dry in air	Suitable for foils. Temperatures must not exceed 65°C when washing and drying	3

		6. Immerse for 10 min at 77 ± 6°C in,
		
		            Parts by wt.
		Sodium metasilicate 30·0
		Sodium hydroxide 1·5
		Sodium pyrophosphate 1·5
		Detergent 0·5
		Water (distilled) 128·0
		Wash in water below 65°C and etch for 10 min at 68 ± 3°C in,
		            Parts by wt.
		Sodium dichromate 1
		Sulphuric acid (93%, sp.gr. 1·84) 10
		Water (distilled) 30
		Rinse in distilled water after washing in tap water and dry in air
			Not recommended for foils. Temperatures must not exceed 65°C when washing and drying.
			
			Alternative proportions of Sodium dichromate Sulphuric acid Water (distilled) which have been used are, 7/2/17, 2/5/15, 2/2/6, 1/2/7
				3, 4

contin

6.1 continued

dherend	Degreasing solvent	Method of treatment	Remarks	Ref. (p.242)
inium inium	Trichloro-ethylene	7. Etch for 3 s at 20°C in, Parts by wt. Nitric acid (67%, sp.gr. 1·41) 3 Hydrofluoric acid (48%) 1 Rinse in distilled water after cold/hot water washing, and dry in air	Suitable for casting alloys with a high copper content. Temperatures must not exceed 65°C when washing and drying	3
lium	Trichloro-ethylene	1. Immerse for 5–10 min at 20°C in, Parts by wt. Sodium hydroxide 20–30 Water (distilled) 170–180 Rinse in distilled water after washing in tap water and oven dry for 10 min at 121–177°C		4
s and ze lso er and er alloys)	Trichloro-ethylene	1. Etch for 5 min at 20°C in, Parts by wt. Zinc oxide 20 Sulphuric acid (93%, sp.gr. 1·84) 460 Nitric acid (67%, sp.gr. 1·41) 360 Rinse in water, below 65°C, and re-etch in the acid solution for 5 min at 49°C. Rinse in distilled water after washing, and dry in air	Temperatures must not exceed 65°C when washing and drying	3
nium	Trichloro-ethylene	1. Abrasion. Grit or vapour blast or 100 grit emery cloth, followed by solvent degreasing	Suitable for general purpose bonding	1
		2. Electroplate with nickel or silver	For maximum bond strength	1

Table 6.1 SURFACE PREPARATION OF METALS

dherend	Degreasing solvent	Method of treatment	Remarks	Ref. (p.242)
omium	Trichloro-ethylene	1. Abrasion. Grit or vapour blast, or 100 grit emery cloth, followed by solvent degreasing	Suitable for general purpose bonding	1
		2. Etch for 1–5 min at 90–95°C in, Parts by wt. Hydrochloric acid (37%) 17 Water 20 Rinse in distilled water after cold/hot water washing, and dry in hot air	For maximum bond strength	1, 8
per and per ys	Trichloro-ethylene	1. Abrasion. Sanding, wire brushing or 100 grit emery cloth, followed by vapour or solvent degreasing	Suitable for general purpose bonding. Use 320 grit emery cloth for foil	5

		2. Etch for 10 min at 66°C in,
		
		                        Parts by wt.
		Ferric sulphate 1·0
		Sulphuric acid (95%) 0·75
		Water
		Wash in water at 20°C and etch in cold solution of,
		                        Parts by wt.
		Sodium dichromate 5
		Sulphuric acid (95%) 10
		Water 85
		Rinse in water, dip in ammonium hydroxide (sp.gr. 0·88) and wash in tap water. Rinse in distilled water and dry in warm air
			For maximum bond strength. Suitable for brass and bronze	5

continued

Table 6.1 continued

Adherend	Degreasing solvent	Method of treatment	Remarks

Copper and Copper alloys	Trichloro-ethylene	3. Etch for 1–2 min at 20 C in,
		
		                                         Parts by wt.
		     Ferric chloride (42% w/w solution) 0·75
		     Nitric acid (sp.gr. 1·42) 1·5
		     Water 10
		 Rinse in distilled water after cold water wash and dry in air stream at 20°C
			Use of a primer X83/65 resin is recommended for maximum bond strength with epoxy adhesives. Alternative proportions of
			     Ferric chloride
			     Nitric acid
			     Water
			 which have been used are, 1/1/10, 1/2/7, 2/4/27, 1/2/13

		4. Etch for 30 s at 20°C in,
		
		                                           Parts by wt.
		     Ammonium persulphate 1
		     Water 4
		 Rinse in distilled water after cold water wash and dry in air stream at 20°C.
			Alternative etching solution to above where fast processing is required

Germanium	Trichloro-ethylene	Abrasion. Grit or vapour blast followed by solvent degreasing	
Gold	Trichloro-ethylene	Solvent or vapour degrease after light abrasion with a fine emery cloth	
Iron		see Steel (mild) p. 226	
Lead and solders	Trichloro-ethylene	Abrasion. Grit or vapour blast, or 100 grit emery cloth, followed by solvent degreasing	
Magnesium and Magnesium alloys	Trichloro-ethylene	1. Abrasion with 100 grit emery cloth followed by solvent degreasing	Apply the adhesive immediately after abrasion

		2. Immerse for 5 min at 70–75°C in,
		
		                                     Parts by wt.
		     Sodium hydroxide 126
		     Water 1,000
		 Wash in cold water and etch for 5 min at 20°C in,
		                                       Parts by wt.
		     Chromium trioxide 100
		     Water 1,000
		     Sodium sulphate (anhydrous) 28
		 Rinse in distilled water after cold water wash and dry in air at 40°C.
			For maximum bond strength. Adhesive should be applied immediately after treatment

		3. Immerse for 3 min at 20°C in,
		
		                                       Parts by wt.
		     Chromium trioxide 16·6
		     Sodium nitrate 20
		     Acetic acid (glacial) 105
		     Water 100
		 Rinse in distilled water after cold water wash and dry in air at 40°C
			Alternative etching solution to (2)

cont

6.1 continued

Adherend	*Degreasing solvent*	*Method of treatment*	*Remarks*	*Ref. (p.242)*
Magnesium Mag-nesium alloys	Trichloro-ethylene	4. Vapour degrease and immerse for 10 min at 70°C in sodium hydroxide (6·3 % w/v). Rinse in water and immerse for 5 min at 55°C in, 　　　　　　　　　　　　　　Parts by wt. 　　Chromium trioxide　　　13·8 　　Calcium nitrate　　　　　1·2 　　Water　　　　　　　　　85 Rinse in distilled water and immerse for 3 min at 55°C in, 　　　　　　　　　　　　　　Parts by wt. 　　Chromium trioxide　　　10 　　Sodium sulphate　　　　0·5 　　Water　　　　　　　　　89·5 Rinse in distilled water and dry in air at 60°C		10
		5. Etch for 30 min at 60–70°C in, 　　　　　　　　　　　　　　Parts by wt. 　　Sodium dichromate　　　10 　　Magnesium sulphate　　　5 　　Manganese sulphate　　　5 　　Water　　　　　　　　　80 Rinse in water and distilled water. Dry in air.	Min. Tech. Specification DTD 911B, III	2
		6. Solvent degrease and immerse for 20 min in boiling solution of, 　　　　　　　　　　　　　　Parts by wt. 　　Ammonium dichromate　　1·5 　　Ammonium sulphate　　　3 　　Sodium dichromate　　　1·5 　　Ammonia (sp.gr. 0·880)　0·3 　　Water　　　　　　　　　93·7 Rinse in warm water and distilled water and dry.	Min. Tech. Specification DTD 911B, III	2
		7. Anodise below 30°C in a solution of ammonium bifluoride (10 % w/v) at 90–120 V A.C. until the current density falls below 0·45 A/m² (area of one electrode). Rinse in water and dry.		14
		8. Etch for 10–15 s at 20°C in, 　　　　　　　　　　　　　　Parts by wt. 　　Sulphuric acid (93 % sp.gr. 1·84)　1 　　Nitric acid (67 % sp.gr. 1·41)　　2 　　Water　　　　　　　　　　　　45 Rinse in cold water and etch in either of the chromic acid solutions (2) or (3).	Useful etching solution for badly oxidised or tarnished surfaces contaminated with burned-on forging or drawing lubricants.	
		9. Vapour degrease and anodise in following solution at 20–30°C, 　　　　　　　　　　　　　　Parts by wt. 　　Potassium hydroxide　　　　12 　　Aluminium (metal)　　　　　0·75 　　Potassium fluoride (anhydrous)　3·4 　　Trisodium phosphate　　　　3·4 　　Potassium permanganate　　1·5 　　Water　　　　　　　　　　80 Apply an alternating voltage up to 85 V to give a current density of 1·1–1·4 A/m² (area of one electrode). Wash in cold distilled water and air dry.		
Steel	Trichloro-ethylene	1. Abrasion with 100 grit emery cloth followed by solvent degreasing	For general purpose bonding	1

continued

Table 6.1 continued

Adherend	Degreasing solvent	Method of treatment	Remarks	R (p.

Nickel	Trichloro-ethylene	2. Etch for 5 s at 20°C in,
		    Nitric acid (67% sp.gr. 1·41)
		    Wash in cold and hot water followed by a distilled water rinse, and air dry at 40°C.
			For general purpose bonding.	1

Platinum	Trichloro-ethylene	Solvent or vapour degrease after light abrasion with a 320 grit emery cloth		1
Silver	Trichloro-ethylene	Abrasion with 320 grit emery cloth followed by solvent degreasing		1
Steel (stainless)	Trichloro-ethylene	1. Abrasion with 100 grit emery cloth, grit or vapour blast, followed by solvent degreasing		1
		2. Vapour blast with a water suspension of 67% garnet grit (200 grade) + 33% garnet grit (400 grade) at a pressure of 55–60 N/cm². After treatment, immerse part in oil (Shell Ensis 252) or isopropanol to prevent rusting. Degrease part for 10 min in trichloroethylene vapour before bonding		1

		3. Immerse for 10 min at 70–85°C in,
		                             Parts by wt.
		    Sodium metasilicate 6·4
		    Sodium pyrophosphate 3·2
		    Sodium hydroxide 3·2
		    Detergent powder (Nansa S)* 1
		    Water 32·1
		Rinse in cold water with a final rinse in distilled water. Dry in air at 93°C.
			Alkaline bath degreasing treatment. General purpose bonding.
			
			*from Marchon Products, 140, Park Lane, London, W.1.
				1

		4. Vapour degrease and immerse for 15 min at 65°C in,
		                           Parts by wt.
		    Sodium metasilicate 5
		    Detergent powder (Empilan NP4)* 9
		    Water 236
		Rinse in hot distilled water and dry at 70°C
			*from Marchon Products.	▶

		5. Etch for 15 min at 50°C in,
		                           Parts by wt.
		    Sodium dichromate (saturated solution) 0·35
		    Sulphuric acid (93% sp.gr. 1·84) 10
		Remove carbon residue with nylon brush while rinsing.
		Rinse in distilled water and dry in warm air, at 70°C
			For maximum peel strength. Alternative etch is, sodium dichromate : Sulphuric acid : water solution in ratio, 7/7/400 parts by wt. Etch for 15 min at 71°C	▶

		6. Vapour degrease and immerse for 10 min at 65°C in,
		                           Parts by wt.
		    Hydrochloric acid (sp.gr. 1·18) 100
		    Formalin (40%) 20
		    Hydrogen peroxide (30%) 4
		    Water 90
		Rinse in distilled water and immerse for 10 min at 65°C in,
		                           Parts by wt.
		    Sulphuric acid (93% sp.gr. 1·84) 100
		    Sodium dichromate 10
		    Water 30
		Rinse in distilled water and dry at 70°C
				1

cont

le 6.1 continued

Adherend	Degreasing solvent	Method of treatment	Remarks	Ref. (p.242)
el (inless)	Trichloro-ethylene	7. Etch for 5–10 min at 65–70°C in, Parts by wt. Hydrochloric acid (37%) 2 Hexamethylene tetramine 5 Water 20 mixed and added to: Hydrogen peroxide 30% Rinse in water with final distilled water rinse. Dry in air below 93°C		1
		8. Etch for 2 min at 93°C in, Parts by wt. Hydrochloric acid (37%) 20 Orthophosphoric acid (85%) 3 Hydrofluoric acid (48%) 1 Rinse in warm water with a final rinse in distilled water. Dry in air below 93°C	Alternative etch is, Hydrochloric: phosphoric: hydrofluoric acids ratio of 24/3·5/1 Etch for 2 min at 82°C.	3, 1

		9. Vapour degrease and etch for 5 min at 50°C in, %w/v
		   Hydrochloric acid (sp.gr. 1·18) 10
		followed by rinsing in,
		   Phosphoric acid (sp.gr. 1·75) 1
		and dry at 70°C
				10, 14

		10. Anodic etch for 90 s at 6 V in, Sulphuric acid, 500 g/l contained in a lead lined tank. Rinse in water and distilled water. Dry in air at 70°C	The part forms the anode and the tank lining the cathode. Treatment may be followed by passivation for 20 min in 5–10% w/v chromic acid (CrO_3) solution	14
		11. Vapour degrease for 10 min and pickle for 10 min at 20°C in, Parts by vol. Nitric acid (67% sp.gr. 1·41) 10 Hydrofluoric acid (40% v/v) 2 Water 88 Rinse in hot distilled water and dry at 70°C	Treatment may be followed by passivation for 20 min in 5–10% w/v chromic acid (CrO_3) solution	14
el (inless)	Trichloro-ethylene	12. Etch for 10 min at 85–90°C in, Parts by wt. Oxalic acid 37 Sulphuric acid (93% sp.gr. 1·84) 36 Water 300 Remove black carbon residue with a nylon brush while rinsing in distilled water and dry in warm air	For maximum heat resistance	1, 6
el (mild) a and ous metals er than nless	Trichloro-ethylene	1. Abrasion. Grit or vapour blast followed by solvent degreasing with water free solvents	Xylene or toluene is preferable to acetone and ketones which may be moist enough to cause rusting	1, 4
		2. Abrasion. 100 grit emery cloth followed by solvent degreasing	For general purpose bonding. 60–70 mesh sand at 50 N/cm² pressure is suitable for grit blast	1, 8

continued

Table 6.1 continued

Adherend	Degreasing solvent	Method of treatment	Remarks
Steel (mild) Iron and ferrous metals other than stainless	Trichloro-ethylene	3. Etch for 5–10 min at 20°C in, Parts by wt. Hydrochloric acid (37%) — 1 Water — 1 Rinse in distilled water after cold water wash and dry in warm air for 10 min at 93°C	Bonding should follow immediately after etching treatment since ferrous metals are prone to rusting. Abrasion is more suitable for procedures where bonding is delayed
		4. Etch for 10 min at 60°C in, Parts by wt. Orthophosphoric acid (88%) — 1 Ethyl alcohol (denatured) — 2 Brush off carbon residue with nylon brush while washing in running water. Rinse with distilled water and heat for 1 h at 120°C	For maximum strength. Alternative procedure to heating at 120°C is to etch electrolytically in sulphuric acid until gas bubbles appear on surface
		5. Immerse for 5 min at 60–65°C in, Parts by wt. Sodium orthosilicate — 30 Sodium alkyl aryl sulphonate — 3 Water — 967 Rinse in hot distilled water and oven dry at 100–105°C	For general purpose bonding
		6. Etch for 10 min at 71–77°C in, Parts by wt. Sodium dichromate — 4 Sulphuric acid (93% sp.gr. 1·84) — 10 Water — 30 Rinse in water and distilled water. Dry at 93°C	Alternative etch is, Sodium dichromate: Sulphuric acid: Water in proportions 2/5/15 for 10 min at 71–77°C
Tin	Trichloro-ethylene	Solvent or vapour degrease after light abrasion with a fine emery cloth (320 grit)	
Titanium and titanium alloys	Trichloro-ethylene	1. Abrasion. Grit or vapour blast, or 100 grit emery cloth, followed by solvent degrease	For general purpose bonding
		2. Immerse in alkaline bath prescribed under Method 2 for stainless steel.	
		3. Etch for 5–10 min at 20°C in, Parts by wt. Sodium fluoride — 2 Chromium trioxide — 1 Sulphuric acid (93% sp.gr. 1·84) — 10 Water — 50 Rinse in water and distilled water. Dry in air at 93°C	
		4. Etch for 2 min at 20°C in, Parts by vol. Hydrofluoric acid (60%) — 84 Hydrochloric acid (37%) — 8·9 Ortho phosphoric acid (85%) — 4·3 Rinse in water and distilled water. Dry in air at 93°C	Suitable for alloys to be bonded with poly-benzimidazole adhesives. Bond within 10 min of treatment

ble 6.1 continued

Adherend	Degreasing solvent	Method of treatment	Remarks	Ref. (p.242)

tanium and tnium oys	Trichloro-ethylene	5. Etch for 2 min at 20°C in,
		
		  Parts by wt.
		  Trisodium phosphate (5% w/v soln) 25
		  Potassium fluoride (2% w/v soln) 10
		  Hydrofluoric acid (2·6%) 15
		Rinse, and soak in water for 15 min at 65°C.
		Rinse with distilled water and air dry.
				9

		6. Etch for 10–15 min at 38–52°C in,
		
		  Parts by vol.
		  Nitric acid (69·8% sp.gr. 1·42) 6
		  Hydrofluoric acid (60%) 1
		  Water 20
		Rinse with water and distilled water. Dry in oven at 71–82°C for 15 min
			Alternative etch for alloys to be bonded with poly-imide adhesives is,
			Nitric : Hydrofluoric : water in ratio 5/1/27 by wt. Etch 30 s at 20°C
				4, 9

ungsten and ngsten oys	Trichloro-ethylene	Etch for 1–5 min at 20°C in,
		
		  Parts by wt.
		  Nitric acid (69% sp.gr. 1·42) 6
		  Hydrofluoric acid (60%) 1
		  Sulphuric acid (93% sp.gr. 1·84) 10
		  Water 3
		Add a few drops of hydrogen peroxide (20%).
		Rinse with water and distilled water. Dry in air at 71–82°C for 15 min
				4

anium	Trichloro-ethylene	Abrasion of the metal in a pool of liquid adhesive (epoxy resins have been used)	Prevents oxidation of metal	7
nc and tc alloys	Trichloro-ethylene	1. Abrasion. Grit or vapour blast or 100 grit emery cloth followed by solvent degreasing	For general purpose bonding	1

		2. Etch for 2–4 min at 20°C in,
		
		  Parts by vol.
		  Hydrochloric acid (37%) 10–20
		  Water 90–80
		Rinse with warm water and distilled water. Dry in air at 66–71°C for 30 min
			Glacial acetic acid is an alternative to hydro-chloric acid	4

		3. Etch for 3–6 min at 38°C in,
		
		  Parts by wt.
		  Sulphuric acid (93% sp.gr. 1·84) 2
		  Sodium dichromate (crystalline) 1
		  Water 8
		Rinse in water and distilled water. Dry in air at 40°C
			Suitable for freshly galvanised metal	10

229

Table 6.2. EFFECT OF PRETREATMENT ON THE SHEAR STRENGTH OF STEEL JOINTS

Pretreatment	Method See Steel (Stainless) p. 226	Martensitic steel FV 448		Austenitic steel S 521		Mild steel S 3	
		M	C	M	C	M	C
Grit blast (No. 40 chilled iron shot at 28 N/cm²) + vapour degreasing		3 540	13·1	2 840	4·7	3 020	7·8
Vapour blast + vapour degreasing	2	4 260	5·6	3 420	7·1	3 320	5·9
Following treatments were preceded by vapour degreasing:							
Cleaning in meta–silicate solution	4	3 020	5·7	2 460	7·8	3 140	6·1
Cleaning in proprietary alkaline solution	3	3 560	8·3	2 220	11·6	2 590	3·3
Acid–dichromate etch	5	4 000	5·8	1 488	22·5	2 820	4·0
Vapour blast + acid dichromate etch	2, 5	4 280	4·1	—	—	—	—
Acid/Formalin/Peroxide etch + acid dichromate etch	6, 5	5 050	9·5	2 560	28·6	1 598	18·3
Anodic etching in sulphuric acid	10	4 540	3·9	2 480	4·6	3 980	6·4
Anodic etching in sulphuric acid + chromic acid passivation	10	4 660	1·7	2 630	7·1	3 800	8·4
Hydrochloric acid etch + phosphoric acid etch	9	2 560	17·9	660	20·2	2 140	20·7
Nitric/hydrofluoric acid etch	11	4 550	7·5	2 220	15·2	2 800	8·4
Nitric/hydrofluoric acid etch + chromic acid passivation	11	4 640	2·0	2 380	15·5	3 120	7·4

M = mean failing load (N/cm²)   C = coefficient of variation (%)

Evaluation of pretreatments

Table 6.2 summarises some early work on the evaluation of pretreatments, on a comparative basis, for three different types of steel bonded with a polyvinyl formal-phenolic based adhesive[14]. The results show that, where optimum strength is required, no single pretreatment is universally applicable.

6.5.   PLASTICS

Plastics materials are described as thermosetting or thermoplastic according to 'their behaviour on exposure to heat, solvent action or stress conditions. Thermosetting plastics are relatively infusible and not subject to flow when heated. These materials are insoluble in solvents although some solvent absorption may occur leading to swelling. The infusibility and insolubility of thermosetting plastics precludes the use of heat sealing or solvent cementing processes as joining methods and restricts their assembly to adhesive bonding. Thermoplastics are soluble in specific solvents (amorphous type) or become so at high temperatures (crystalline type) and under heat or stress conditions these materials are subject to flow. Heat sealing and solvent cementing are bonding techniques applicable to many thermoplastics. The solvent cementing of thermoplastics is dealt with elsewhere in this book (Table 9.3) and may be regarded as a supplementary technique to adhesive bonding.

Pretreatment is mandatory for many plastics which are among the most difficult materials to bond.

Apart from substrate roughening by abrasion to improve mechanical adhesion, surface treatments for plastics are basically concerned with the removal of release agents and the formation of higher energy surfaces with a greater affinity for polar adhesives. Release agents present on the substrate reduce the interfacial contact between adhesive and adherend and limit adhesion. They include the mould release materials usually found on the surfaces of plastics laminates and mouldings, oils, water, plasticisers exuded from flexible plastics, and other contaminants which constitute a weak boundary layer at the joint interface. Their removal may be achieved by various means, e.g. solvent action, abrasion, gentle heat treatment of the plastic to promote syneresis, or by their absorption into an adhesive with which they are compatible. In general, it is inadvisable to bond plastics materials having incorporated release agents such as stearates, since this can lead to disruption of an assembly when subsequent agent migration to the interface takes place. On the other hand, certain adhesive systems are unaffected by migrant plasticisers. Thus, nitrile-phenolic and polyurethane based materials are suitable for flexible vinyl plastics containing plasticisers.

Plastics substrates have lower surface energies than hydrophilic materials such as glass, metals, and oxides and minerals, and this can prevent the adhesive from wetting the adherend surface sufficiently for good adhesion. Untreated polyethylene and fluorinated polymers, and other non-polar materials, present particular bonding difficulties because their surface energies are below those of solvents or adhesives. Chemical and physical treatments for these materials are designed to promote adhesion by converting the substrate to a higher energy surface. The surface modification may be brought about by diverse means such as:

abrasion to remove weakly adherent surface material (of low molecular weight) to expose more adherent bulk material (high molecular weight),

reagents which chemically modify and/or etch the substrate,

bombardment with ions in a gaseous discharge,

exposure to oxidising flames.

Foamed plastics are sufficiently porous to permit the easy absorption of adhesive and rarely require pretreatment. Spongy materials are primed with the adhesive (possibly diluted) which is then partially cured. This technique forms a rigid glue-line foundation and avoids the danger of joint starvation by absorption. Rigid foams are usually cut to expose fresh surface and freed from cutting debris with an air jet. The interlocking of the glue-line with the adherend provides adequate fastening for foamed polymers which are not often subject to high loading.

The modification of raw polymer materials with fillers, pigments, and plasticisers, etc., has resulted in a variety of plastics displaying significant differences in adhesion properties. Consequently, a particular surface pretreatment which is suitable for a selected plastics material may not be the best one for the same material in modified form. The user must here resort to the evaluation of several pretreatment processes and joint tests to be certain of the adequacy and long-term performance of a bonded assembly in service. Again, a reassessment of the pretreatment adopted is called for where new adhesive materials are employed for an existing process.

Table 6.3. SURFACE PREPARATION FOR PLASTICS

Adherend	Degreasing solvent	Method of treatment	Remarks	Ref. (p.242)
rylonitrile adiene ene	Acetone	1. Abrasion. Grit or vapour blast or 220 grit emery cloth followed by solvent degreasing		10
		2. Etch in chromic acid solution for 20 min at 60°C	Recipe 2 for methyl-pentene (p. 232)	10
LULOSICS ellulose ellulose cetate ellulose cetate-utyrate ellulose itrate ellulose ropionate hyl ellulose	Methanol, Isopropanol	1. Abrasion. Grit or vapour blast or 220 grit emery cloth followed by solvent degreasing	For general bonding purposes	
		2. After procedure (1), dry the plastic at 100°C for 1 h and apply adhesive before the plastic cools to room temperature	For maximum strength	1, 8
llylphtha-ate llyliso-hthalate	Acetone, Methyl ethyl ketone	Abrasion. Grit or vapour blast or 100 grit emery cloth, followed by solvent degreasing	Steel wool may be used for abrasion	1, 8
xy resins	Acetone, Methyl ethyl ketone	Abrasion. Grit or vapour blast or 100 grit emery cloth, followed by solvent degreasing	Sand or steel shot are suitable abrasives	1, 8
ylene l acetate	Methanol	Prime with epoxy adhesive and fuse into the surface by heating for 30 min at 100°C		10
ane	Acetone, Methyl ethyl ketone	Abrasion. Grit or vapour blast or 100 grit emery cloth, followed by solvent degreasing		10

Table 6.3 continued

Adherend	Degreasing solvent	Method of treatment	Remarks	R (p.
Ionomer	Acetone, Methyl ethyl ketone	Abrasion. Grit or vapour blast or 100 grit emery cloth, followed by solvent degreasing	Alumina (180 grit) is a suitable abrasive	10
Melamine resins	Acetone, Methyl ethyl ketone	Abrasion. Grit or vapour blast or 100 grit emery cloth, followed by solvent degreasing		4,
Methyl pentene	Acetone	1. Abrasion. Grit or vapour blast or 100 grit emery cloth, followed by solvent degreasing	For general purpose bonding	10
		2. Immerse for 1 h at 60°C in, Parts by wt. Sulphuric acid (93% sp.gr. 1·84) 26 Potassium chromate 3 Water 11 Rinse in water and distilled water. Dry in warm air		12

		3. Immerse for 5–10 min at 90°C in,
		    Potassium permanganate (saturated soln)
		    acidified with sulphuric acid (93% sp.gr. 1·84).
		Rinse in water and distilled water. Dry in warm air
				12

		4. Prime surface with lacquer based on urea formaldehyde resin diluted with carbon tetrachloride	Coatings (dried) offer excellent bonding surfaces without further pre-treatment	1
Phenolic resins Phenolic– melamine resins	Acetone, Methyl ethyl ketone, Detergent	1. Abrasion. Grit or vapour blast or abrade with 100 grit emery cloth, followed by solvent degreasing	Steel wool may be used for abrasion. Sand or steel shot are suitable abrasives. Glass fabric decorative laminates may be degreased with detergent solution	1
		2. Removal of surface layer of one ply of fabric previously placed on surface before curing. Expose fresh bonding surface by tearing off the ply prior to bonding		6,
Polyamide	Acetone, Methyl ethyl ketone, Detergent	1. Abrasion. Grit or vapour blast or abrade with 100 grit emery cloth, followed by solvent degreasing	Sand or steel shot are suitable abrasives	1
		2. Prime with a spreading dough based on the type of rubber to be bonded in admixture with isocyanate	Suitable for bonding polyamide textiles to natural and synthetic rubbers	1
		3. Prime with resorcinol formaldehyde adhesive	Good adhesion to primer coat with epoxy adhesives in metal-plastic joints	1
POLY-CARBONATES Allyl diglycol carbonate	Methanol, Isopropanol, Detergent	Abrasion. Grit or vapour blast or 100 grit emery cloth followed by solvent degreasing	Sand or steel shot are suitable abrasives	1
FLUOROCARBONS Polychloro-trifluoro-ethylene Polytetra-fluoro-ethylene	Trichloro-ethylene	1. Etch for 10–30 s in, 1% sodium in ammonia (sp.gr. 0·880) solution until surface is light brown. Rinse with water and distilled water. Dry in warm air.	Sodium treated surfaces must not be abraded before use. Hazardous etching solution requiring skilful handling. Proprietary sodium-etch reagents are available	1
		2. Etch for 15 min in following prepared solution, Naphthalene (128 g) dissolved in Tetrahydrofuran (1 litre,		7

conti

e 6.3 continued

dherend	Degreasing solvent	Method of treatment	Remarks	Ref. (p.242)
		denatured), to which is added, Sodium (23 g of pea size pellets) during a stirring period of 2 h. Rinse with water and distilled water after an acetone rinse. Dry in warm air	commercially (refs. A, B) Commercial treatment of PTFE materials also undertaken (refs. C.D.E.)	
		3. Prime with epoxy adhesive and fuse into the surface by heating for 10 min at 370°C followed by 5 min at 400°C		7
		4. Reflux for 30 min in, Parts by wt Sodium hydroxide 10 Diallyl melamine 8		7, 10
		5. Immerse in fused potassium acetate for 30 min at 295°C. Rinse with water and distilled water and dry in warm air		10

		6. Expose to one of the following gases activated by corona discharge,
		    Air (dry) for 5 min
		    Air (wet) for 5 min
		    Nitrous oxide for 10 min
		    Nitrogen for 5 min
			Bond within 15 min of pretreatment	10

		7. Expose to electric discharge from a Tesla coil (50,000 V A.C.) for 4 min	Bond within 15 min of pretreatment	10
'ESTERS ethylene ·hthalate	Detergent, Acetone, Methyl ethyl ketone	1. Abrasion. Grit or vapour blast or 100 grit emery cloth, followed by solvent degreasing	For general purpose bonding	1, 8
		2. Prime the surface with an anhydrous 2% solution of tetrabutyl titanate in petroleum ether (or chlorinated hydrocarbon). Allow to hydrolyse in moist atmosphere to give a continuous film of primer coat	Reagents available commercially (ref. F)	7, 10
		3. Immerse for 10 min at 70–95°C in, Parts by wt. Sodium hydroxide 2 Water 8 Rinse in hot water and dry in hot air	For maximum bond strength. Suitable for linear polyester films	1, 7
rinated ether	Acetone, Methyl ethyl ketone	Etch for 5–10 min at 66–71°C in, Parts by wt. Sodium dichromate 5 Water 8 Sulphuric acid (93% sp.gr. 1·84) 100 Rinse in water and distilled water. Dry in air	Suitable for film materials such as Penton	4, 8
ethylene ethylene lorinated) ethylene ephtha- e (see, lyesters)	Acetone, Methyl ethyl ketone	1. Solvent degreasing	Low bond strength applications	1
		2. Expose surface to gas burner flame (or oxy-acetylene oxidising flame) until the substrate is glossy		1, 10
		3. Etch for up to 90 min at 20°C in, Parts by wt. Sodium dichromate 5 Water 8 Sulphuric acid (93% sp.gr. 1·84) 100	For maximum bond strength	1, 8

continued

Table 6.3 continued

Adherend	Degreasing solvent	Method of treatment	Remarks
		4. Expose for 15–30 s to hot solvents or their vapours, Toluene, perchloroethylene or trichloroethylene are suitable	Hydrocarbons etch or swell to give surfaces with improved adhesion properties
		5. Prime the surface with an anhydrous 2% solution of tetrabutyl titanate in petroleum ether. Allow to hydrolyse in moist air to give a continuous adherent film of primer coat	Titanium esters are available commercially for priming polythene films (ref. F)
Polypropylene	Acetone, Methyl ethyl ketone	1. Abrasion. Grit or vapour blast or 100 grit emery cloth, followed by solvent degreasing	For general purpose bonding
		2. Etch for 1–2 min at 66–71°C in, Parts by wt. Sodium dichromate 5 Water 8 Sulphuric acid (93% sp.gr. 1·84) 100 Rinse in water and distilled water. Dry in air	Alternative etch solution is, Sodium dichromate: Water: Sulphuric acid in ratio 1·7/20/30. Etch for 15 min at 20°C
		3. Prime with 'Propathene' primer prior to bonding with cold curing (preferred) adhesives	Suitable for film material. Available commercially (ref. G) in brush or spray form.
		4. Prime after dichromate etch (method 2) with epoxy-polyamide adhesive coat	Suitable for rigid plastic. Epikote 825 + Versamid 125 in ratio 20 : 21 parts by wt. is effective. (ref. H)

		5. Expose to following gases activated by corona discharge
		  Air (dry) for 15 min
		  Air (wet) for 5 min
		  Nitrous Oxide for 10 min
		  Nitrogen for 15 min
			Bond within 15 min of pretreatment

		6. Expose to electric discharge from a Tesla coil (50,000 V A.C.) for 1 min	Bond within 15 min of pretreatment
Polyformaldehyde (acetal polymers)	Acetone, Methyl ethyl ketone	1. Abrasion. Grit or vapour blast or 100 grit emery cloth, followed by solvent degreasing	For general purpose bonding
		2. Etch for 10–20 s at 20°C in, Parts by wt. Sodium dichromate 5 Water 8 Sulphuric acid (93% sp.gr. 1·84) 100 Rinse in water and distilled water. Dry in air	

		3. Etch for 5 min at 120°C in,
		  Para-toluene sulphonic acid 1 g
		  Dioxane 10 ml
		  Perchloroethylene 200 ml
		Rinse in cold/hot water and dry in hot air
			Stresses in the plastic due to moulding should be relieved by heat treatment prior to acid etching. (e.g. 1 h at 120°C)

Polyimide	Acetone	Etch for 1 min at 60–90°C in Parts by wt. Sodium hydroxide 5 Water 95 Rinse in cold water and dry in hot air	Adhesive X33/1180 with resin: catalyst ratio of 1 : 1 recommended (ref. Q)

con

6.3 continued

therend	*Degreasing solvent*	*Method of treatment*	*Remarks*	*Ref. (p.242)*
nethyl-thacry-acrylate adiene rene	Acetone, Methyl ethyl ketone, Detergent, Methanol, Trichloro-ethylene, Isopropanol	Abrasion. Grit or vapour blast or 100 grit emery cloth, followed by solvent degreasing	For maximum strength relieve stresses by heating plastic for 5 h at 100°C	1
phenylene	Trichloro-ethylene	Abrasion. Grit or vapour blast or 100 grit emery cloth, followed by solvent degreasing		10
pheny-e oxide	Methanol	Solvent degrease	Plastic is soluble in xylene and may be primed with adhesive in xylene solvent	10
tyrene	Methanol, Isopropanol, Detergent	Abrasion. Grit or vapour blast or 100 grit emery cloth, followed by solvent degreasing	Suitable for rigid plastic	1
phone	Methanol	Vapour degrease		10
urethane	Acetone, Methyl ethyl ketone	Abrade with 100 grit emery cloth and solvent degrease		8
inyl oride inyl-ne oride	Trichloro-ethylene, Methyl ethyl ketone	1. Abrasion. Grit or vapour blast or 100 grit emery cloth, followed by solvent degreasing	Suitable for rigid plastic. For maximum strength, prime with nitrile-phenolic adhesive	1 1
inyl ride		2. Solvent wipe with ketone	Suitable for plasticised material	1
ne ylo-ile	Trichloro-ethylene	Solvent degrease		10
nalde-e	Acetone, Methyl ethyl ketone	Abrasion. Grit or vapour blast or 100 grit emery cloth, followed by solvent degreasing		10

Mechanical, chemical and physical pretreatment of plastics

Table 6.4 shows some of the results of recent research at Sira into various surface pretreatments for plastics which are not easily bonded without preparation. Pretreatment processes have not been described in detail since these are still being investigated. Polypropylene (PP), polytetrafluoroethylene (PTFE) and 4-methylpent-l-ene (TPX) have been selected to illustrate that adhesion is considerably improved by a variety of surface preparation methods. Peel specimens were prepared with a nitrile rubber based adhesive; shear specimens were bonded with an epoxy-polyamide adhesive which was heated at 100°C for 40 min. Bonding pressures were 96·5 N/cm² (Peel specimens) and 2·0 N/cm² (Shear specimens).

The results show that chemical oxidation and glow discharge treatments are generally superior to abrasion for increasing the shear and peel strengths of these plastics. Combination of treatments is also an effective method of improving the adhesion obtained with a single treatment.

/

Table 6.4 ADHESION OF PLASTICS TO PLASTICS OR TO STEEL (STAINLESS)

Method of Treatment / Adherends	Peel strength ASTM D1876–61T (N/cm)			Tensile shear strength ASTM D897–49 (N/cm²)		
	PP to PP Failing loads Mean (Min–Max)	PTFE to PTFE Mean (Min–Max)	TPX to TPX Mean (Min–Max)	PP to steel Failing loads Mean (Min–Max)	PTFE to steel Mean (Min–Max)	TPX to Mean (Min–M
None	0	0	0	0	0	0
Degreased with acetone				(40–60)	(30–40)	(30–4
Acetone degreased and following treatments applied:						
Abraded (0 grade emery) degreased	3 (2–6)	1 (0–5)	2 (0–4)	(40–60)	(30–40)	(30–4
Abrasion (0 grade) + chromic acid etch	—	n/a	25 (10–40)			
Abrasion + acid permanganate etch	18 (10–22)	n/a	17 (15–20)			
Abrasion + primer (trade product)	12 (8–15)	n/a	n/a	(200–300)	n/a	n/a
Chromic acid etch	25 (20–30)	n/a	17 (10–25)	(500–600)	n/a	(200–3(
Acid permanganate etch	30 (20–45)	n/a	30 (20–45)	(300–400)	n/a	(200–3(
Aminosilane primer application after						
(a) Acid permanganate etch	—	n/a	—			
(b) Chromic acid etch	30 (20–40)	—	40 (30–50)	—	n/a	—
(c) Abrasion	7 (5–12)	—	35 (20–45)	(150–200)	n/a	(150–2(
Glow discharge in dry air or helium for 300 s	7 (4–15)	2 (0–4)	2 (0–4)	(400–500)	(500–600)	(500–6(
Sodium–ammonia etch	n/a	10 (8–15)	n/a	n/a	(600–700)	n

n/a = not applicable   0 = specimen hand-broken

6.6. RUBBERS (NATURAL AND SYNTHETIC)

Table 6.5. SURFACE PREPARATIONS FOR RUBBERS

Adherend	Degreasing solvent	Method of treatment	Remarks	
Natural rubber	Methanol, Isopropanol	1. Abrasion followed by brushing. Grit or vapour blast or 280 grit emery cloth, followed by solvent wipe	For general purpose bonding	
		2. Treat the surface for 2–10 min with sulphuric acid (93% sp.gr. 1·84) at room temperature. Rinse thoroughly with cold water/ hot water. Dry after rinsing in distilled water. (Residual acid may be neutralised by soaking for 10 min in 10% ammonium hydroxide after hot water washing).	Adequate pretreatment is indicated by the appearance of hair line surface cracks on flexing the rubber. Suitable for many synthetic rubbers when given 10–15 min etch at room temperature. Unsuitable for use on butyl, polysulphide, silicone, chlorinated polyethylene, and polyurethane rubbers	
		3. Treat surface for 2–10 min with paste made from sulphuric acid (93%) and barium sulphate. Apply paste with stainless steel spatula and follow the procedure described in method 2		
		4. Treat surface for 2–10 min in, Parts by vol. Sodium hypochlorite 6 Hydrochloric acid (37%) 1 Water 200 Rinse with cold water and dry	Suitable for those rubbers amenable to treatments (2) and (3)	
Butyl	Toluene	1. Solvent wipe	For general purpose bonding	
		2. Prime with butyl rubber adhesive in an aliphatic solvent	For maximum strength	
Butadiene styrene	Toluene	1. Abrasion followed by brushing. Grit or vapour blast or 280 grit emery cloth, followed by solvent wipe	Excess toluene results in swollen rubber. A 20 min	

cor

e 6.5 continued

dherend	Degreasing solvent	Method of treatment	Remarks	Ref. (p.242)
			drying time will restore the part to its original dimensions	
		2. Prime with butadiene styrene adhesive in an aliphatic solvent		1
		3. Etch surface for 1–5 min at room temperature following method 2 for Natural rubber.		3
diene trile	Methanol	1. Abrasion followed by brushing. Grit or vapour blast or 280 grit emery cloth, followed by solvent wipe		1, 3
		2. Etch surface for 10–45 s at room temperature following method 2 for Natural rubber		3
ro- phonated lythene	Acetone or Methyl ethyl ketone	Abrasion followed by brushing. Grit or vapour blast or 280 grit emery cloth, followed by solvent wipe	General purpose bonding	10
lene ylene	Acetone or Methyl ethyl ketone	Abrasion followed by brushing. Grit or vapour blast or 280 grit emery cloth, followed by solvent wipe	General purpose bonding	10
ro- icone	Methanol	Application of fluorosilicone primer (A4040) to metal where intention is to bond unvulcanised rubber	Primer available from Dow Corning International Ltd., Castle Chambers, 3/9, Sheet St, Windsor (ref. I)	
chloro- ene	Toluene, Methanol, Isopro- panol	1. Abrasion followed by brushing. Grit or vapour blast or 100 grit emery cloth, followed by solvent wipe	Adhesion improved by abrasion with 280 grit emery cloth followed by acetone wipe	1, 3
		2. Etch surface for 5–30 min at room temperature, following method 2 for Natural Rubber		3
acrylic	Methanol	Abrasion followed by brushing. Grit or vapour blast or 100 grit emery cloth, followed by solvent wipe	General purpose bonding	10
butadiene	Methanol	Solvent wipe	General purpose bonding	10
sulphide	Methanol	Immerse overnight in strong chlorine water, wash and dry		
- ane	Methanol	Incorporation of an aminosilane into the adhesive elastomer system. 1 % w/w is usually sufficient	Aminosilane A1100 is available commercially. (ref. J). Addition to adhesive eliminates need for priming and improves adhesion to glass, metals. A1100 may be used as a surface primer	10
ne	Acetone or Methanol	1. Application of primer to substrate other than silicone rubber. MS 602 in solvent (dries in ½–2 h) for glasscloth MS 2402 in solvent (dries in ½–1 h) for use with silicone adhesive DP 2401. Chemlok 607 in solvent (dries 10–15 min for assemblies below 200°C. Silcoset Primer OP in toluene dried in air or 3 min at 100°C	Primer available commercially. MS602, MS2402 (ref. K) Chemlok 607 (ref. L) Silcoset OP (ref. G) Selected from a range of products	
		2. Incorporation of an integral bonding additive into the uncured adhesive system ESP.2437	Obviates need for a primer. (ref. K)	

6.7. FIBRE PRODUCTS

Fibre products are based on a wide range of organic and inorganic materials and are used extensively in filament or fabric form for the manufacture of textiles, plastics-laminates, and resin composites. For many applications the mechanical adhesion obtained by saturating the adhesive into the fabric interstices is sufficient. Measurements on textiles reveal that the relationship between the size of the interstices between weft and warp yarns and the adhesion obtained for a specified material and adhesive coating, is a linear one. Adhesive penetration of the yarn filaments is another factor which favours adhesion to low twist rather than high twist yarns. Certain fabrics fail to give adequate mechanical adhesion regardless of their construction, and chemical or physical treatments are then necessary to reinforce specific adhesion at the interface.

Natural fibres, such as cotton wool and cellulosic materials, contain sufficient polar sites for adhesives to be employed satisfactorily, whereas synthetic fibre materials based on polyacrylonitriles, polyamides, polyesters, and polyolefins, tend to have low surface energies which may limit the adhesion obtainable. The surface finish of many synthetic materials is irregular and not conducive to bonding and chemical modification of the substrate is necessary whenever performance requirements are critical, e.g. for fabric-elastomer applications such as conveyor belting and tyre cords.

Reinforced composites and laminates generally have a thin surface layer of the polymer component so that adhesion to a fibrous substrate is not involved; the adherend is treated as it would be for an unreinforced polymer material. Special pretreatments have been devised to promote specific adhesion in polyester fibre in which no easily reactive site is

Table 6.6. SURFACE PREPARATIONS FOR FIBRE PRODUCTS

Adherend	Degreasing solvent	Method of treatment	Remarks
Cotton, Wool, Silk, Felt	Acetone, Methanol	Dip in a primer based on isocyanate containing a small amount of the rubber (natural or synthetic) to be bonded	For maximum strength bonding of textiles to rubber substrates. Isocyanate primers are usually available from suppliers of rubber based adhesives
Leather	Acetone	1. Abrasion with glass paper followed by solvent degreasing	For general purpose bonding
		2. Abrasion by brush machine. Material is roughened from the centre outwards in one direction and then rotated through 180° so that the remaining half is roughened in the opposite direction	Procedure employed in the Footwear industry to achieve maximum strength. Details of brush scouring machines and processes from SATRA (ref. M)
Paper Cardboard	Methanol	Treatment is usually unnecessary since these cellulose fibre materials are sufficiently polar and porous to provide good adhesion with any adhesive material. Certain papers are treated with adhesive, moisture proofing primer coats, such as silicones, and a solvent treatment (toluene) will be required for bonding areas	
Synthetic textiles based on polyesters, polyamides, polyacrylics		Dip-coat the material into a dilute solution of a suitable primer, e.g. Chemlok 607.	Primer available commercially (ref. L)
Inorganic fibres. Asbestos, glass	Trichloroethylene	Prime with aminosilane coating prior to bonding	Method of improving adhesion of laminated composites of these materials with polymers. Primers available commercially (refs. G, I, J, K)
Carbon fibre	Trichloroethylene	Treat with nitric acid (67% sp.gr. 1·41) for 30 min. Dry, and bond immediately	

available. The most widely used system for polyester fibre involves a two-stage process. The fibre is first treated with an adhesive primer based on isocyanates, epoxies, polyvinylchloride and amines in appropriate combinations. This is followed by the application of an adhesive based on a resorcinol-formaldehyde-latex system (R.F.L.) which is then dried. The end-use requirements determine the extent of curing given to the treated fabric. An alternative one-stage process involves the addition of an isocyanate compound to the aqueous R.F.L. adhesive. The composition and solids content of the R.F.L. system are varied to suit particular applications of fibres and rubbers. The treatment is applicable for adhesion to natural, neoprene, butyl, acrylonitrile, styrene butadiene, reclaim, and other types of rubber.

Some of the treatments recommended for rigid polymers (p. 244) may be used to advantage with these materials in woven fabric form. The treatment of fibrous substrates with activated species of inert gases, such as helium, has been employed to effect the direct bonding of rubber to nylon. High bond strengths are not usually required of textile bonds since the fabric strength is generally low and large bonding areas are involved and consequently, pre-treatments often need not be elaborate. The selection of adhesives for these materials is more concerned with the retention of fabric flexibility and durability when cleaned with solvents and detergent washes.

Paper consists of a random arrangement of short cellulosic fibres. The high porosity of paper combined with its weak but relatively rigid structure makes it a suitable adherend for almost any adhesive without pretreatment, provided it has not been subjected to previous waterproofing treatment with waxes or silicones. Processing and end-use generally determine the choice of adhesive for paper.

Non-woven fabrics resemble paper, although the fibres are larger in size and can be considered as a reinforcement for the selected bonding materials. The selection of the bonding material is important to the final performance of the fabric although most non-woven fabrics are bonded with adhesive systems based on aqueous latices and polymeric rubber and plastics dispersions.

Inorganic fibres are extensively used as the constituents of reinforced composites. Thus, chlorosilane primer pretreatment has been used to increase the specific adhesion of asbestos, quartz, and glass fibres, where these are used for plastics laminate fabrication. Carbon fibres, formed by decomposing organic filaments (such as rayon) at high temperatures, are really composites that consist of high modulus graphite crystals in a carbon-graphite matrix. Pretreatment in concentrated nitric acid (60–70%) followed by immediate coating with the adhesive polymer, is an effective means of improving interfacial adhesion.

6.8. INORGANIC MATERIALS

Ceramics and glasses do not usually possess weak surface layers so that substrate modification is not required. Hot acid-chromate solutions (60°C) provide a more effective means of removing surface contamination than solvents. Glass for optical and electronic assemblies may require special treatment where the performance reliability of a finished assembly is important, and which is frequently dependent on the preparation of its component parts. A reproducibly 'clean' surface is produced by the cleaning procedure outlined below (Glass methods); the method has proved to be suitable for the preparation of glass substrates intended for subsequent thin-film metal plating by electroless or vacuum deposition techniques, and would appear to be worth consideration where adhesive bonding to the metal film on glass is envisaged. Chlorosilane priming of glass substrates considerably improves adhesion and is widely employed as a pretreatment for glass-fibre fabrics in plastics laminate composites. For other inorganic materials or minerals, solvent degreasing treatments are sufficient. Silicate minerals such as mica, quartz, and gem stones may be treated like glass.

Table 6.7. SURFACE PREPARATION FOR INORGANIC MATERIALS

Adherend	Degreasing solvent	Method of treatment	Remarks	Ref. (p.242)
asbestos (rigid)	Acetone	1. Abrasion. Abrade with 100 grit emery cloth remove dust and solvent degrease	Allow the board to stand for sufficient time to allow solvent to evaporate off	1, 6, 8
		2. Prime with diluted adhesive or low viscosity rosin ester		
bitumenised surfaces	Acetone	Abrasion. Abrade with 100 grit emery cloth and solvent degrease	Applicable to bitumenised pipework	6
carbon, graphite	Acetone	Abrasion. Abrade with 220 grit emery cloth and solvent degrease after dust removal	For general purpose bonding	1

continued

Table 6.7 continued

Adherend	Degreasing solvent	Method of treatment	Remarks	Re (p.2
Ceramics (glazed, 2, 3, 4). Ceramics (unglazed, 1, 4). Clays (fired 5) Stonework, (5) Plaster, masonry, (6)	Acetone	1. Abrasion. Grit blast with carborundum and water slurry and solvent degrease	Suitable for unglazed ceramics such as alumina, silica, etc	1
		2. Solvent degrease or wash in warm aqueous detergent. Rinse and dry	For glazed ceramics such as porcelain	
		3. Immerse for 15 min at 20°C in, Parts by wt. Sodium dichromate 7 Water 7 Sulphuric acid (93 % sp.gr. 1·84) 400 Rinse in water and distilled water. Oven dry at 66°C	For maximum strength bonding of small ceramic (glazed) artefacts	1
		4. Prime with 2·4 % w/w solution of A–174 silane in alcohol, allow to dry in moist air and bond	For bonding polyester gaskets to ceramic pipes and tiles. A–174 (ref. J)	
		5. Solvent degrease, abrade with wire brush and solvent degrease after dust removal. Prime with diluted adhesive or low viscosity rosin ester	Non-glazed fired materials and masonry	1
		6. Abrade with fine emery cloth and remove dust		
Concrete Granite, Stone	Perchloro-ethylene, Detergent	1. Abrasion. Abrade with a wire brush, degrease with detergent and rinse with hot water before drying	For general purpose bonding	
		2. Etch with 15 % hydrochloric acid until effervescence ceases. Wash with water until surface is litmus neutral. Rinse with 1 % ammonia and water. Dry thoroughly before bonding	Applied by stiff-bristle brush. Acid should be prepared in a polythene pail. 10–12 % hydrochloric or sulphuric acids are alternative etchants. 10 % w/w sodium bicarbonate may be used instead of ammonia for acid neutralisation	1 6
Ferrite	Acetone	Solvent degrease		10
Glass	Acetone, Detergent	1. Abrasion. Grit blast with carborundum and water slurry and solvent degrease. Dry for 30 min at 100°C. Apply the adhesive before the glass cools to room temperature	For general purpose bonding. Drying process improves bond strength	1, 6
		2. Immerse for 10–15 min at 20°C in, Parts by wt. Sodium dichromate 7 Water 7 Sulphuric acid (93 % sp.gr. 1·84) 400 Rinse in water and distilled water. Dry thoroughly	Alternative solution is, chromium trioxide and water in the ratio 1 : 4. Unsuitable for optical glass applications	1, 4
		3. Detergent, such as 2 % Quadralene in water, in an ultrasonic cleaning bath	Quadralene available as a powder (ref. P). Suitable for some optical glass assemblies	1, 4
		4. Prime with 2–5 % vinyl aminosilane in alcohol, allow to dry in moist air, and bond	Primer available as A1100 (ref. J)	1

continu

Table 6.7 continued

Adherend	Degreasing solvent	Method of treatment	Remarks	Ref. (p.242)
		5. Swab surface with a filtered detergent/cerium oxide polishing powder slurry. Rinse in tap water and swab with filtered detergent. Wash in tap water for 5 min followed by distilled water. Finally, flush the substrate with hot isopropanol in a Soxhlet apparatus. Tazlanised cloth should be used for swabbing processes	'Teepol' is a suitable detergent for the slurry. Cerrirouge E powder may be recommended (ref. N). Tazlanised cloth is supplied as 'Tez' (ref. O). Reliable cleaning process for glass intended for metal film deposition by chemical or vacuum means. Further information available	11
Magnesium fluoride	Acetone	Solvent degrease		10
Mica and Quartz	Acetone	1. Solvent degrease and prime with a 2–5% vinyl aminosilane in ethanol, allow to dry in moist air and bond	Primer available as A1100 (ref. J)	
		2. Prime the surface with an adhesive coating	Silicone resins are useful bonding agents for mica plates. (ref. K).	
Silicon carbide	Trichloro-ethylene	1. Solvent degrease	For general purpose bonding	
		2. Prime the surface with an adhesive coating		
Sodium chloride	Acetone	Degrease with water-free solvent		10
Tungsten carbide	Acetone	1. Abrasion. Grit or vapour blast or abrade with emery cloth, followed by solvent degreasing	Carborundum abrasive is suitable	
		2. Etch for 10 min at 80–90°C in, Parts by wt. Sodium hydroxide 3 Water 7 Rinse in cold/hot water and distilled water. Dry in hot air	For maximum strength	1

6.9. WOOD AND ALLIED MATERIALS

Wood surfaces are often contaminated with dust, dirt and exuded natural oils and resins which have an adverse effect on the wettability by adhesives. Improvements are effected when the surface is sanded or planed. Sanded surfaces, with an increased surface bonding area derived by roughening, do not always produce glued joints of high strength, since the surface imperfections are filled with wood dust which may prevent good contact between adhesive and substrate. Planed surfaces bonded with thin glue-lines produce the highest strength joints, but with care sawn joints can be as strong as planed ones. The method effectively removes the top surface and any contaminant material which may have been absorbed below the surface. The porous, cellular structure of wood limits the degree of smoothness which is attainable. Some species of timber absorb adhesive solvents to the extent that starved glue-lines result; primers or high viscosity adhesives may be used to overcome the drainage problems occurring with porous or high moisture woods. It should be noted that 'ink-bottle' pits in the surface also prevent good contact between glue and wood. This type of defect arises where the ends of cells have become burred over during surface preparation. Occluded air in glue-lines from score marks, grooves, pits, open-ended cells, is another cause of poor interfacial contact and leads to the presence of stress centres which can promote bond failure on loading.

Contamination of a prepared surface can proceed rapidly and wood should be glued as soon as possible after preparation to ensure adequate wetting. Manufacturing standards require that bonding should take place within 48 h of preparation or within 12 h if the wood has been pretreated with fire-retardants or preservatives. Hydrophilic (polar) adhesives, such as the urea and phenolic resins make good joints with wood by virtue of the chemical affinity between them

Table 6.8.   SURFACE PREPARATION FOR WOOD AND ALLIED MATERIALS

Adherend	Degreasing solvent	Method of treatment	Remarks	R (p.
Cork		Brush coat with adhesive as a primer or pore sealant		
Wood Plywood		1. Abrasion. Dry wood is smoothed with a suitable glass paper. Sand plywood along the direction of the grain. The moisture content should be adjusted to suit the adhesive employed, e.g. Mouldrite Phenolic plywood glues are applied to woods with 6 to 12% moisture content	For general purpose bonding	1
		2. Plane the surface and lightly sand with glass paper complying with Grade 1 of the BS 871. (Abrasive papers and cloths for general purposes.) Sand with double strokes parallel to the grain direction and avoid rounding of the edges	For maximum strength adjust moisture content to $12 \pm 2 \cdot 5\%$ on oven dry weight	1

which leads to good wetting properties. High quality glue-lines are more easily achieved with softwoods (e.g. ash, aspen, beech, birch, elm) than with hardwoods (e.g. cypress, cedar, fir, hemlock, larch, pine, redwood, spruce) which are less easily wetted. Wetting properties can be affected where timber has been chemically pretreated with preservatives and fire-retardant materials and further treatment with chemicals such as caustic soda, detergents or solvents, may be required before satisfactory glue-lines can be made.

Moisture content is another factor which affects the quality and reliability of glued timber structures. Unsatisfactory bonding is often a consequence of inattention to the drying of the wood beforehand. Where timber structures are concerned, the moisture content of the wood at the time of bonding, should lie within 3% of the average equilibrium moisture content possible in service, providing this falls within a range of 7–15%. Generally, adhesives give optimum results when the timber moisture content is within the range 8–12%, although for some applications satisfactory joints may be formed within a range of 2–25%.

6.10. PAINTED SURFACES

In general, the application of an adhesive to a painted surface is not to be recommended, since the adhesive bond will be only as strong as the bond of the paint to the substrate. Moreover, the migration of adhesive constituents into some types of paint film can lead to early failure at the paint-substrate interface. For improved performance it is desirable to remove the paint film, by abrasion or solvent action, and apply the adhesive to the exposed substrate after a suitable pretreatment. Adhesive bonding of foamed rubbers and plastics to painted metal surfaces with contact adhesives is widely practised in certain industries (e.g. automobile). These applications usually require a low adhesive strength well below the bond strength of the paint-

metal bond. Pretreatment in such applications need be no more than cleaning the paint surface with detergent solution followed by abrasion with a medium emery cloth. A final wash with detergent removes the fine dust.

REFERENCES

Pretreatment methods

1. Document D65/1123, B.S.I. London, January (1965).
2. Ministry of Aviation Specification, DTD 915B, H.M.S.O., London.
3. SHARPE, L. H., 'Adhesives', *Mach. Des.*, **39** (14), June (1967).
4. GUTTMAN, W. H., *Concise Guide to Structural Adhesives*, Reinhold, New York, (1961).
5. *Handbook of Adhesives*, Cyanamid International Corporation (Bloomingdale Dept.).
6. Tech. Data Sheet A15, CIBA (ARL) Ltd., Duxford, Cambridge.
7. MARTIN, J. T., 'Surface treatment of adherends', *Adhesion and Adhesives*, Vol. 2 (Ed. Houwink R., and Salomon, G.), Elsevier, Amsterdam, (1967).
8. *Assembly and Fastener methods*, January (1965).
9. HENDERSON, A. W., 'Pre-treatment of surfaces for adhesive bonding', *Aspects of Adhesion*, Vol. 1 (Ed. Alner, D. J.), Univ. Lond. P., (1965).
10. Sira evaluation.
11. HUNT, P. G., *A method of cleaning glass surfaces*, SIRA *Review*, **9** (4), (1968).
12. *Adhesives for TPX Polymers*, Tech. Data I.C.I. (Plastics Div.), Ltd., (1968).
13. CHUGG, W. A., *Glulam: the manufacture of glued laminated structures*, Benn, London, (1964).
14. Unclassified Reports of Mintech. contract work at B.A.C. (Operating) Ltd. Filton, Bristol, (1963–1966).

Other References

ROGERS, N. L., 'Surface Preparation of Metals for Adhesive Bonding, *Applied Polymer Symposia*, No. 3, Interscience/ Wiley, New York, (1966).
SMITH, R., 'How to prepare the surface of metals and non-metals for adhesive bonding', *Adhes. Age*, March (1967).
MUCHNICK, S. N., 'Treatment of metal surfaces for adhesive bonding', *W.A.D.C. tech. Rep.*, 58–87, April (1956).

EICKNER, H. W., 'Adhesive bonding properties of various metals as affected by chemical and anodizing treatments of the surfaces', *Forest Prod. U.S.,* Nos. 1842, April (1954) and 1842–A, Feb. (1955).

SNOGREN, R. C., 'Surface treatment of joints for structural adhesive bonding', *ASME Paper* 66–MD–39, May (1966).

JACKSON, L. C., 'Preparing plastics surfaces for adhesive bonding', *Adhes. Age,* **4** (2), 30, (1961).

GRAY, C. L., Jr., MACCARTHY, H. L. and MCLAUGHLIN, T. F., Jr., 'Adhesion promoter for polyethylene', *Mod. Packag.,* **34**, 143, (1961).

WECHSBERG, H. E., and WEBBER, J. B., 'Surface treatments of polythene film by electrical discharge', *Mod. Plast.,* **36** (11), 101, (1959).

VINCENT, G. G., 'Bonding polyethylene to metals', *J. appl. Polym. Sci.,* **11** (8), 1553, (1967).

NELSON, E. R., KILDUFF, T. J. and BENDERLEY, A. A., 'Bonding of Teflon', *Ind. Engng Chem.,* **50**, 329, (1958).

British Standards Institution, *Cleaning and preparation of metal surfaces,* CP 3012: 1972.

ASTM, *Tentative recommended practice for preparation of surfaces of plastics prior to adhesive bonding,* ASTM D2093–62T (1962).

JACKSON, L. C., 'How to select a substrate cleaning solvent', *Adhes. Age,* **17** (12), 23–31, (1974).

WHITE, M. L., *Clean surfaces: their preparation and characterization for interfacial studies,* G. Goldfinger (Ed), Marcel Dekker: New York, (1970).

CAGLE, C. V. and LEE, H., 'Using the stereo-scanning microscope for adhesive and surface studies', *Adhes. Age,* **14** (5), 40–44, (1971).

FOWKES, F. M., 'Surface chemistry', in *Treatise on Adhesion and Adhesives,* Vol. 1, 325–349, Marcel Dekker: New York, (1967).

JENNINGS, C. W., 'Surface roughness and bond strength of adhesives', *J. Adh.,* **4** (1), 25–38, (1972).

CASSIDY, P. E., and YAGER, B. J., 'A review of coupling agents as adhesion promoters', *Tracor Inc. NASA Contract Report 24073,* September, (1969).

CASSIDY, P. E., JOHNSON, J. M., and ROLLS, G. C., 'Coupling agents for adhesive systems', *Ind. Eng. Prd.,* **11**, 170–174, (1972).

BESSIN, R. L., 'How to obtain strong bonds via plasma treatment', *Adhes. Age,* **15** (3), 37–40, (1972).

EVANS, J. M., 'Nitrogen corona activation of polyethylene', 1–7; 'Influence of oxygen on the nitrogen corona treatment of polyolefins', 9–16, in *J. Adh.,* **5** (1), (1973).

AYRES, R. L., and SHOFNER, D. L., 'Preparing polyolefin surfaces for inks and adhesives', *Spe. J.,* **28** (12), 51–55, (1972).

MORRISS, C. E. M., 'Adhesive bonding of polypropylene', *J. App. Poly.,* **15**, 501–505, (1971).

WATSON, C. A., 'Bonding Plastics', *Eng. Mach. Des.,* **15** (9), 721–724, (1972).

BRAGOLE, R. A., 'Adhesion of EPDM', *Rubber Age,* **106** (1), 53–56, (1974).

DELOLLIS, N. J., and MONTOYA, O., 'Bondability of RTV Silicone rubber', *J. Adh.,* **3** (1), 57–67, (1971).

PETTIT, D., and CARTER, A. R., 'Behaviour of urethane adhesives on rubber surfaces', *J. Adh.,* **5** (4), 333–349, (1973).

BASCON, W. D., and PATRICK, R. D., 'The surface chemistry of bonding metals with polymer adhesives', *Adhes. Age,* **17** (10), 25–32, (1974).

BIJLMER, P. F. A., 'Influence of chemical treatments on surface morphology and bondability of aluminium', *J. Adh.,* **5** (4), 319–331, (1973).

MINFORD, J. D., 'Effect of surface preparations on adhesive bonding of aluminium', *Adhes. Age,* **17** (7), 24–29, (1974).

ALLEN, K. W., ALSALIM, H. S., and WAKE, W. C., 'Bonding of titanium alloys', *J. Adh.,* **6** (1 and 2), 153 156, (1974).

PAUL, R. D., and MCGIVERN (JR.), J., 'Electrochemical characterization and control of titanium surfaces for adhesive bonding', *Adhes. Age,* **17** (12), 41–50, (1974).

ALLEN, D. R., and ALLEN, S. A., 'Testing titanium surface stability', *Adhes. Age,* **14** (10), 41–42, (1971).

Trade Sources

A Welwyn Electric Ltd., Strain Measurement and Equipment Division, 70, High Street, Teddington, Middlesex.

B Marshall-Howlett Ltd., 44, Tower Hill, London, E.C.3.

C Fluorocarbon Ltd., Fluorocarbon House, Caxton Hill, Hertford, Herts.

D Hydralon Ltd., Princes Street, Northam, Southampton.

E Tygadure Ltd., Littleborough, Lancs.

F British Titanium Intermediates Ltd., 10, Stratton Street, London, W.1.

G I.C.I. Ltd., Plastics Division, Welwyn Garden City, Herts.

H Cornelius Chemical Co., Ibex House, Minories, London, E.C.3.

I Dow Corning International Ltd., Castle Chambers, 3/9, Sheet Street, Windsor.

J Union Carbide Ltd., P.O. Box 111, 8, Grafton Street, London, W.1.

K Midland Silicones Ltd., Cardiff Road, Barry, Glamorgan.

L Durham Raw Materials, 1–4, Gt. Tower Street, London, E.C.3.

M SATRA, Satra House, Rockingham Road, Kettering, Northants.

N Thorium Ltd., Widnes, Lancs.

O Takdust Products Ltd., Hayes Lane, Lye, Stourbridge, Worcs.

P Fisons Scientific App. Loughborough, Leicester.

Q CIBA (A.R.L.) Ltd., Duxford, Cambridge.

CHAPTER 7

The Bonding Process

7.1. STORAGE

Many adhesives require to be stored in the dark or in opaque containers while others should be stored at low temperatures (e.g. 5°C) to prolong shelf life. Base resins and curing agents should be kept apart so that accidental container breakage will not lead to contamination problems. Containers for solvent-based adhesives should generally be sealed immediately after use to prevent solvent loss or the escape of toxic or inflammable vapours.

Since adhesive materials differ in their storage requirements, the conditions of storage should follow the manufacturers' recommendations closely.

7.2. PREPARATION OF THE ADHESIVE

Adhesive preparation requires careful attention. After cold storage the adhesive must be warmed to the correct temperature for application. Usually this is room temperature but in some cases (e.g. hot-melt adhesives) the application temperature may be considerably higher. Where component mixing is concerned it may be important to measure the proportions correctly if optimum properties are required. This is particularly the case with catalysation reactions (e.g. amine curing agents for epoxy resins) where insufficiency of catalyst prevents complete polymerisation of the binder resin while too much catalyst can lead to brittleness in the cured material.

Excess unreacted curing agents may cause corrosion of metallic adherends. Some two-component adhesives have less critical mixing ratios (e.g. epoxy polyamides) and component volumes may often be judged by eye without too adverse an effect on the ultimate bond strength properties. As for storage precautions, the manufacturers' directions for adhesive preparation will regulate the procedures used.

7.3. METHODS OF ADHESIVE APPLICATION

The method of application of the adhesive material to an adherend, following a proper surface preparation, requires consideration where optimum performance from an adhesive bond is required. Improper application of an otherwise suitable adhesive to a well-designed joint often results in inadequately bonded assemblies. Suitable methods of adhesive application distribute the material as a uniform film of the correct thickness. The requirement is met by a number of methods, the employment of which is determined by such factors as the physical properties of the adhesive, the shape and dimensions of the bonding surface and the existing production facilities. Thus, for film and encapsulated adhesives, the form limits the application technique to placement of the adhesive between the bonding surfaces. Adhesives in liquid form may be applied by the methods described here.

7.3.1. BRUSHING

This method is suited to adhesives application to selected areas of a surface without the use of masks or stencils, or to complicated shapes. The control of adhesive film thickness is limited and films are often uneven and blobbed. The technique is generally not suited to rapid assembly work. Stiff brushes provide the most satisfactory coatings.

7.3.2. FLOWING

This method is useful for applying adhesives to flat surfaces. Pressure-fed flow guns are commonly used to apply the adhesive through a nozzle or flow brush attachment which promotes liquid spreading over extended areas. The technique lends itself to fast assembly work. Superior control over adhesive coating thickness is obtained compared with the brushing technique.

7.3.3. SPRAYING

This technique employs equipment of the type used for paint spraying and is suited to the coverage of large surface areas of uneven contour. Even coating depends on the strict control of adhesive consistency; film thickness uniformity is better than for coats produced by brushing or flowing. The presence of a possible health hazard arising from solvent spray mists should always be considered and adequate ventilation ensured.

7.3.4. ROLL COATING

This technique is based on the transference of adhesive material from a trough, by means of a pick-up roller partially immersed in it, to a contacting transfer roller sheet. Material is continuously coated with adhesive when fed between the transfer roller and a pressure roller which is adjusted to determine the coat thickness. This method is excellent for the adhesive coating of flat sheets and films of large width. Of all the methods considered it provides the fastest production rate and most uniform adhesive coverage.

7.3.5. KNIFE COATING

This method employs an adjustable knife-blade, bar, or rod to control the deposition of adhesive flowing on to a sheet moving under the blade. The distance between the blade-edge and the adherend surface determines the adhesive coating thickness.

7.3.6. SILK SCREENING

This technique is used to ensure that only selected areas or patterned spaces are adhesively coated. Printing equipment, masks and stencils may be used to advantage to apply adhesives to a required pattern.

7.3.7. MELTING

The term refers to the heating of adhesives, of the hot-melt type, to reduce them to a fluid state for application by flowing. The hot liquid adhesive in reservoirs is dispensed by suitable nozzles to the work piece. The use of rollers and knife blades with hot-melt adhesives is a feature of some laminating processes.

7.4. METHODS OF ADHESIVE BONDING

There are several methods for assembling bonded joints which are open to modification to suit particular production needs. Bonding techniques must fulfil certain requirements which are essential to successful joint assembly. The important stages concern:

Liquefaction of the adhesive during the process to ensure wetting of the adherend surface and to promote contact

Removal of unwanted components of the adhesive (e.g. organic solvents, water or volatile by-products from a curing reaction) from the joint to prevent the occurrence of voids, vapour locks and faults in the glue-line. Solvents are present in an adhesive to provide a suitable consistency for application and must be removed for the adhesive to take effect

Application of pressure to the joint while the adhesive cures to maintain the assembly without displacement of the adherends.

Bonding techniques commonly employed are as follows.

7.4.1. WET BONDING

The adhesive is applied to one adherend surface and the other surface (which may also be coated) is united to the wet coating. The method is suitable for non-porous adherends only when the adhesive contains no solvents. The occlusion of solvents would lead to fissure formation within the adhesive layer on curing. Where one adherend is porous, the method may be used with any adhesive that can wet the surface. The time for which bonding pressure is applied to the assembly is reduced for solvent-based adhesives by allowing the adhesive coat to dry partially, for a period (open-assembly time), to a tacky state. The open assembly time is usually specified by the manufacturer and depends on environmental temperature, humidity, and other variables. The coating of both adherends can result in stronger bonds but may extend the open-assembly time.

7.4.2. REACTIVATION BONDING

This refers to methods for reactivating adhesive coatings which have been applied to non-porous materials. These methods are usually unsuitable for porous adherends being mainly intended for large bonding areas where the control of adhesive drying is difficult. Two commonly employed reactivation techniques involve:

Solvent reactivation
This is achieved by moistening the dried adhesive

film with a suitable fast-drying solvent prior to joining the surfaces. This process is convenient for bonding completely dried adhesive surfaces which may have been stored for some time under dust free conditions. The method applies only to adhesives amenable to solvent activation such as thermoplastic and rubber based formulations in a solvent vehicle.

Heat reactivation

This process is confined to adherends which are resistant to the heat involved. The parts are assembled after adhesive coating and drying, and are heated during the application of sufficient bonding pressure to induce adhesive fusion. The bond is formed when the adhesive cools. It is important to allow time for the adhesive to reach a fusion temperature and to bear in mind that adherends often act as heat insulators. Sometimes the coated adherends are heat activated before assembly.

7.4.3.   PRESSURE-SENSITIVE BONDING

The method is restricted to adhesive materials which retain their tack, when dry, for long periods. Joint assembly is completed when the surfaces (one of which is coated) are brought together and clamped to form a bond. Where both surfaces are adhesive coated, the method is called 'Contact Bonding' and refers to the self-adhesive property of some adhesives. Adhesive backing papers are invariably used to protect pressure sensitive adhesives from aerial contamination.

7.4.4.   CURING

Certain adhesives need to be cured to develop joint strength. These materials exist in film and liquid forms; the latter usually relying on catalytic reaction to effect a cure. Application of a pre-mixed liquid adhesive-catalyst system must be completed within its working life if spreading and wetting are to be adequate. Following its application, an adequate time allowance must be made for the adhesive to cure. Room temperature curing adhesives often require some hours to set during which time the assembly must be jig-supported. Heat curing adhesives, which set within minutes, are available and suitable for application to adherends able to withstand the heat. It may be essential to heat cure certain metallic adherends, subject to oxidation, in an inert atmosphere, such as nitrogen, if weak interfacial oxide layers are to be avoided. One component adhesives exist which rely on heat activation to initiate the action of a constituent catalyst. Some adhesives contain volatile components to improve their consistency so that processing may involve an intermediate liquid removal stage before bonding (i.e. evaporation by heat or even force drying). Other forms are supplied as films, which may or may not be supported on carrier cloths; these types avoid the problems of solvent removal and tackiness associated with many liquid adhesives. The film is easily cut to shape and applied to the joint area. Structural adhesive films generally require high bonding pressures to be sustained during hot-curing schedules. Another recent approach to rapid curing at room temperature concerns the use of ultra-violet radiation (e.g. see p. 102).

7.4.5.   OTHER METHODS OF BONDING

Other bonding techniques involve combinations of the methods described above. Examples of mixed processes are the following:

The coating of different adherends with different adhesives which join to form a bond. Thus the bonding of nylon to steel is effected by priming the nylon with a resorcinol formaldehyde adhesive and bonding to an epoxy coated steel.

Pre-coating an adherend surface and bonding it, when dried, to another adherend with an undried coat.

Application of several adhesive coats to highly porous adherends to overcome the absorption problems responsible for starved glue-lines.

Adhesive application to a primed adherend where bonding directly to the adherend is difficult.

7.5.   INADEQUATE BONDING

The poor execution of an otherwise suitable bonding technique is often reponsible for unsatisfactory joints. Among the reasons for inadequate bonding are:

Starved joints—the deficiency of adhesive material on the adherend surface.

Dry joints—the adhesive is overdried to the extent that there is non-fusion with another overdried adhesive film on bonding.

Wet joints—the incomplete removal of component solvent from an adhesive or the incomplete curing of liquid adhesive.

Joint misalignment—due to insufficient fixation of the work piece, displacement of the assembly parts occurs during adhesive curing.

Inadequate surface preparation—often responsible for non-wetting surfaces and poor adhesive contact; contamination of cleaned surfaces is a similar hazard.

Incomplete processing—in which there is insufficient heat, pressure or assembly time, etc., to permit the adhesive to adhere to surfaces or fully cure.

Adhesive deterioration—arising from the migration of plasticisers from flexible plastics adherends, into the adhesive film, which diminishes its strength properties.

Void formation—the production of fissures, air/vapour holes and other faults within the glue-line; caused by incomplete solvent removal or faulty adhesive mixing or curing: air entrapment during adhesive application is also a danger.

Non-uniform glue-line—a consequence of unsatisfactory process control leading to incorrect or uneven adhesive film thickness inconsistent with maximum strength.

7.6. METHODS OF BOND CURING

The majority of adhesives need to be cured to ensure that the full strength of the assembled joint is realised. The curing conditions will be determined by the nature of the adhesive material and the adherends under consideration. For solvent adhesives and porous adherends it is sufficient to allow solvent loss, at ambient temperature, to set the adhesive. Many structural adhesives require the application of heat with or without pressure to effect curing. Pressures may vary from 7 to 1 050 kN/m² and curing temperatures from 20–350°C. With ceramic based adhesives, dependent on a sintering action for adhesion, the processing temperatures can reach 1 800°C. Adhesive manufacturers are usually able to specify the curing schedule(s) giving optimum adhesion. Faster production generally results from heat curing procedures. A number of methods are used to apply heat and pressure either separately or together to bonded assemblies.

7.6.1. DIRECT HEAT CURING

Where direct heat is employed for curing, it is essential that the glue-line is maintained at the specified temperature for the requisite time. The bond temperature usually increases more slowly than the ambient temperature and should be checked (e.g. with a thermocouple) during the heating of a specimen assembly. This procedure ensures that the correct heating schedule is adopted and avoids under-curing of the adhesive due to insufficient heat exposure. Hot curing techniques are associated with the heating equipment discussed below.

Ovens
These are the commonest means of heating bonded assemblies since they are readily adapted for use with pressurising appliances. Temperature control is poor unless the oven is of the circulating air type; recent models realise temperatures of 450 ± 1°C with design features such as forced air fans and transistorised thermostats. Heat transfer, by convection, to assemblies is usually slow.

Liquid baths
Various liquids are employed to provide rapid heat transference by conduction. Water is commonly used but for higher curing temperatures mineral or silicone oils are required. The silicone fluids are useful, inert and non-toxic heating media for temperatures up to 300°C. Direct contact between the bond and the heating medium should be avoided; the method depends on heat conduction through the adherend to cure the adhesive.

Hot presses or platens
Such equipment relies on electrical resistance heaters or steam to provide heat to the platens compressing the bonded assembly. The highest temperatures are obtained with electrical heating elements and these are amenable to relay control (i.e. where the curing cycle involves various temperature-time stages). Steam heating is a faster process and it is often advantageous to circulate cooling water through the piping after curing. The technique is an effective one for cooling bonded structures under pressure conditions.

7.6.2. RADIATION CURING

This technique, involving infra-red radiation heaters, produces an increase in the heat transference rate exceeding that of oven heaters. Infra-red lamps provide a useful way of removing solvents from contact adhesives prior to bonding, and are suited to the rapid heating of localised areas of a substrate.

7.6.3. ELECTRIC HEATERS

A conductor strip of metal is embedded in the adhesives to act as an internal heater. The heating of the bond is achieved by the passage of current through a metallic adherend or conductor within or adjacent to the glue-line for non-metallic adherends.

More recently, graphite has been used as an internal electric resistant heater for curing structural adhesives. The graphite is available in various physical forms such as felts, yarns, woven fabrics and tapes. It can be utilised as a heating element over a wide temperature range; up to 360°C in air and beyond 2 800°C in an inert atmosphere. A negative coefficient of resistance with temperature prevents current surges during heating. Rapid heating and cooling of the cloth results from the low thermal mass and high emissivity of radiation per unit area of graphite fabric. This internal heat source method with graphite resistance elements provides bond strengths of joints which compare favourably with similar joints prepared by oven curing. Advantages which the method offers over conventional external heating methods are the following:

Rapid attainment of curing temperature since the adhesive is directly heated; provision is usually made for heat losses to adherends and the environment.

Easy application of heat to localised areas of an assembly.

Fabrication of assemblies with high temperature curing adhesives without the risk of distortion; uniform heating of fabric eliminates hot spots.

Closer control of glue-line temperatures with consequent realisation of maximum adhesion performance; fabric functions as a glue-line spacer and ensures uniform thickness of adhesive layer.

Restriction of heating to the glue-line thus avoiding unnecessary heating of adherends; reduced expenditure on large assemblies following a lower power consumption.

Simplicity of process excluding the need for ovens or platens; ready on site repair of damaged assemblies with transportable power equipment.

Realisation of improvements in design and on site modifications to structural units, in a room temperature environment.

The technique may be used to advantage with hot-melt adhesives to achieve easier processing. Conventionally, assemblies are heated up to the melting point of the hot-melt (applied as a film or powder between the adherends) and then cooled. By impregnating graphite fabric with the hot-melt and passing a current for a short period to liquefy the adhesive, the need to heat up and cool down the entire assembly is eliminated. Processing times may be considerably reduced.

Electrical heating—other methods
Other methods for electrical heating of bonded assemblies make use of wrap-around electrical heating tapes or resistance elements within the jig supporting the jointed structure. It is generally difficult to attain uniform heating with these methods.

7.6.4. HIGH FREQUENCY (RADIO FREQUENCY) DIELECTRIC HEATING

The curing of glue-lines by heat conduction from hot platens is inefficient where thick non-conductive adherends are involved. High frequency dielectric heating has been developed as a curing means for bonds based on organic polar materials which are poor conductors (unlike metals for which inductive heating is preferred) or insulators (e.g. polystyrene). High frequency heating processes are particularly effective with the thermosetting resins employed for woodworking applications, e.g. urea, melamine, resorcinol (and with phenol) formaldehydes, polyvinyl acetate (and with urea formaldehyde), and to a lesser extent, animal and casein glues.

The process is based on the absorption of energy by the adherend material (or dielectric) when it is placed in an alternating electric field. At high fre-

quencies, from (10 to 15) 10^6 Hz, molecular vibration (resonance) occurs which leads to heat generation within the material provided the material has an appropriate loss factor at that frequency. High loss factors favour rapid heating. [Loss factor, (dielectric constant × power factor) determines the heating rate of a material. Thus, an acetal plastic (L.F. = 0·016) is 5 000 times more difficult to heat by dielectric means than water (L.F. = 80).]

Circuit arrangements usually conform to one of the following.

Glue-line heating
This arrangement, Fig. 7.1, is favoured for assembly work. Electrodes are adjacent to the glue-line so that the electric field is parallel to it. Heat is generated

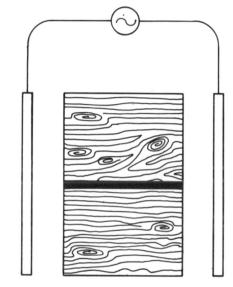

Fig. **7.1**. *Glue-line heating*

in the glue-line and setting times can lie within 10–60 s. For wood, about 0·06 m² bonding area is cured in 1 min for 1 kW power consumption.

Transverse heating
This is employed for bonding laminates and shapes, Fig. 7.2. Electrodes are on opposite sides of the glue-line so that the electric field is at right angles to it. Heat is generated within the material and is conducted to the glue-line.

For wood, about 1 kW of power heats 0·45 kg by 38°C in 1 min. Longer setting times (exceeding 3 min) are recommended to avoid heat damage to the wood.

Stray field heating
This is the least efficient heating method and may be regarded as a combination of glue-line and transverse heating, Fig. 7.3. It is suited to the glueing of small and unsymmetrical assemblies. Setting times are intermediate (e.g. allied to the typical cases above,

the process would set about 0·03 m²/kW/min). Heating up to a material depth of 5 cm is feasible.

Recently, 'stray field platens', in which the electrodes are embedded in the platens, have been introduced which avoid the need for special jigs and

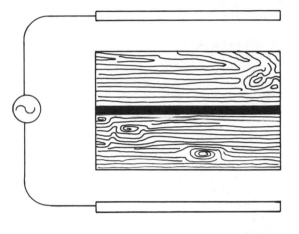

Fig. **7.2**. *Transverse heating*

Fig. **7.3**. *Stray field heating*

electrode arrangements. Heat is generated in any glue-line close to the platens and only component location jigs are required.

7.6.5. INDUCTION HEATING

This is a similar method to dielectric heating; electric power is used to generate heat in a conducting material. Hence the technique is applicable to metal adherends or to adhesive materials filled with metal powder. If one adherend is conductive and the other non-conductive, either dielectric or induction heating may be applied.

The possibility of heat charring the adhesive, where rapid heat curing is involved, should be recognised and care taken to control the heat input adequately. Rapid curing usually results in less than optimum strength for bonded joints and it is preferable to maintain slow heating and cooling rates for heat curing processes.

7.6.6. LOW-VOLTAGE ELECTRIC HEATING (L.V.H.)

This is employed for a variety of wood bonding applications, e.g. scarf joint manufacture, boat-building hull work. The method is cheaper to instal than dielectric heating but is not as efficient for wood joints where adherend thickness exceeds a few millimetres. The method employs an A.C. step-down transformer to give low-voltages (from 3–12 V with currents 500–1 000 A are commonly used) which are applied to a metal platen (e.g. steel) pressed against the adherend panel. Approximately 1 kW is required to heat 0·3–0·4 m² with platen temperatures varying between 70 and 200°C. Temperature uniformity with large panels can sometimes be a problem.

7.6.7. ULTRASONIC ACTIVATION

This method of curing the adhesive is based on the transmission of mechanical vibrations from an ultrasonic transducer to the adhesive at the interface between the mating parts. It is used most efficiently where a bead or film adhesive can be incorporated into the design. Heat produced by the absorption of ultrasonic energy melts or sets the adhesive. The adhesive layer may be a dried solution or emulsion coating applied to one or both adherends or in film form. Viscoelastic adhesives are particularly suited to the process whereas slow-curing liquid thermosetting types do not respond to ultrasonic activation. Some of the adhesive types amenable to the technique are listed in Table 9.5 (p. 281). Ultrasonic activation usually increases the bond strength and reduces curing times where it can be employed as an alternative to conventional thermal or drying processes.

7.7. BONDING PRESSURE

In addition to time, joints often require the application of pressure during bond formation. The need for pressure during bonding will be determined by the adhesive employed. The application of pressure may fulfil several of the purposes described here:

Maintains the integrity of a jointed assembly during adhesive cure

Ensures close contact between joint parts and the adhesive material. Contact pressure induces the flow of the adhesive over the bonded areas and thus promotes wetting

Prevents the evolution of volatile components during the heat curing of certain adhesives (e.g. phenolic resins) which would otherwise result in porous glue-lines of low strength

Causes hot-curing film adhesives to flow before finally setting. Pressures of 350–1 400 kN/m² are usually required to spread adhesive systems, based on polyaromatic materials, during an intermediate curing stage. Flow is one of the most important properties of an adhesive; too little flow produces patchy bonds containing voids and low strength areas; excess flow results in a variable glue-line thickness and non-reproducible strengths in metal-to-metal bonding. For honeycomb core bonding, the flow requirements are less critical and the objective is the formation of full fillets with the core edges. Honeycomb cores in an unbonded state usually limit bonding pressures to 350 kN/m².

The pressure applied during the cure should never exceed that specified for the adhesive. Too much pressure produces internal stresses in a joint and results in either a decrease in its bond strength or early failure. Pressure may be applied, according to the joint type and the bonding area concerned, by the pressure equipment discussed below which must apply the correct amount of pressure and sustain it uniformly over the surface. A pressure pad of a soft material, such as rubber or asbestos board, interposed between the pressure plate and the adherend surface ensures a uniform pressure distribution. Constant bonding pressure must be arranged to allow for changes in assembly dimensions during curing (e.g. adhesive shrinkage may occur). The type of equipment able to satisfy these requirements is various and includes the following.

Hydraulic presses

These maintain an easily controlled continuous pressure. Large area joints require pressure pads to distribute the pressure uniformly. The pressure plates may incorporate electrical heating elements for hot curing.

Hydraulic pads

The pads are similar to the presses and provide good pressure distribution. Pressure is supplied to an adherend by inflating an adjacent pad with compressed air or fluid. The technique has been used to bond flexible interliner materials to piping and cylinder tubes.

Weight loading

The simplest method of applying pressure. Usually sand or shot bags are feasible where flat adherend surfaces requiring low bonding pressures are involved.

Clamps

These are generally screw-toggle arrangements which require pressure pads to distribute the pressure evenly. Spring loading provides for follow-up pressure where assemblies undergo dimension changes during cure. Screwed wedges represent another form of pressure applicator.

Vacuum bag application

The vacuum bag is applied to the component being bonded thus providing a maximum bonding pressure of 103 kN/m² (atmospheric pressure). Greater pressures are achieved with the use of pressure transmission bars between the bag wall and the contained joint.

Autoclave vessels

These are pressurising chambers, sometimes used with steam or hot-air to meet the need for heat curing under pressure. Hot air autoclaving provides a heat transfer rate which is air-density dependent; heating rates increase with pressure.

Autoclave bonding generally involves high pressures which can create assembly distortion problems not significant in other bonding equipment. For example, variations in glue-line pressure, due to non-conformity of assembly parts, is often a problem with autoclave bonding. The concentration of bonding pressure at adherend edges causes thinning of the glue-lines at the edge, which may lead to distortion of the structure.

The requirement for sustained bonding pressure is not so critical for certain structural adhesives which set at short curing cycles involving high temperatures. Thus, some modified phenolic adhesives will cure at 250°C in a matter of minutes; maintenance of pressure is superfluous after curing.

Curing methods requiring the application of both heat and pressure are usually based on combinations of the equipment and techniques already described. It is often expedient to design equipment for a specific assembly problem.

7.8.   EQUIPMENT FOR PROCESSING ADHESIVES

The application and processing of adhesives demands equipment which will satisfy requirements for:

provision of an adequate quantity of adhesive on a specified bonding area with control over the glue-line thickness,

proportioning and mixing of two or more component adhesives,

handling of different forms of adhesive in small or large quantities, for batch or assembly-line operation,

removal of solvents and volatile by-products of adhesive curing reactions,

application and maintenance of heat and pressure.

A variety of jigs, ovens, presses, roller-coaters, etc. have been developed or adapted for adhesives technology. A list of currently available adhesives processing plant and equipment will be found below, together with the manufacturers and their addresses.

7.8.1. PLANT AND PROCESSING EQUIPMENT: TRADE SOURCES

The equipment is followed by the trade source (for the key to Code No. see Section 7.8.2.)

Adhesive binding machinery, 24, 33, 56
Applicators, 2, 4, 9A, 18, 22, 38A, 47, 53, 54, 56, 66, 73
Assembly presses, 66
Autoclaves, 4
Bagmaking machinery, 17, 38, 64
Bag 'neck-sealing' machinery, 5, 63
Bundlers, 33
Carton gluing equipment, 2, 4, 9, 19, 24, 25, 30, 33, 38, 40, 51, 54, 64, 73
Carton sealing equipment, 2, 5, 25, 38, 39, 51, 54, 73
Case sealing equipment, 2, 4, 9, 25, 30, 38, 39, 54, 73
Caulking guns, 11, 18
Clamps and jigs, 35, 63
Coating machinery, curtain, 40
    double sided coating 9A, 21, 24, 26, 40
    edge coating, 9A, 2, 21, 26, 37, 41, 46, 60
    hot melt, 21, 24, 26, 66
    polyvinyl chloride, 46, 60
    roller, 9, 9A, 21, 24, 26, 40, 45, 47
    reverse roller, 21, 24, 26, 45, 47
    waxes, 9, 9A, 21, 24, 26
Compressors, air cooled, 2, 18, 38
    water cooled, 18, 38
Dispenser equipment, resins, 2, 5, 38, 53, 54, 66, 73
    glue, 2, 4, 22, 54, 73
    gummer paper, 73
Edge jointing equipment, 5, 9, 35, 40, 45, 59
Edge veneering machinery, 35, 40, 45, 59
Filling equipment, 26, 30, 36, 38, 73
Folding/gluing machinery, 2, 9, 19, 24, 25, 33, 38, 64
Glue mixing equipment, 3, 35, 40, 67, 72
Glue pots, 7
Glue spreaders, curtain coating, 9A, 33, 60
    roller coating, 9A, 26, 33, 36, 46, 52, 57, 60
    for abrasive papers and cloths, 46
Hand rollers, 19, 40
Heating equipment, glue kettles, 7
    high frequency, 8, 35, 59
    hot spraying, 4
    heat sealing, 26, 38, 47
    low voltage, 35, 39, 59
    ovens, 7, 16, 26, 36, 38, 40, 65
    RF, 8, 35, 59

Hot-melt applicators, 2, 4, 21, 25, 30, 33, 40, 47, 54, 56, 65, 66, 71, 73
Labelling machinery, 5, 14, 25
Laminating equipment, 9, 21, 24, 26, 33, 35, 45, 59
Metering/dispensing equipment, 36, 41A
Mills, 26, 38, 67
Mixing/metering equipment, 3, 6, 67, 72
Mixing/metering/dispensing equipment, 3, 6, 67, 72
Mixers, dough, 6, 12, 36, 38, 67, 72
    high temperature, 3, 6, 26, 67, 72
Moisture meters, 7, 26
Photo-electric process controllers, 26, 33, 43
Presses, cold, 32, 40
    hot, 32, 40
    rotary, 17, 26, 32, 33, 40
Pressure reducing valves, 4, 32
Pumps, 3, 4, 19, 32, 36, 38, 56
Self-adhesive tape dispensers, 5, 11, 14, 53, 73
Spray booths, 7, 18, 20, 22
Spray equipment, automatic, 2, 4, 18, 20, 22, 45
    manual, 4, 18, 20, 22
Stencil gluing equipment, 9, 19, 24, 33
Strip bonding machinery, 9, 33, 59
Strut bonding machinery, 54
Temperature monitoring and control equipment, 26
Viscometers, 26

7.8.2. TRADE SOURCES AND ADDRESSES

Code No.   Manufacturer

1   AEROSTYLE LTD,
    Sunbeam Road, London NW10.
    01-965 3464
2   AIR INDUSTRIAL DEVELOPMENTS LTD
    Aidspray Works, Shenstone, Lichfield, Staffs.
    0543 480341
3   A.P.V. OSBORNE CRAIG LTD
    Glenburn Road, College Milton North, East Kilbride, Glasgow, Scotland G74 5BJ.
    03552 25461
4   ARO CORPORATION (UK) LTD
    Caernarvon, North Wales.
    0286 3551
5   AVERY-MONSON LTD
    Walkinstown Avenue, Long Mile Road, Dublin 12, Ireland.
    Dublin 507755
6   BAKER PERKINS LTD
    Wewswood Works, Peterborough.
7   BARLOW-WHITNEY LTD
    Watling Street, Bletchley, Bucks.
    0908 23571
8   BELLOW MACHINE CO LTD
    Ellerby Lane, Leeds 9.
9   BERRY EDE & WHITE LTD
    Esplanade, Rochester, Kent.

9A BONE BROTHERS LTD
Manor Farm Road, Wembley,
Middlesex.
01-997 9555

10 BOSTIK LTD
Ulverscroft Road, Leicester LE4 6BW.
0533 50015

11 BRISSCO EQUIPMENT LTD
Cater Road, Bishopsworth,
Bristol BS13 7TX.
0272 664216–0

12 CASCO MIXING MACHINES LTD
Lloyd's Bank Chambers, 125 Oxford
Street, London W1.

13 CEFMOR-BREHMER LTD
Tariff Road, London N17.
01-808 4577

14 CHARTPAK LTD
Station Road, Didcot, Berks.
023581 2607

15 CHEKMIAN SILKS LTD
45 Poland Street, Manchester 4.
061-205 2565

16 CHURCHILL INSTRUMENT CO LTD
Walmgate Road, Perivale, Greenford,
Middlesex.
01-998 3322

17 COBDEN CHADWICK LTD
Century Works, Havelock Street,
Oldham, Lancs.
061-625 1200

18 COLOUR SPRAYS LTD
Albion Works, North Road, London
N7 9HP.
01-607 6091

19 COOPER PRINTING EQUIPMENT CO LTD
New Barnes Mill, Cottonmill Lane,
St Albans, Herts.
St Albans 50920 and 59622

20 DEVILBISS CO LTD
Ringwood Road, Bournemouth, Hants.
0202 3131

21 DIXON & CO LTD, T H
Works Road, Letchworth, Herts.
04626 5101

22 DUNLOP CHEMICAL PRODUCTS DIVISION
Chester Road, Erdington,
Birmingham
B35 7AL.
021-373 8101

23 EAGLE PACKAGING & PRINTING CO LTD
Compsall Mill, Compsall, Cheshire.

24 EDLON MACHINERY LTD
Edlon House, 17–19 Barter Street,
London WC1.
01-242 2341

25 EMHART (UK) LTD
Crompton Road, Wheatley, Doncaster,
Yorks.
0302 65226

26 ENGELMANN & BUCKHAM LTD
William Curtis House, Alton, Hants.
0420 82421

27 EVODE LTD
Industrial Adhesives Division, Common
Road, Stafford.
0785 2241

28 EX-CELL-O CORPORATION (STANDE
DIVISION)
3 Noble Corner, Great West Road,
Hounslow, Middlesex.
01-570 9659

29 FJELLMANS AD, R T
Mariestad, Sweden 13830.

30 FLEXILE METAL CO LTD
Bessemer Drive, Stevenage, Herts.
0438 51491

31 FLOWER-FARADAY LTD
Eclipse Works, Wimborne, Dorset.
0201-25 2251

32 FOGG & YOUNG ENGINEERING LTD
Norfolk Road, Gravesend, Kent.
0474 63521

33 FRIEDHEIM LTD, OSCAR
246 Waterloo Road, London SE1.
01-928 1363

34 FUSSELL'S RUBBER CO LTD (LARS FOSS KEMI
LTD)
Knightstone Mills, Worle, Weston-
Super-Mare, Somerset.
0934 3573

35 GIBBS CONSTRUCTOR (1961) LTD
First Avenue, Edmonton, London N18.
01-807 4789

36 GILBERT (INDUSTRIAL) LTD, G & R
Restmor Way, Hackbridge Road,
Hackbridge, Wallington, Surrey.
01-669 1173/6

37 HALLEY & SONS LTD
Farm Street, West Bromwich, Staffs.
021-553 2141

38 HART & SONS (LONDON) LTD, W J
212/4 Putney Bridge Road, London
SW15.
01-788 9295/7

38A INDUSTRIAL SCIENCE
Leader House, Snargate Street, Dover,
Kent.
0304 2656

39 INTERTHERM LTD
Blenheim Gardens, Brixton Hill,
London SW2 5EY.
01-674 9531

40 INTERWOOD LTD
362 Old Street, London EC1.
01-739 9866

41 JACOB, WHITE & CO LTD
Westminster Mill, Horton Kirby,
Dartford, Kent.
Farningham 3511

41A KANE GROUP LTD, DOUGLAS
　　Swallowfields, Welwyn Garden City,
　　Herts.
　　Welwyn Garden 21261

42 KANE LTD, C FREDK
　　Coronation Road, Park Royal,
　　London NW10.
　　01-965 5555

43 KAPPA JANES LTD
　　27 Stewart Avenue, Shepperton,
　　Middlesex TW17 0EQ.
　　Chertsey 2772

44 LIQUID CONTROLS CORPORATION (WADE
　　ENGINEERING)
　　Crowhurst Road, Brighton, Sussex
　　BN1 8AJ.
　　0273 506311

45 LOHMANN & CO LTD
　　The Mill Trading Estate, Acton Lane,
　　London NW10.
　　01-965 8981

46 LOTZ, ABBOTT & CO LTD
　　928 High Street, North Finchley,
　　London N12.
　　01-445 0256

47 MARTIN & CO LTD, T G
　　Gilmar Works, Hyde, Stockport,
　　Cheshire SK14 5RN.
　　061-368 2123

48 MATHER & PLATT LTD
　　Radcliffe Works, Radcliffe,
　　Nr Manchester.
　　061-723 2641

49 MAYHALL CHEMICAL (UK) LTD
　　1 Boundary Road, New Ferry,
　　Bebington, Wirral, Cheshire L62 5AL.
　　051-645 8220/9

50 MORANE PLASTIC CO LTD
　　Gresham Road, Staines, Middlesex.
　　Staines 61985

51 MORGAN FAIREST LTD
　　Fairway Works, Carlisle Street,
　　Sheffield S4 7LP.

52 NEWMAN LABELLING MACHINES LTD
　　Queen's Road, Barnet, Herts.
　　01-449 9666

53 NORPRINT LTD
　　Horncastle Road, Boston, Lincs.
　　0205 5501

54 PAFRA LTD
　　Bentalls, Basildon, Essex.
　　0268 280606

55 PETERS PACKAGING LTD
　　Windsor Works, Mill Street, Slough,
　　Bucks SL2 5DG.
　　Slough 34511

56 PLANAX BINDING SYSTEMS LTD
　　15–17 Cheam Road, Sutton, Surrey.
　　01-643 4441

57 PRACTICAL MACHINES CO (A T GADSBY) LTD
　　St George's Works, 303 Camberwell
　　New Road, London SE5.
　　01-703 5204/5

58 PURDY MACHINERY CO LTD
　　41/2 Prescot Street, London E1.
　　01-481 8401

59 PYE THERMAL BONDERS LTD
　　28 James Street, Cambridge.
　　0223 57590

60 SCHUBERT LTD, H
　　Imperial Works, Wenlock Road,
　　London N1.
　　01-235 1772

61 SESSIONS LTD, WILLIAM
　　The Ebor Press, Huntingdon Road,
　　York.
　　0904 59224

62 SKERMAN & SONS LTD, C
　　10 Parson's Green, London SW6.
　　01-736 6402

63 SMITH & CO, RONALD
　　106 Eldon Street, York.
　　0904 22361

64 SOAG MACHINERY LTD
　　Transport Avenue, Great West Road,
　　Brentford, Middlesex.
　　01-560 5181

65 STABILAG ENGINEERING LTD
　　Tremix Works, Blagden Street,
　　Sheffield S2 5QS.
　　0742 77921

66 STANDARD ENGINEERING CO LTD
　　Ebbington Valley Road, Leicester.
　　0533 736575

67 STEELE & COWLISHAW LTD
　　Cooper Street, Hanley, Stoke-on-Trent,
　　Staffs.
　　0782 23333

68 THOMASONS LTD, G C
　　820a Green Lanes, London N21.
　　01-360 8700

69 TREMCO LTD
　　27 St George's Road, London SW19 4DY.
　　01-947 3451

70 WANTZEN LTD, A
　　68 Coleman Street, London EC2.
　　01-606 3928

71 WERNER & PFLEIDERER (UK) LTD
　　72a Compstall Road, Romiley,
　　Stockport, Cheshire SK6 4DE.
　　061-494 1135

72 WINKWORTH MACHINERY LTD
　　65 High Street, Staines, Middlesex.
　　Staines 55951/3

73 WIX OF LONDON LTD
　　41 Standard Road, Park Royal,
　　London NW10.
　　01-965 1255

REFERENCES

PAGE, W. D., 'Using conduction heating to cure plywood components', *Adhes. Age,* **14** (1), 36–39, (1971).

MYNOTT, T. I., 'Adhesives for radio-frequency heating', *J. Inst. Wood Sci.,* **973** (33), 2–07, (1973).

ANON., 'Adhesives set sights on process equipment', *Chem. Engng.,* **75** No 11, 78–80, (1968).

NYSTROM, R. G., 'Applying adhesives automatically', *Automation,* **14** No 12, 60–62, (1967).

TELLER, W., 'New heating and cooling systems control application temperatures', *Adhes. Age,* **17**(1), 36–37, (1974).

CHAPTER 8

Physical Testing of Adhesives

8.1. INTRODUCTION

A detailed description of the various test methods that have been developed for adhesive bonds is beyond the scope of this handbook. A short outline, in chart form, of commonly used test specimens follows the remarks on the evaluation of adhesive strength. Non-destructive test methods and the effects of adverse service conditions on bond strength are dealt with in subsequent paragraphs and the section concludes with a list of titles of widely accepted standard test methods.

8.2. STRENGTH PROPERTIES

Specialised testing methods are required for the evaluation of the strength properties of adhesives. In addition to joint strength determination these methods provide a means for checking the efficacy of the processes used to make the bonds. The joint strength is invariably dependent on bonding technique factors such as adhesive application, adherend pretreatment and the adhesive curing conditions. Bonding conditions also determine the reproducibility of test results and complete information on a number of variables is, therefore, necessary before undertaking an adhesive evaluation. The following particulars are essential for the fabrication of reliable test specimens.

Instructions for the preparation of the adhesive

Adherend surface pretreatment procedures recommended for the adhesive under consideration. (Special treatments may be involved where certain environmental tests are envisaged)

Adhesive application and processing before bonding. (Attention to coating thicknesses and their control, or drying conditions is often important)

Manufacturers' specified conditions for joint assembly (temperatures, humidities or times)

Adhesive curing conditions relating to bonding temperatures, pressures and times.

Test specimens required to give reproducible failing strengths need to be carefully designed and prepared; unsatisfactory bonds will result from the faulty execution of any stage in the assembly process. Most of the standard test methods employ test specimens of definite shape and size which have to be machined to specified tolerances. Most methods specify the number of specimens to be tested in order to obtain a reliable result because testing factors and slight differences between adhesive batches prepared under identical conditions lead to joint strength variations. Ten or more specimens may be required to give meaningful data. The equipment used for testing is important and will influence the reliability of strength values obtained. Some variations in the performance between machines of the same type are to be expected. Often, test machine accuracy is greatest over a limited working range of the loading capacity (usually 10–90%), and test specimens should fail at loads within this span. The rate at which test specimens are stressed is another factor influencing the strength values obtained for adhesive bonds. Standard test methods generally specify the testing rate although it may be better to adopt a rate which more closely resembles the stressing rate that an actual assembly is likely to experience.

The data obtained from test methods is useful for comparing the performance of several adhesives prior to the selection of one for a particular assembly job. It must be emphasised that the test specimen rarely simulates the actual configuration of an assembly and that the test data cannot therefore, be relied upon to predict the performance of the assembly in service. The same limitation applies to test specimens removed from an assembly; these are unlikely to represent the behaviour of the whole structure. Short of testing an assembly under service conditions it is necessary to adopt a test specimen and method which simulate the assembly and its working environment as closely as is practicable. The testing procedure finally employed must produce results that are likely to show a good correlation with the results that would be obtained in tests on assemblies. In this respect, selected standard test procedures are frequently employable without modification.

Table 8.1. SUMMARY OF STRENGTH PROPERTIES AND TEST SPECIMENS

Strength Property	Test Specification	Design of Test Specimen (Direction of Stress)		Adherend materials	Remarks
Shear	ASTM D1002–64	1		Rigid materials such as metals, wood, reinforced plastics	Shearing stresses act in the pl∂ of the glue-line. Specimens are tension loaded. Generally unsuitable for weak flexible materials such as ther moplastic laminates, rubbers ∂ cloth fabrics, which are likely fail adhesively before bond rupture. Joint 1 may be used these materials provided the overlap area is minimal. Joint is also suitable for bonding lo strength materials by sandwic between metal adherends. Join is designed to minimise deflec ing of adherends and peel stresses during loading. Specir is not so easy to prepare as types 1 and 2 and consequent is less frequently used. Optim adhesive glue line thickness exists; bond strengths decreas for very thin or thick films. Joint strength does not vary directly with overlap area.
Shear		2			
Shear		3			
Shear	ASTM D906–47 BS 1203: 1954	4		Plywood	Factors of importance which affect the results obtained include, length of overlap, alignment of bonded area an∂ alignment in the tensometer, the rate of loading.
Shear	ASTM D905–49	5		Wood and similar materials	Loading by compression. Hi∂ shear strengths are obtained tension loaded specimens to improved distribution of stresses.

conti

Table 8.1 continued

Strength Property	Test Specification	Design of Test Specimen (Direction of Stress)		Adherend materials	Remarks
ensile	ASTM D897–49	6		6, Wood 7, Metals	Breaking forces are applied at right angles to the plane of the glue-line. Reduction of bonded area in Joint 6 reduces the risk of wood failure. Weak materials such as leather, rubber, plastics, can be sandwiched into the joint (6 or 7) to determine their bond strengths. Technique may be extended to determine composite joints such as glass-rubber, by using the test pieces as backing plates. Alternatively, for brittle materials (e.g. glass) it is preferable to use cross-lapped rectangular test pieces. (ASTM D1344–57.) Joint 6 is subject to cleavage and lower strengths are obtained with non-rigid adhesives compared to Joint 7. Modified types 7 and 6 are also used for impact strength measurements.
ensile		7	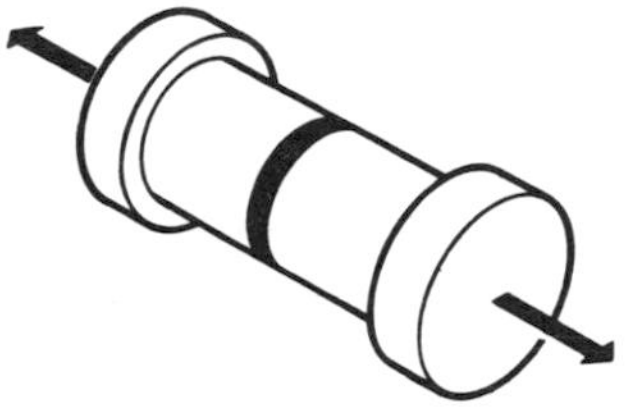		
ensile		8	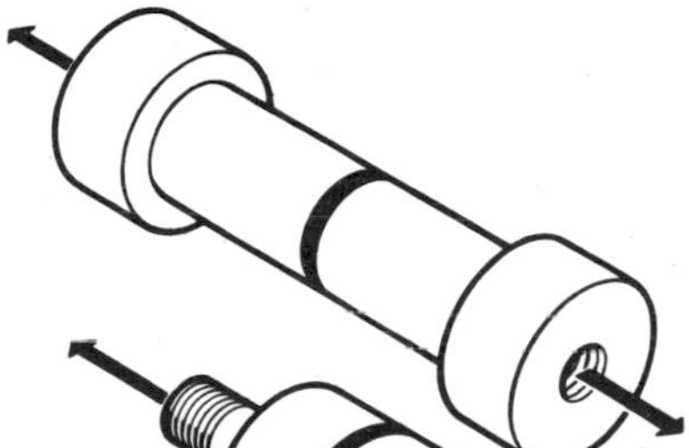	8 and 10, Metals 9, Rubber to Metal	Joints 8, 9, 10, are loaded in tension although the method of application differs for types 8 and 10. Joint 8 is also used for fatigue testing.
ensile	ASTM D429	9	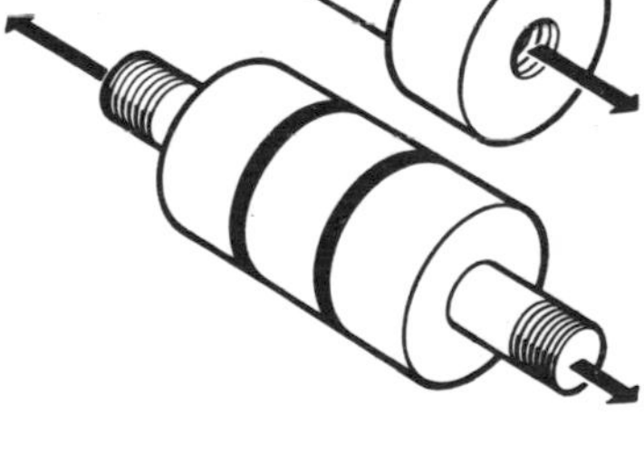		
ensile	ASTM D2095–62T	10	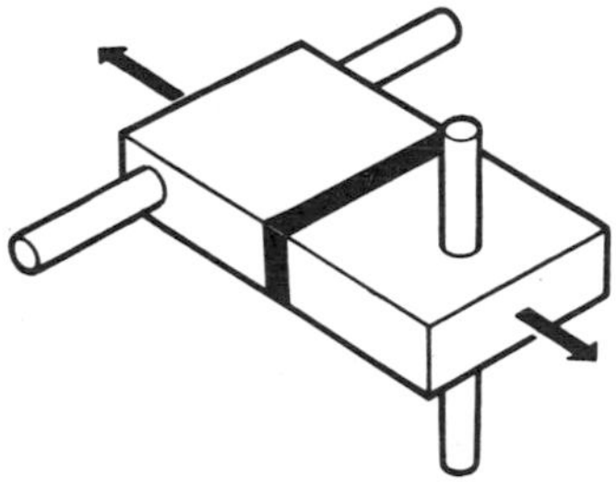		

continued

Table 8.1 continued

Strength Property	Test Specification	Design of Test Specimen (Direction of Stress)		Adherend materials	Remarks
Impact	ASTM 950–54	11 		Metals, Wood	Impact strength of an adhesive is the least force needed to break the bond in a single blow. The determination is made by shear loading the specimen. The lower block is secured in a grip and the upper block impacted with a pendulum hammer swinging at a known velocity along an axis parallel to the glue line.
Impact 1		12			Joint 12 is tension loaded for impact strength testing. The specimen is essentially the same type as is used for butt-jointing testing of tensile strength (ASTM D2095–62T). Joint 13 is employed for repeated impact tests. An annular ring is dropped repeatedly, from a known height, on to the shoulder of the bonded specimen. The impact fatigue resistance of the adhesive is obtained by noting the number of impacts at given heights and the weights causing joint failure (ASTM Bulletin 141, 42 (1946).)
Impact fatigue 2		13			
Cleavage	ASTM D1062–51	14 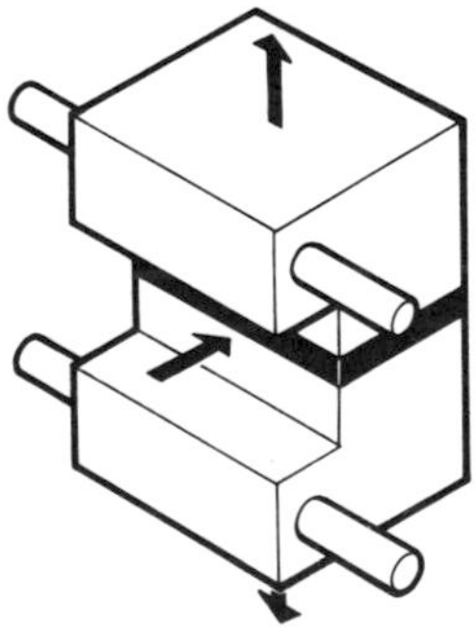		Rigid materials such as metals, reinforced plastics, hard woods. Unsuitable for flexible, extensive adherends	Tensile loading is applied to spl the bonded adherends apart. The stress distribution is complex; the cleavage forces are affected by the adherend mechanical properties, the test specimen dimensions and the adhesive strength properties. Joint is much weaker than in shear or tension.

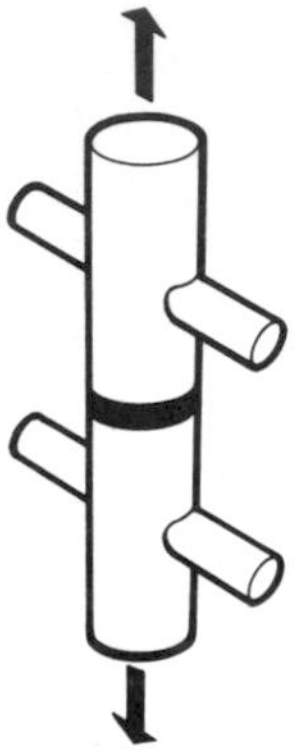

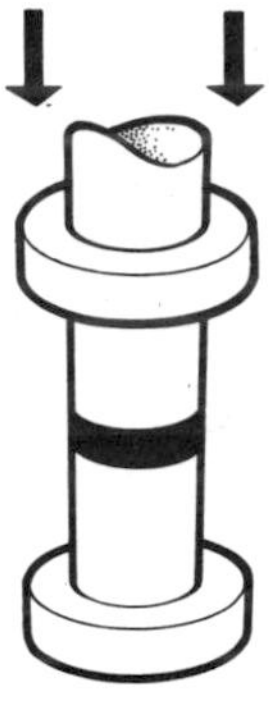

continued

Table 8.1 continued

Strength Property	Test Specification	Design of Test Specimen (Direction of Stress)	Adherend materials	Remarks
Shear		15 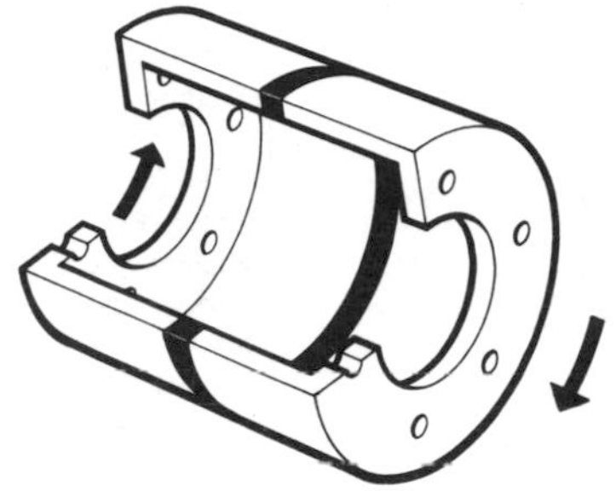	Rigid materials	Pure shear strength and shear modulus are of interest to the design engineer but are not easily determined. Simple lap–shear joints, when stressed parallel to the glue-line produce a shear in this plane, together with additional stresses due to stretching and bending of the adherends. The 'Napkin-Ring' type of test specimen (Joint 15) shown here relies on an even distribution of the almost pure shear resultant when the joint is subjected to torsion as the bonded cups are twisted to failure. Some tensile stresses are applied to the test specimen. Glue-line thickness is difficult to control during joint fabrication.
Shear Shear		16 17	Rigid materials	The specimen shown as Joint 16 is used in the Shawbury Torsion test equipment developed by the Rubber and Plastics Research Association (RAPRA). When test specimens are carefully machined and prepared torsional shear strengths are reproducible to within 2–3% (to at least 12 000 N/cm^2). The technique is potentially one for the measurement of the fundamental adhesive strength and is particularly useful for comparative evaluation work with adhesives. An alternative test specimen (Joint 17) consists of washers bonded together. Shear modulus is determined by taking advantage of the extra movement gained from multiple glue-lines. Care is needed to control glue-line thickness and a sensitive extensometer is essential for the determination of shear modulus.

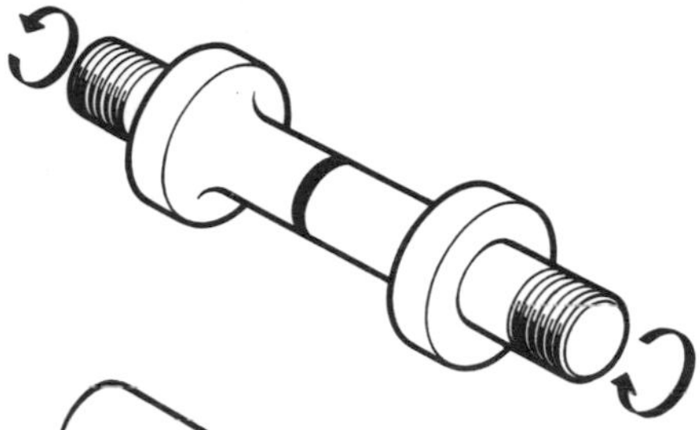

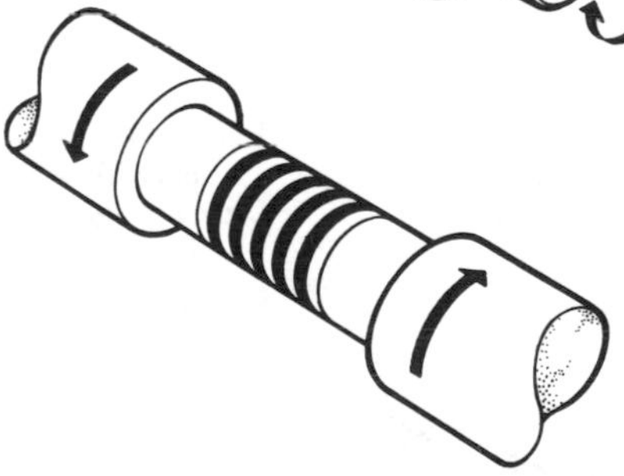

continued

260

Table 8.1 continued

Strength Property	Test Specification	Design of Test Specimen (Direction of Stress)	Adherend materials	Remarks
Peel	ASTM D903–49	18 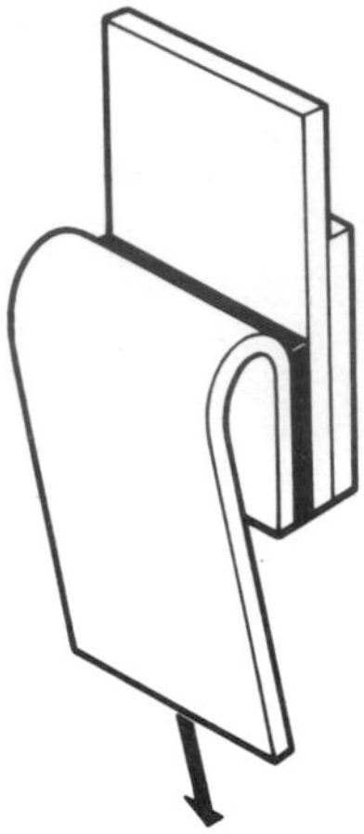	Flexible materials fastened to rigid or flexible materials. Plastics, rubbers, fabrics, leather, etc., to themselves or metals, wood, glass	The flexible member is pulled at an angle of 180°C from the other adherend which may be supported by a rigid backing plate. The test is useful for flexible materials requiring only a small stress to fold them back on themselves with a minimum radius at the bend point.
Peel	ASTM D1876–61T	19 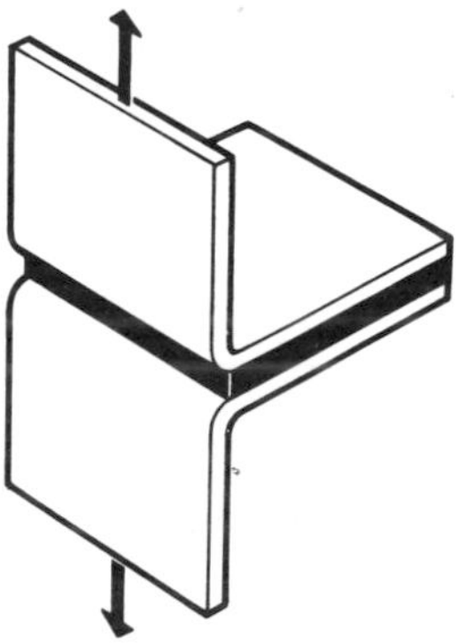	Materials slightly less flexible than those employed for 180° peeling. Plastic–rubber composites, leather, metal foils	Referred to as the 'T' peel test. Tensile loading is applied to separate two adherend strips bonded along three-quarters of their length. The unbonded ends are bent through 90° to the glue-line plane. The test is particularly suited to high peel strength adhesives. Peeling rates (for 2·54 cm wide specimens) range from 1–50 cm/min but 25·4 cm/min is commonly employed.
Peel	ASTM D773–47	20 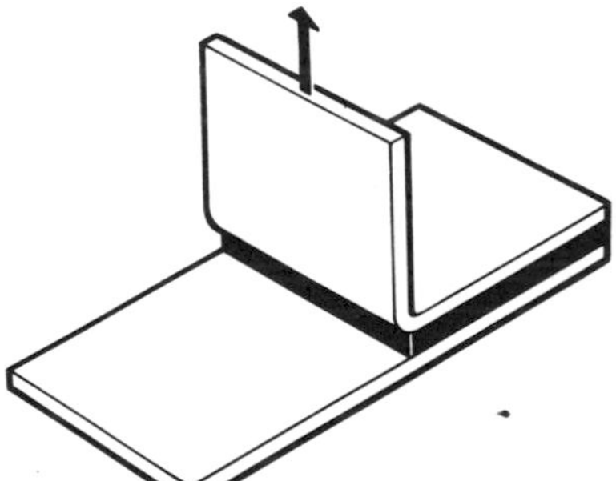	Rubber from metal or glass plates	The flexible member is peeled, by tensile loading, from a rigid adherend at a 90° angle. The test is often employed for pressure-sensitive tapes evaluation.

continued

Table 8.1 continued

Strength Property	Test Specification	Design of Test Specimen (Direction of Stress)	Adherend materials	Remarks
Peel	ASTM D1781–62	21	Laminated materials having a member flexible enough to bend over the drum and resistant to peeling forces. Metal laminates, honeycomb sandwich structures where the core is bounded by metal skins	Referred to as the 'climbing drum' technique. The peeling torque is the resultant of the bond breaking stresses and the forces responsible for bending the flexible adherend. Peeling rates vary from 7–13 cm/min.
Flexural	ASTM D1184	22	Laminates based on metals, wood, reinforced plastics. Honeycomb sandwich structures where the core is bounded by metal skins	The laminated or sandwich specimen is supported at, or near, the ends and loaded from the top of the centre (sometimes with 2-point loading).

8.3. ASSESSMENT OF DURABILITY AND STRENGTH PARAMETERS

8.3.1. FATIGUE

Fatigue testing refers to the repeated application of a specified load or deformation on a bonded specimen. Tests may be conducted under static or dynamic conditions (or both separately if necessary) according to the data required to evaluate an adhesive under service conditions.

Static fatigue properties are determined by measuring the maximum loading sustained by an adhesive over a given time. Various weight loadings applied to shear or tensile specimens provide a measure of the time required for bond failure.

Dynamic fatigue properties are measured by cycling test specimens with specified minimum to maximum stress loading for a given period or number of cycles or until failure. Cycle frequencies usually vary within the range 5 000 to 10^7 Hertz. In addition to frequency, fatigue life is determined by amplitude, temperature and mode of stressing; these variables must be specified along with the extent of loading. These tests do not determine the damping properties or elastic moduli of adhesives.

8.3.2. CREEP

There is no standard test for measuring the distortion or dimensional change in a bonded specimen under sustained loading (Creep). Deformation of the adhesive is generally measured by noting the dimensional change occurring when a bonded specimen is subjected to a constant load for a specified time and temperature. Room temperature creep is known as 'cold flow'. Higher temperatures usually increase the rate of creep significantly.

Creep tests are often carried out to determine joint deformation when stressed below the failing load required to break the bond. Joints may be loaded by springs (ASTM D2294–64T) or dead weights (MIL-A-5090E) to maintain constant loading in a specified environment. Optical measurement of the shift in scribed reference lines on a lap joint edge is a useful method of creep assessment. Alternatively, the relaxation, or the ability of an adhesive to restore to its former state, may be optically determined on removal of stress.

Rigid thermosetting adhesives display little or no creep under stress in contrast to thermoplastic or plasticised adhesives. Prolonged stressing of thermoplastic adhesives always reduces bond strength.

8.3.3. FLEXURAL STRENGTH

The shear strength of beams composed of adhesive laminated strips may be determined by flexural loading. The load is applied to the mid-span to develop maximum shear stress and delamination in the centre layer of adhesive. The method gives higher shear strengths than are obtained with tensile or compressive shear specimens because the resistance of the adhesive to shear failure is increased by the compressive loading normal to the glue-line.

8.3.4. PEEL STRENGTH

Peel tests involve complex stress distributions. Peel strengths vary with the speed of testing (particularly with low modulus adhesives) and the forces needed to start and sustain peeling action are determined by the physical properties of the adherends, test specimen geometry (adherend thickness and width), and the adhesive strength characteristics. Peel strength increases with adherend thickness and adhesive thickness but decreases with adhesive modulus of elasticity. Steel adherends give higher peel strengths than aluminium adherends of similar thickness. Low peel strengths are usually a feature of brittle adhesives with high tensile strengths. Peeling rates of 15·2 cm/min for adherend widths of 2·54 cm are commonly specified.

Extension of a flexible member during peeling leads to inaccurate results; the stretching forces are included in the test readings where the line of contact (at which bond failure occurs) becomes reduced. Nevertheless, the imperfect peeling action may closely simulate an actual assembly under stress, and provide a more meaningful test result.

N.B. Tensile strength and shear strength are discussed on pages 257 and 258 respectively. Shear stresses are also covered in Chapter 2, page 19.

8.3.5. DURABILITY

The test specimens described previously may be used to determine the effects of various adverse environments on an adhesive bond although no single test (or series) exists which will enable the user to predict its service life. A suitable test will provide information on the permanence of the bond when it is exposed to deteriorating circumstances such as temperature changes leading to oxidation, thermal degradation or softening of the adhesive. Other destructive hazards include low temperatures, sunlight and radiation, water, chemical reagents, oils and biodeterioration. Test specimens and procedures should be selected to simulate the type of service conditions envisaged for the bonded assembly. Consideration must be given to the adherend material for certain tests, e.g. for the evaluation of the acid resistance of an adhesive, certain metals would be unsuitable as adherends. Some of the environments which often provide the basis for unfavourable long-term conditions of exposure for adhesive bonds are discussed here.

Temperature

Adhesives and adherends are affected by high, low and varying temperatures. Elevated temperatures may decompose adhesive materials by oxidation or thermal degradation. Long exposure to moderate temperatures often leads to polymerisation changes in adhesives. Displacement of bonded surfaces occurs where high or low temperatures accentuate differences between the thermal coefficients of expansion for adherends and adhesives; stresses are set up at the interface which influence bond strength. Low temperatures embrittle many adhesives causing a reduction in their peel and cleavage strengths.

Test chambers with heating or cooling units can be employed for environment simulation. Adhesive durability is best determined at the service temperature; higher test temperatures should be regarded for their comparative value only. Destructive temperature effects may become apparent in hours or years.

Weathering

The long term ageing or weathering properties of bonded structures are difficult to predict since there are no standard short term permanence tests. Actual long time weathering is usually a reliable guide to adhesive durability although variations in exposure conditions over long test periods can make data difficult to interpret. Rainfall, humidity and temperature vary widely with locality. Accelerated weathering tests designed to reduce the long exposure periods are useful if the results can be correlated with actual weathering. Several tests have been adopted in the U.K. for providing a uniform testing procedure for military equipment. Referred to as I.S.A.T. (Intensified Standard Automating Trials), these tests are claimed to be equivalent to prolonged storage under existent weather conditions. A typical cycle test representative of 9 wk continual tropical storage is the following:

I.S.A.T. Series B Cycle

    46°C, 95% R.H., 2 d; 60°C, 60% R.H., 1 d;
    Cooled to ambient temperature 1 d; 75°C,

without moisture addition 8 h; Cooled to ambient temperature 16 h; 46°C, 95% R.H., 1 d; Cooled to ambient temperature 1 d.

The 1 wk cycle is repeated as often as desired.

Other cyclic tests have been developed which incorporate temperature, sunlight and moisture changes. One such cycle which resembles weather cycling without the duplication of any specific weather cycle is described here,

Light, corresponding to sunlight, 2 h; Rain 5 min; Dry heat at 38°C to complete 24 h cycle.

Many factors influence test results and these must be assessed before testing. Adherends, adhesive and exposure period need to be appraised before cyclic testing. Thus, thick porous adherends may require longer exposures than thin ones where moisture absorption is the adhesive hazard. For metal adherends the adhesive layer may not be exposed to moisture; a cyclic test involving moisture would be of dubious value.

Chemical

The permanence of a bond may be affected by exposure to external chemical agents or by the latent chemical reactivity of the adhesive for an adherend.

Several tests have been specified to evaluate adhesive bond strength on exposure to reagents such as acids, alkalies, water, sea water, petrol, organic solvents and lubricating oils. The deterioration in adhesion is sometimes dependent on reagent concentration. Temperature and exposure period should also be considered as test factors. Other tests are concerned with the effects of atmospheric constituents which are known to cause adhesive deterioration, e.g. salt spray or ozone.

Chemical constituents in the adherends, such as plasticisers, can migrate into the adhesive and destroy adhesion. Additionally, the by-products of an adhesive curing reaction may attack the adherend at the interface and cause loss of adhesion.

Biological

Certain adhesive types based on natural products such as casein, cellulose, dextrine or protein, etc., are subject to attack by bacteria, fungi, insects and rodents. Tests are available to check the effectiveness of preservative agents for adhesive formulations otherwise subject to biodeterioration.

Radiation

The effects of light, either artificial or natural, on bonded glass or optical assemblies, involving transparent or translucent materials, may be important. Adverse effects include the loss of adhesive strength and the discoloration of the glue-lines following photochemical changes in the adhesive. Light is unlikely to present a hazard for impervious adherend structures but is often an important factor with glass and transparent or translucent plastics.

Nuclear radiation is known to effect structural changes in high molecular weight polymers but has been scarcely studied for adhesive systems. The advent of nuclear technology and space research can be expected to produce test methods for this type of environment soon. An extensive literature on radiation induced changes in polymers provides a basis for studying adhesive performance. Some selected references appear at the end of the section.

8.3.6. NON-DESTRUCTIVE TESTING

The objective of these methods is to determine whether an adhesive bond is satisfactory without destroying it. Such tests do not determine the joint strength directly, but rely on a previously established correlation between bond strength and some property of the bonded system to indicate whether a given bond is adequate.

Properties such as dielectric constant, dynamic moduli, viscous moduli, thermal conductivity, and the ability to transmit sound or radiation, may be measured by non-destructive techniques and used to differentiate between bonds of high and low strength. The methods are of value in indicating potential joint performance and have the advantage that complete inspection of bonded assemblies can be carried out. Non-destructive testing procedures are the subject of much recent research and many of the techniques still need to be established as reliable means of checking bond performance. Some of the methods adopted for the inspection of assemblies are as follows.

Sonic testing

Sometimes referred to as tapping, the method relies on the variation in the audible sound emitted when bonded areas are tapped. Areas containing voids emit a different sound.

Ultrasonic testing

Similar to sonic testing except that the methods employ high frequency sound with instrumental detection of frequency changes.

Visual inspection

The appearance of the bond and whether there is a sufficiency of adhesive at the joint edges, or between adherends (transparent), determines its adequacy. The technique might include spot checks with a micrometer on glue-line thickness over a bonded area. More recently, photoelastic properties of adhesives and joints have been studied by observing stress patterns in transparent bonds with polarised light. Variations in the colour pattern produced by liquid cholesterol crystals on a substrate are dependent on the heat transmission properties of the joint. Certain colour areas correspond to deficiency of adhesive and may be detected visually.

Dielectric constant
Change in the electrical capacitance of the adhesive as the test specimen is stressed, has been used to determine dielectric constant.

Proof loading
The bonded structure is subjected to a greater load than it will sustain in service to determine its reliability. The proof loadings are restricted to values which do not cause structural deformation.

Core sampling
The method is not strictly non-destructive in that it depends on the removal of small annular sections from a bonded assembly for destructive testing. The

test sections are too small to affect the structural integrity of the remaining assembly.

Thermal infra-red inspection (TIRI)
Detection of internal voids and flaws in the glue-line is carried out by radiometer measurement of infrared emission from adjacent adherend surfaces, following the injection of thermal energy into the bond. Variations in volume heat diffusivity in the region of flaws, give rise to a time-varying temperature gradient sensed as an infra-red emission from the surface.

8.3.7   STANDARD TEST METHODS

Table 8.2.   INDEX TO STANDARD TEST METHODS

Property under test	Title	Test Specification Number		
		ASTM	*BSI*	*Other*
MECHANICAL STRENGTH Shear	Method of test for shear strength and shear modulus of structural adhesives.	E229–63(T)		
	Method of test for adhesion of vulcanised rubber to wire cord.	D2229–63(T)		
	Method of test for static adhesion of textile to rubber (H-pull test).	D2138–67		
	Methods of tests for adhesion of vulcanised rubber to single-strand wire.	D1871–67(T)		
	Method of test for bond strength of ceramic tile to Portland cement mortar.	C482–64(T)		
	Method of test for wet adhesion of thermal insulating cements.	C383–58		
	Recommended practice for determining strength development of adhesive bonds.	D1144–57		
	Method of conducting shear-block test for quality control of glue bonds in scarf joints.	D1759–64		
	Method of test for strength properties of metal-to-metal adhesives by compressive loading (disc shear).	D2182–63(T)		
	Method of test for strength properties of adhesives in two-ply wood construction in shear by tension loading.	D2339–65(T)		
	Testing of sole adhesives			*DIN 53273: 1953
	Determination of the bond strength of single shear lap joints under tensile stress.			†DIN 53283
	Method of test for comparing concretes on the basis of the bond developed with reinforcing steel.	C234–62		
	Method of test for strength properties of adhesives in plywood type construction in shear by tension loading.	D906–64		
	Method of test for strength properties of adhesives in shear by tension loading (metal-to-metal).	D1002–64		
	Method of test for strength properties in shear by compression loading.	D905–49		
	Testing of wood adhesives and glued wood joints (determination of breaking strength of lap joints under longitudinal shear.)			†DIN 53254
	Shear test in flatwise plane of flat sandwich construction of sandwich cores.	C273–61		
	Test for shear fatigue of sandwich core materials.	C394–62		

*English translation from SATRA, Kettering   †English translation from BSI, London   *continued*

Table 8.2 continued

Property under test	Title	Test Specification Number		
		ASTM	*BSI*	*Other*
Tensile	Method of test for tensile properties of adhesive bonds.	D897–49		
	Method of testing cross-lap specimens for tensile properties of adhesives.	D1344–57		
	Spec. for adhesive for acoustical materials.	D1779–65		
	Method of test for tensile strength of adhesives by means of bar and rod specimens.	D2095–62(T)		
	Testing of wood adhesives and glued wood joints.			†DIN 53257: 1964
	Determination of tensile strength of butt joints.			†DIN 53288
	Testing of adhesives for metals and of bonded metal joints.			
	Preparation and testing of the adhesion of vulcanised rubber to metal where the rubber is assembled to two metal plates.			ISO DR615: 1963
	Method of test for adhesion of dried thermal insulating cement.	C353–56		
	Spec. and tests for adhesives for fastening gypsum wallboard to wood framing.	C557–67		
	Methods of test for adhesion of vulcanised rubber to metal.	D429–64		
	Tension test of flat sandwich constructions in flatwise plane.	C297–61		
Peel	Preparation and method of test of the adhesion of vulcanised rubber to metal where the rubber is assembled to one metal plate.			ISO DR614: 1963
	Testing of metal adhesives and bonded metal joints. T-peel test with angle test piece.			DIN 53282: 1965
	Testing of sole adhesives.			*DIN 53274: 1953
	Testing for evaluating peel strength of shoe sole-attaching adhesives.	D2558–66(T)		
	Method of test for pressure-sensitive tack of adhesives.	D1878–61(T)		
	Method of test for peel resistance of adhesives. (T-peel test.)	D1876–61(T)		
	Testing of adhesives for metals and of bonded metal joints.			†DIN 53289
	Climbing drum peel test for adhesives.	D1781–62		
	Method of test for peel or stripping strength of adhesive bonds.	D903–49		
	Methods of test for adhesion of vulcanised rubber (friction test).	D413–39		
	Method of test for adhesion ratio of polyethylene film.	D2141–63(T)		
	Determination of adhesion of vulcanised natural or synthetic rubbers to textile fabrics.		BS 903: Pt A12: 1958	ISO/R36: 1957
	Method of determination of rubber-to-metal bond strength.		BS 903: Pt A21: 1961	
	Test for delamination strength of honeycomb type core material.	C363–57		
	Test for adhesiveness of gummed tape.	D773–47: 1961		
Cleavage	Method of test for cleavage strength of metal-to-metal adhesive bonds.	D1062–51		
Impact	Method of test for impact strength of adhesive bonds.	D950–54		

*English translation from SATRA, Kettering   †English translation from BSI, London   *continued*

Table 8.2 continued

Property under test	Title	Test Specification Number		
		ASTM	BSI	Other
Compression	High alumina cement. Method of test for cold bonding strength of air-setting refractory mortar (wet type).	C198–47	BS 915 : 1947	
	Portland cement (ordinary and rapid hardening).		BS 12 : 1958	
	App. C. Test for compressive strength of cement using mortar cubes (test m/c as BS 1881).			
	App. D. Test for compressive strength of cement using concrete cubes.			
	App. H. Optional test for 1-day tensile strength of rapid hardening Portland cement.			
	Portland – blastfurnace cement.		BS 146 : 1958	
	App. C. Test for compressive strength using mortar cubes.			
	Compression test using compression testing m/c complying with Clause 58, BS 1881 'Methods of testing concrete'.			
	App. D. Test for compressive strength of cement using concrete cubes (as above).			
	Tests for flatwise compressive strength of sandwich cores.	C-365–57		
	Test for edgewise compressive strength of flat sandwich constructions.	C-364–61		
Flexural	Method of test for shear strength of adhesive bonds on flexural loading	D1184–55		
	Test for bond strength of electrical insulating varnishes by the helical coil test.	D2519–16(T)		
	Test for bond strength of mortar to masonry units.	E149–66		
	Test for flexure-creep of sandwich constructions.	C480–62		
	Flexure test of flat sandwich constructions.	C393–62		
Creep	Method of test for creep properties of adhesives in shear by compression loading (metal-to-metal).	D2293–64(T)		
	Recommended practice for conducting creep tests of metal-to-metal adhesives.	D1780–62		
	Method of test for creep properties of adhesives in shear by tension loading (metal-to-metal).	D2294–64(T)		
Surface coatings	Method of testing films deposited from bituminous emulsions.	D466–42		
	Method of test for adhesion of coatings of paint, varnish lacquer and related products (scrape-adhesion).	D2197–67(T)		
	Recommended practice for scratch-adhesion testing of cathode coatings.	F32–63(T)		
Various strength tests	Methods of testing rubber cements. (Prepn. surfaces, shear, tension, stripping method (friction test), cold brittleness (bending.)	D816–55		
	Testing veneer, plywood, and other glued veneer constructions. (Compression, tension and shear strengths.)	D805–63		
	Methods of testing adhesives for brake lining and other friction materials. (Prepn. of surface, tension, shear, shelf life, ultrasonic non-destructive adhesion test.)	D1205–67		
	Methods of test for wood chipboards and other particle boards. (Flexural, creep, impact, tensile.)		BS 1811 : 1961	
	Testing of sole adhesives. (Characteristics of materials to be bonded, adhesives and adhesive procedure.)			*DIN 53272 : 1953

*English translation from SATRA, Kettering

continued

Table 8.2 continued

Property under test	Title	Test Specification Number		
		ASTM	BSI	Other
DURABILITY Weathering	Adhesives for structural laminated wood products for use under exterior (wet use) exposure conditions.	D2559–66(T)		
	Method of test for bonding permanency of water – or solvent – soluble liquid adhesives for automatic machine sealing top flaps of fibreboard specimens. (Peel)	D1713–65		
	Method of test for bonding permanency of water – or solvent-soluble liquid adhesives for labelling glass bottles.	D1581–60		
	Recommended practice for atmospheric exposure of adhesive-bonded joints and structures. (Tentative)	D1828–61(T)		
	Method of test for resistance of adhesives to cyclic laboratory ageing conditions.	D1183–61(T)		
	Standing atmospheres for conditioning and testing materials.	E171–63		
Ageing	Methods of test for working life of liquid or paste adhesives by consistency and bond strength.	D1338–56		
	Method of test for storage life of adhesives by consistency and bond strength. (Strength as D906 and D1002.)	D1337–56		
	Test for laboratory ageing of sandwich constructions.	C481–62		
Temperature	Testing of sole adhesives			*DIN 53271: 1960
	Test for strength properties of adhesives in shear by tension loading in temperature range −267·8 to −55°C (−450 to −57°F).	D2557–66(T)		
	Method of test for strength properties of adhesives in shear by tension loading at elevated temperatures (metal-to-metal).	D2295–64(T)		
	Method of test for effect of moisture and temperature on adhesive bonds.	D1151–61		
Water	Testing of wood adhesives and glued wood joints.			†DIN 53258: 1964
	Test for integrity of glue joints in structural laminated wood products for exterior use.	D1101–59		
	Plywood manufactured from tropical hardwoods.		BS 1455: 1963	
	Method of test for water absorptiveness of fibreboard specimens for adhesives.	D1714–65		
	Method of test for water absorptiveness of paper labels.	D1584–60		
Radiation	Recommended practice for exposure of adhesives specimens to high-energy radiation.	D1789–61(T)		
	Recommended practice for determining the effect of artificial (carbon-arc-type) and natural light on the permanence of adhesives.	D904–57		
Chemical reagents	Method of test for resistance of adhesive bonds to chemical reagents.	D896–66		
	Method of test for bond strength of chemical-resistant mortars.	C321–64		

*English translation from SATRA, Kettering   †English translation from BSI, London

continued

Table 8.2 continued

Property under test	Title	Test Specification Number		
		ASTM	*BSI*	*Other*
Biodeterioration	Method of test for permanence of adhesive-bonded joints in plywood under mould conditions.	D1877–61(T)		
	Method of test for effect of bacterial contamination on permanence of adhesive preparations and adhesive bonds.	D1174–55		
	Method of test for effect of mould contamination on permanence of adhesive preparations and adhesive bonds.	D1286–57		
	Test for susceptibility of dry adhesive films to attack by roaches.	D1382–64		
	Test for susceptibility of dry adhesive films to attack by laboratory rats.	D1383–64		
ELECTRICAL PROPERTIES	Methods of testing adhesives relative to their use as electrical insulation.	D1304–60		
	Method of test for bond strength of plastics and electrical insulating materials. (Ply adhesion of sheet plastic.)	D952–51		
SPECIFIC ADHESIVE TESTS	Methods of sampling and testing glues (bone, skin, and fish glues).		BS 647: 1959	
	Synthetic resin adhesives (phenolic and aminoplastic) for plywood.		BS 1203: 1963	
	Glue size for decorators' use.		BS 3357: 1961	
	Methods of sampling and testing vegetable adhesives.		BS 844: 1965	
	Low heat Portland cement (compressive strength).		BS 1370: 1958	
	Methods of test for polyvinyl acetate adhesives for wood.		BS 3544: 1962	
	Adhesives based on bitumen or coal tar. (Flexural, Peel strengths for Kraft paper.)		BS 3940: 1965	
	Paper-hanging pastes and powders. (Peel test for wallpaper.)		BS 3046: 1958	
	Cold-setting casein glue for wood. (Shear tests.)		BS 1444: 1948 (Confirmed 1957)	
MISCELLANEOUS Preparation of test specimens	Recommended practice for preparation of bar and rod specimens for adhesion tests. (Prepn. of adherends.)	D2094–62(T)		
	Testing of adhesives for metals and bonded metal joints. Sheet 1 – Preparation of surfaces to be joined. Sheet 2 – Method of manufacture. Sheet 3 – Indication of data relating to bonding process.			†DIN 53281
	Methods of sample preparation for physical testing of rubber products.	D15–66(T)		
Non-destructive tests	1967 Book of ASTM Standards, Part 16, pp 869–881. Proposed method of inspection of adhesive bonded structures utilising the Fokker Bond Tester. (Void detection in adhesive bond lines and adherends by ultrasonic testing.)			
Additional tests	Method of test for penetration of adhesives.	D1916–61(T)		
	Methods of test for ply adhesion of paper.	D825–54		
	Method of test for ply separation of combined container board (visual).	D1028–59		

†English translation from BSI, London

continued

Table 8.2 continued

Property under test	Title	Test Specification Number		
		ASTM	*BSI*	*Other*
	Method of test for density of adhesives in fluid form.	D1875–61(T)		
	Method of test for applied weight per unit area of dried adhesive solids.	D898–51		
	Method of test for applied weight per unit area of liquid adhesive.	D899–51		
	Method of test for tack-free time of oil and resin-based caulking compounds.	D2377–65(T)		
	Method of test for flow properties of adhesives (flow during curing or setting).	D2183–63(T)		
	Method of test for consistency of adhesives.	D1084–63		
	Method of test for filler content of phenol, resorcinol and melamine adhesives.	D1579–60		
	Methods of test for non-volatile content of aqueous adhesives.	D1489–60		
	Method of test for non-volatile content of phenol, resorcinol and melamine adhesives.	D1582–60		
	Method of test for hydrogen ion concentration of dry adhesive films.	D1583–61		
	Wall tiling. (Cohesive and shear strengths.)		CP 212	
	Method of test for measuring the volume resistivity of conductive adhesives.	D2739–68(T)		

ADDITIONAL STANDARDS

Animal glue for wood, BS 745:1969.
Adhesives for packaging, BS 1133, Section 16, (1968).
Synthetic resin adhesives (phenolic and amino plastic) for wood, BS 1204, Part 1: 1964, *Gap filling adhesives*, Part 2: 1965, *Close-contact adhesives.*
Cleaning and preparation of metal surfaces, CP 3012: 1972.
BS *Methods of test for adhesives* (in course of preparation)

    Method 1.1   *Adherend preparation*
    Method 2.4   *Pot life*
    Method 2.8   *Determination of solids content*
    Method 3.5   *Longitudinal shear*
    Method 3.7   *Sustained loading and creep resistance*
    Method 5.2   *Guide to statistical analysis*

British Standards are obtainable from: BSI, Sales Department, 101 Pentonville Road, London N1 9ND.

ASTM Standards may be obtained from: Heyden and Sons Ltd., Spectrum House, Alderton Crescent, London NW4.

Adhesives: methods of testing, U.S. Federal Standard 175a. (A review of applicable ASTM methods but includes additional test methods, e.g. longitudinal shear fatigue test.)
Adhesives, heat resistant, airframe structural, metal-to-metal, MMM–A–132.
Adhesives, epoxy resins, metal-to-metal structural bonding, QPL–MMM–A–134–2.

REFERENCES

GENERAL TESTING

DELAY, L. J., 'Testing methods to evaluate adhesive performance', *ASME Paper* 66–MD–46, May (1966).

NEUSS, W. H., (Ed), *Testing of Adhesives*, Tappi Monograph Series, No. 26, (1963).

YURENKA, S., 'How to test structural adhesives', *Adhes. Age*, **4** (40), 36, (1961).

VAN RAALTE, A. O., 'Testing of adhesives and bonded joints', *Sh. Metal Inds*, **42** (462), 743, (1965).

MARKIN, YO. I. and VOYUTSKII, S. S., *Ind. Lab.*, **29** (7), 867, (1963).

FORT, R. J. and SHELDON, R. P., 'Apparatus for the testing of low strength adhesives', *J. scient. Instrum.*, **40** (5), 264, (1963).

BULL, R. F., *et al.*, 'Measuring tack using rotating drum technique', *Adhes. Age*, **11** (5), 20, (1968).

POLYAKOV, YO. N., 'A method for determining the adhesion of cast plastics and foam plastics to various building materials', *Ind. Lab.*, **31** (10), 1564, (1966).

VOYUTSKII, S. S. and MARKIN, YO. I., 'Measurement of polymer—metal adhesion', *Ind. Lab.*, **28** (10), 1280, (1963).

HARDY, A., 'Peeling adhesion', *Aspects of adhesion 1st Proc. Conf.*, 47, (1963).

GRIMALDI, A. C., 'Dynamic testing of bonded joints for use under severe vibrational stress', *ASME Paper* 67—Vibr-38, (1967).

HUMPIDGE, R. T. and TAYLOR, B. J., 'Apparatus for measuring the shear strength of adhesive joints at high temperatures and methods for constructing the adhesive joints', *J. scient. Instrum.*, **44** (6), 457, (1967).

RUTHERFORD, J. L., *et al.*, 'Capacitance methods for measuring properties of adhesives in bonded joints', *Rev. scient. Instrum.*, **39** (5), 666, (1968).

SMITH, D. F. and CAGLE, C. V., 'Ultrasonic testing of adhesive bonds using the Fokker bond tester', *Mater. Evaluation*, **24** (7), 362, (1966).

KUTZSCHER, E. W., ZIMMERMAN, K. H. and BOTKIN, J. L.,

'Thermal and infrared methods for non-destructive testing of adhesive-bonded structures', *Mater. Evaluation*, **26**, 143, (1968).

BROWN, S. P., 'Cholesteric crystals for non-destructive testing', *Mater. Evaluation*, **26**, 163, (1968).

SCHLIEKELMANN, R. J., 'Non-destructive testing of adhesive bonded metal structures', Pts. 1 and 2, *Adhes. Age*, May, **7**, (5), 30, 33, (1964).

THOMPSON, W. and GOSLING, A. T., 'A non-destructive test for adhesive bonded joints', Min. of Aviation Aeronautical Inspection Directorate Rept. AID/NDT 1712, 12, (1964).

GONZALEZ, H. M. and CAGLE, C. V., 'Non-destructive testing of adhesive-bonded joints. Symposium on Adhesion', *ASTM Spec. Tech. Publ.*, No. 360, (1964).

EIRICH, F. R. and DIETZ, A. G. H., (Ed.), *High-Speed Testing: Symposium* Vol. 1–5, Interscience, (1960–66).

GARDON, J. L., 'The variables and interpretation of some destructive cohesion and adhesion tests', *Treatise on Adhesives and Adhesion*, (Patrick, Ed.), E. Arnold, (1967).

BENSON, N. K., 'Mechanical testing of bonded joints', *Adhesion and Adhesives*, (Houwink and Salomon, Eds.), Vol. 2 Elsevier, (1967).

LUNSFORD, L. R., 'Bonded metal-to-metal shear testing in Symposium on Adhesion and Adhesives', *ASTM Spec. Tech. Publ.*, 271 and in 'Symposium on Shear and Torsion Testing', *ASTM Spec. Tech. Publ.*, 289, (1961).

YETTITO, P. R., 'A Thermal, infrared inspection technique for bond-flow inspection', *Applied Polymer Symposia* No. 3, 435 Interscience, New York, (1966).

MIXER, R. Y., 'Adhesives', *Radiation Effects on Organic Materials* (Bolt and Carroll, Eds.), Ch. 10, Academic Press, New York, (1963).

ASTM Stand., 'Structural Sandwich Constructions', *Wood and Adhesives*, Pt 16, (1968).

U.S. Military Standards MIL-STD-401A. *Sandwich Construction and Core Materials, General Test Methods*.

CLARKE, T. C., 'Non-destructive testing of adhesively bonded metal assemblies', (Survey article), *Mater. Eval.*, **29A**, 25, (1971).

GLEDHILL, R. A. and KINLOCH, A. J., 'Environmental failure of structural adhesive joints', *J. Adh.*, **6** (4), 315–330, (1974).

DUKES, W. A. and KINLOCH, A. J., 'Non-destructive testing of bonded joints—adhesion science viewpoint', *Non-Destr. T.*, **7** (6), 324–326, (1974).

SCHLIEKELMANN, R. J., 'Non-destructive testing of adhesive bonded metal-to-metal joints', *Non-Destr. T.*, **5**, 79–86, (1972), Part (1); **5**, 144–150, (1972), Part (2).

THOMPSON, D. O., THOMPSON, R. B. and ALERS, G. A., 'Non-destructive measurement of adhesive bond strength in honeycomb panels', *Materials Evaluation*, **32** (4), 81, (1974).

NORRISS, T. H., 'Non-destructive testing of bonded joints—control of adhesive bonding in production of primary aircraft structures', *Non-Destr. T.*, **7** (6), 335–339, (1974).

MACDONALD, N. C., 'Standard test methods for adhesives', *Adhes. Age*, **15** (9), 21–26, (1972).

MADELEY, G. G., 'Improved test for measuring adhesion of steel cord to rubber', *Text. I. Ind.*, **11** (9), 236–237, (1973).

WOLFF, R. V. and STOUT, R. J., 'Tapered overlap shear specimens used to determine bond strength', *Adhes. Age*, **14** (6). 34–37, (1971).

DELOLLIS, N. J., 'High strength versus stress relief in a structural bond', *Adhes. Age*, **14** (4), 22–25, (1971).

LEWIS, A. F., 'Stress endurance limit of lap shear adhesive joints', *Adhes. Age*, **15** (6), 38–40, (1972).

SWANSON, F. D. and GREGOSNIK, N. W., 'Evaluating adhesives by underground exposure in extreme environments', *Adhes. Age*, **14** (10), 18–25, (1971).

TILMANS, A. and KROKOSKY, E. M., 'Effect of radiation on adhesive properties of epoxy-metal and epoxy-glass adhints', *J. Mats.*, **6**, 482–486, (1971).

DURABILITY TESTING

SHARPE, L. H., 'Some aspects of permanence of adhesive joints'. *Appl. Polymer Symp.*, **3** (353), (1966).

KERR, C., et al., 'Effect of certain hostile environments on adhesive joints', *J. appl. Chem.*, **17** (3), 62, (1967).

WEGMAN, R. F., et al., 'How weathering and ageing affect bonded aluminium', *Adhes. Age*, **10** (10), 22, (1967).

CARTER, G. F., 'Outdoor durability of adhesive joints under stress', *Adhes. Age*, **10** (10), 32, (1967).

BARBARISI, M. and HALL, J. R., 'Effect of relative humidity on adhesive bonding of aluminium alloy', *Feltman Res. Labs. Tech. memo* PA-TM-1785 (19), Picatinny Arsenal, Dover, N.J., (1967).

FALCONER, D. J., et al., 'Effect of high humidity environments on strength of adhesive joints', *Chem. and Ind.*, **32**, 1230, (1964).

KINDERMAN, E. M., RADDING, S. B., et al., 'Nuclear Radiation effects on Structural Plastics and Adhesives Parts 1 and II'. *Wright Air Dev. Cen.*, Literature Survey, 230 (71), (1957).

GALLANT, P. E. and SWAFFER, C. S., *The assessment of structural bonds by destructive methods*, presented at City University, London, April, (1967).

'Adhesives for Durability and Permanence, Symposium on Testing', *ASTM Spec. Tech. Publ.*, 138, (1952).

SMITH, T., 'NDT techniques for prediction of adhesive failure loci prior to bonding', *Mater. Eval.*, **33** (5), 101, (1975).

PATRICK, R. L. and DOYLE, M. J., 'Energy criteria test methods for adhesives', *Abstract Paper ACS 1975* (169), 26, (1975).

MUNIHSAV, P., 'Rational test procedure for determining static and dynamic adhesion of brass-plated steel cord in rubber', *Kaut Gum* KU 29–31, Jan. 28 issue, (1975).

MINFORD, J. D., 'Evaluating adhesives for joining aluminium', *Metal Eng. Q.*, **12**, 48–53, (1972).

CLAD, W., 'Testing of adhesives for assembly gluing, *Holz Roh WE*, **31** (9), 329–337, (1973).

BEATTY, J. R., 'New dynamic rubber-to-metal adhesion tester, *Rubber Age*, **105** (9), 55–57, (1973).

QUERIDO, R. J., REINKS, K. J. and SCHLIEKELMANN, R. J., 'Quality control of adhesive bonded metal structures by means of holographic interference methods', *Plastica*, **25** (11), 477–483, (1972).

AZRAK, R. G., JOESTEN, B. L. and HALE, W. F., 'Physical property performance correlations in contact adhesive systems', *Abstract Paper ACS 1975*, (169), 10, (1975).

MITTAL, K. L., 'Surface chemical criteria for maximum adhesion and their verification against adhesive strength values', *Abstract Paper ACS 1975*, (169), 36, (1975).

IWAMOTO, M., et al., 'Observation of fracture and debonding under fatigue of adhesive joints', *Kobunsh Ron*, **32** (3), 137–141, (1975).

CHANG, M. D., et al., 'Fracture mechanics analysis of adhesive failure of a lap shear joint', *Exp. Mech.*, **12**, 34, (1972).

MOSTOVOY, S., 'Fatigue failure of adhesive joints', *Abstract Paper ACS 1975*, (169), 23, (1975).

MCCARVILL, W. T. and BELL, J. P., 'Torsional test method for adhesive joints', *J. Adh.*, **6** (3), 185–193, (1974).

LEWIS, A. F., *et al.*, 'Long term strength of structural adhesive joints', *J. Adh.*, **3** (3), 249–257, (1972).

YALOF, S., 'Tracking adhesive behaviour with dynamic dielectric spectroscopy', *Adh. Age*, **18** (4), 23–31, (1975).

ANON, 'How to test adhesive properties', *Mater. Des. Engng.*, **25**, 60–66, (1972).

More detailed information on testing methods may be found in the publications of the American Society for Testing and Materials and in the U.S. Federal Test Method Standard No. 175, *Adhesives: Methods of Testing*. Additional information on the use of statistical methods to determine the degree of correlation between sets of test results is available in the ASTM Manual on Quality Control of Materials STP–15C, March (1951).

CHAPTER 9

Adhesives Tables

9.1. INTRODUCTION

The following tables are presented to assist in the preliminary selection of suitable adhesive types for any particular application. They have been compiled from trade product data and other published information. It should be realised, however, that tabulations of adhesive types and properties are subject to limitations as a means of selection as invariably they fail to represent the interdependence between all essential properties. Furthermore, commercial adhesives based on a particular chemical class are subject to wide variations in properties and also there is a considerable overlap of properties between formulations in different classes. The designer must beware, therefore, of associating a definite set of properties with a single chemical composition and recognise that an alternative chemical type may be more suitable for other reasons. The tables have been included for their value where simple applications are concerned and for eliminating unsuitable types for more difficult problems. In the latter situation reference to the section on 'Adhesive Materials and Properties' (p. 30) together with the more specific data of representative materials in the 'Adhesive Products Directory' (p. 80) will assist in ultimate selection of adhesives for evaluation trials. A realistic analysis of adhesives performance data requires considerable experience and in cases of uncertainty the designer is strongly recommended to take advantage of advisory services offered by manufacturers and other organisations with staff qualified to give guidance on adhesives (see 'Sources of Information on Adhesives', p. 330). Any such consultation will be facilitated if the bonding problem is completely defined at the outset. The 'Check List of Factors for Adhesive Selection' (p. 29) will provide a basis for this and help to anticipate the technical questions the consultant may ask.

Table 9.1 shows whether adhesion is obtainable to a particular adherend by a code number system for the different types of adhesives listed in Table 9.2. For bonding one material to another the adhesive selection is made by locating the adhesive code number common to both materials. For example, a comparison of the rows for mica and glass indicates that two different adhesive types are feasible as bonding materials. In addition to adhesives materials, several solvents (or mixtures of these) are suitable for bonding plastics—these are listed in Table 9.3. The succeeding tables should then be used to make further selections on the basis of processing requirements and other physical properties.

Tables 9.8 to 9.26 present in histogram form the generalised bulk properties of the various adhesive types. These histograms will assist in indicating in a rough quantitative manner the advantages and limitations of each type. This is useful especially where the adhesive is required to have some secondary property such as electrical resistance, thermal conductivity etc. Again the possibility of modifying such properties (e.g. by the addition of fillers to enhance thermal conductivity) should not be overlooked.

9.2. PHYSICAL PROPERTIES OF ADHESIVE MATERIALS

Brief descriptions are given of the physical properties depicted in Tables 9.8 to 9.26 in histogram form, together with a reference table of factors for the conversion of non-metric and other units of measurement to SI units. ASTM (American Society for Testing and Materials) numbers given are from *Modern Plastics Encyclopaedia* McGraw Hill New York (1968) or from *ASTM Standards,* Part 16 June 1968.

Tensile strength ASTM D638–61T
Related to modulus of elasticity; it is force per unit area required to break a standard specimen; plastics specimen typically 3·2 mm thick by 12·5 mm wide. In tensile testing, the elongation of a specimen is the increase in test specimen length, after rupture, referred to the original length. It is reported as percentage elongation. The elastic modulus (tensile modulus or modulus of elasticity) is the ratio of stress to strain below the proportional limit of the material.

Compressive strength ASTM D695–63T
The force required to produce a given percentage decrease in height of a specimen under compressive load. The specimens are usually either:
  (a) Prisms 12·5 mm wide, 25·4 mm high, or
  (b) right cylinders 12·5 mm diameter, 25·4 mm high.

Flexural strength ASTM D790–63
The force in Newtons required to bend a specimen such that the elongation of the outermost face of the arc is 5%. The specimen is typically 127 mm long by 12·5 mm wide by 3·2 mm thick; the force is applied at the centre of the specimen.

Izod impact strength ASTM D256–56
Calculated from the loss of energy of a pendulum after breaking a standard sample. The sample is typically 50 mm by 12·5 mm by 3·2 mm and a notch is cut in the narrow face to localise the impact so that the break occurs through the notch. The results are calculated on the basis of a sample 25·4 mm (1 inch) thick rather than the 3·2 mm usually used. *N.B.* some substances are 'notch' sensitive and register erroneously low impact strengths.

Tensile impact strength (Charpy) ASTM D1822–61T
A relatively new test where samples are swung under tension on the end of a pendulum to strike an anvil. The loss of energy of the pendulum is used to calculate the impact strength and, as no notch is used, realistic figures are obtained for all materials.

Heat distortion temperature/deflection temperature ASTM D648–56
Samples are 127 mm long by 12·7 mm wide with thickness in range 3·2–12·7 mm. The sample is placed on two supports 100 mm apart and a load of either 45·4 or 182 N/cm^2 applied to its centre. The temperature of the sample is increased gradually until the centre of the sample sags by 0·2 mm. This temperature is the deflection temperature.

Water absorption ASTM D570–63T
Disc specimens of diameter 50·8 mm and 3·17 mm thickness are dried 24 h in an oven at 50°C, cooled in a desiccator and weighed immediately. Water absorption is determined as a percentage gain in weight after specimen immersion for 24 h or longer in water at 23°C. Appropriate corrections are made for materials which lose soluble matter during immersion.

Thermal conductivity ASTM C177
Usually measured by Lee's disc method with samples 0·5–5·0 mm thick, it is the quantity of heat in joules passing through opposite faces of a one metre cube when the temperature difference between them is one degree Celsius.

Thermal expansion ASTM D696
This is usually recorded as the increase in length of a sample divided by the original length for each one degree Celsius of temperature rise.

Continuous heat-resistance temperature range
The continuous heat resistance temperature is the maximum temperature to which a material may be exposed continuously without a significant loss in its properties or performance in service (e.g. extreme heat can cause melting, thermal or oxidative degradation of the material and adversely affect adhesion). The chart also includes the sub-zero service temperature range for continuous exposure conditions, so that the histograms represent the temperature limits within which the materials may be usefully employed for adhesive purposes.

Dielectric constant and dissipation (power) factor ASTM D150–64T
This is determined by applying electrodes of known areas to either side of a thin plastics sheet and the capacitance and conductance measured on an A.C. bridge. The dimensions of the electrodes, sheet thickness and uniformity are important.

$$C \propto \frac{AE}{T}$$

Where C = capacitance, A = areas of electrode. E = dielectric constant and T = sheet thickness.

$$\tan \delta \propto \frac{G}{C}$$

where G = conductance
$\tan \delta$ = tangent of loss angle, or dissipation factor.

Dielectric strength ASTM D149–64
This is the voltage expressed in kV/cm required to pass a large current (exceeding 0·1 A) through a unit area of thin sheet.

Volume resistivity ASTM D257–61
This is usually measured by placing electrodes on either side of a flat sheet. It is the ratio between the applied voltage and the current passing with the values corrected to unit separation and unit areas of the electrodes.

Table 9.1 ADHESION OBTAINABLE TO PARTICU[LAR]

Adherend	Adhesive
METALS, ALLOYS	
Aluminium	18 19 20 21 22 . 24 25 26 27 . 30 31 3
Brass	22 . 25 . 28 . 32 . . 3
Bronze	22 . 25 . 28 . 32 . 3
Cadmium	
Chromium	21 . 25 . 31 . 3
Copper	22 . 25 . 27 28 . 31 32 . . 3
Germanium	
Gold	21 22 . 25 . 31 . 34 .
Lead	
Magnesium	18 19 20 21 22 . 24 25 26 27 28 29 . 31 . 34 . 3
Nickel	21 . 28 . 3
Platinum	
Silver	28 .
Steel (mild), iron	9 10 . 12 . 18 19 20 21 22 . 24 25 26 27 28 29 30 31 32 . . 3
Steel (stainless)	9 10 . 12 . 18 19 20 21 22 . 24 25 26 27 28 29 30 31 32 . 3
Tin	2 . 21 22 . 24 25 26
Titanium	2 . 21 22 . 24 25 26 27 28 . 31 .
Tungsten	22 .
Zinc	18 19 20 21 22 . 24 25 26 27 28 29 . 31 . 34 .
Uranium	
PLASTICS	
Acrylonitrile butadiene styrene	19 . 21 22 . 25 . 28
Allyl diglycol carbonate	
Cellulose	9 . 12 .
Cellulose acetate	9 10 . 12 . 16 . 21 22 . 28 29 30 31 . 34
Cellulose acetate—butyrate	9 10 . 12 . 16 . 21 22 . 28 29 30 31 . 34
Cellulose nitrate	12
Cellulose propionate	9 10 . 12 . 16 . 21 22 . 28 29 30 31 . 34
Diallyl isophthalate	24 . 28 .
Diallyl phthalate	24 . 28 .
Epoxy resins	19 20 21 22 . 25 . 28
Ethyl cellulose	12
Ethylene vinyl acetate	31 .
Furane	
Ionomer	
Methacrylate butadiene styrene	24
Melamine formaldehyde	28 29 .
Methyl pentene	22
Phenol formaldehyde	28 29 .
Phenolic—melamine	19 20 21 22 . 24 25 . 28 . 30 .
Phenoxy	
Polyamide	22 . 25
Polycarbonate	25 .
Polychlorotrifluorethylene	21 . 27 . 31
Polyester (fibre composite)	19 20 21 22 . 25 . 28 . 31 .
Polyether (chlorinated)	22 . 28 .
Polyethylene	22 . 31 .
Polyethylene (film)	21 22 . 30 31 .
Polyethylene (chlorinated)	22 . 28 .
Polyethylene terephthalate	21 22 . 29 30 31 . 34 .
Polyformaldehyde	22 .
Polyamide	
Polymethylmethacrylate	22 . 27 28 29 . 34 .
Polyphenylene	31 .
Polyphenylene oxide	21 22 . 25 . 28 .
Polypropylene	22 . 31 .
Polypropylene (film)	22 . 30 31 .
Polystyrene	
Polystyrene (film)	29 . 34
Polysulphone	22 .
Polytetrafluorethylene	21 . 27 . 31 .
Polyurethane	22 . 28 .
Polyvinyl chloride	21 22 . 25 . 28 29 . 33 34 .

e Numbers

```
.    . 42 43  . 45 46  .   . 49 50  .   . 53 54 . 56  . 58  .   .   .   . 63 64 65 66 67 68 69 70  .   .   .   .   .   .   .   .   .   .   .   . 85  .   .   . 89
.    .   . 43  . 45 46  .   . 49  .   .   .   . 54  .   .   . 58  .   .   .   .   . 64 65  . 67 68 69 70  .   .   .   .   .   .   .   .   .   .   .   .   .   .   .
.    .   . 43  . 45 46  . 48 49  .   .   .   . 54  .   .   . 58  .   .   .   .   . 64 65  . 67 68 69 70  .   .   .   .   .   .   .   .   .   .   .   .   .   .   .

.    .   .   .   .   .   .   .   .   .   .   .   .   .   .   .   .   .   .   .   .   . 65 66  .   .   .   .   .   .   .   .   .   .   .   .   .   .   .   .   .
.    .   .   .   .   . 46  .   .   . 49  .   .   . 53  . 55 56  .   .   .   .   .   . 65 66 67 68 69  .   .   .   . 75  .   .   . 78 79 80 81  .   .   .   . 85  .   .
.    .   . 43  . 45 46  .   . 49 50  .   . 53 54 55  .   . 58  .   .   .   .   . 64 65 66 67 68 69 70 71  .   . 75  .   . 78 79 80 81  .   .   .   .   .

.    .   .   . 44  .   .   .   .   .   .   .   . 53  . 55  . 58  .   . 63  .   . 65  . 67  .   .   .   .   .   .   .   .   .   .   .   .   .   .   .   .
.    .   .   .   .   .   .   .   .   .   .   .   .   . 55 56  .   .   .   .   .   .   . 65  . 67  . 69  .   .   .   .   .   .   .   .   .   .   .   .   .

.    .   .   . 45 46  .   . 49 50  .   . 53  .   .   . 58  .   .   . 63 64 65 66 67 68 69  . 71  .   . 75  .   . 78 79 80 81  .   .   .   .   .
.    .   .   .   .   .   .   . 49  .   .   . 53  .   .   . 58  .   .   .   .   . 65 66 67 68 69  .   .   .   . 72  .   .   .   .   .   .   .   .   .   .
.    .   .   .   .   .   .   .   .   .   . 53  . 56  . 58  .   .   .   .   . 65 66 67 68 69 70  .   .   .   .   .   .   .   .   .   .   .   .   .   .
.    .   . 43  . 45 46  .   . 49 50  .   . 53 54 55 56  . 58  .   .   .   . 64 65 66  . 68 69 70 71  .   . 74 75  .   . 78 79  . 81  .   . 85 86 87 88  .
.    .   . 43  . 45 46  .   . 49 50  .   . 53 54 55 56  . 58  .   .   .   . 64 65 66  . 68 69 70 71  .   . 74 75  .   . 78 79  . 81  .   . 85 86 87 88  .

.    .   . 43  .   .   .   .   .   .   .   . 53  .   .   .   .   .   .   .   . 65 66 67 68 69 70  .   .   .   .   .   .   .   .   .   .   .   .   .
.    . 42  .   . 46  .   . 49  .   .   . 53 54 55  . 58  .   .   .   . 63 64 65 66 67 68 69 70 71  .   .   .   .   . 78 79 80 81  .   .   .   .
.    .   .   .   .   .   .   .   .   .   .   .   .   .   .   .   .   .   .   .   . 65 66  .   .   .   .   .   .   .   .   .   .   .   .   .   .   .   .

.    . 42  . 45 46  .   . 49  .   .   . 54 55 56  . 58  .   .   .   . 64  . 66 67 68  . 70  .   .   .   .   .   . 78  .   .   .   .   .
.    .   .   .   .   .   .   .   .   .   . 55  .   .   .   .   .   .   .   .   .   . 67  .   .   .   .   .   .   .   .   .   .   .   .

.    .   .   . 45 46  .   .   . 53  . 55 56  .   .   .   . 65  . 67 68  .   .   .   .   .   . 78  .   .   .   .   .
.    .   .   . 45  .   .   .   .   .   .   .   .   .   .   .   .   .   .   .   .   .   .   .   .   .   .   .

.    . 43  . 46  .   . 50  . 53  . 55  .   .   .   .   .   .   .   .   .   . 78  .   .   .   .   .
.    . 43  . 46  .   . 50  . 53  . 55  .   .   .   .   .   .   .   .   .   . 78  .   .   .   .   .

.    . 43  . 46  .   . 50  . 53  . 55  .   .   .   .   .   .   .   .   . 78  .   .   .   .   .
.    .   .   .   .   .   . 57  .   .   .   .   . 65  .   .   .   .   .   . 78  . 80  .   .
.    .   .   .   .   .   . 57  .   .   .   .   . 65  .   .   .   .   .   . 78  . 80  .   .

.    .   .   .   .   . 53  . 55  . 58  .   .   . 64 65  . 67 68  . 70  .   .   .   . 78  .   .
.    .   .   .   .   .   .   .   .   .   .   .   .   .   .   .   .   .   .   .   .   .   .

.    .   .   .   .   .   .   .   .   .   .   .   .   .   .   .   .   .   .   . 76  .   .
.    .   .   .   . 48  .   .   .   .   .   .   .   .   .   .   .   .   .   .   .   .   .

.    .   .   .   .   .   .   . 57  .   . 61  .   . 65  . 67  .   .   .   .   .   . 78  . 81  .
.    .   .   .   .   .   .   .   .   .   .   .   . 65  .   .   .   .   .   .   .   .   .
.    .   .   .   .   .   .   . 57  .   . 61  .   . 65  . 67  .   .   .   .   .   . 78  . 81  .

.    .   .   . 49  .   . 53  . 55  .   . 58  . 61  . 63 64 65 66 67 68 69 70 71  .   . 76  .   . 79  . 81  .
.    .   .   .   .   .   . 53  .   .   .   .   .   .   .   .   .   .   .   .   .   .   .
.    . 43  .   . 49  .   .   . 55  .   .   . 61 62  .   . 67  .   .   .   .   .   . 78  .   .

.    .   . 46  .   .   .   . 55 56  .   .   .   . 65  . 67  .   .   .   .   .   .
.    .   .   .   .   . 53  .   .   .   .   . 64 65 66 67 68 69  .   .   . 78  .
.    . 43  . 45  .   .   . 55 56  .   .   . 65  . 67 68  . 70  .   .   .   . 78  .

.    . 43  .   .   . 50  . 53  .   .   . 65  .   .   .   .   . 78  .
.    . 43  .   .   . 50  . 53  . 55  .   . 64  .   .   . 74  . 78  .

.    .   .   .   . 55  . 58  .   . 64  .   .   .   .   .   .   .
.    . 43 44  .   . 50  . 54 55  . 58  .   . 64  .   .   . 74 75  .   .
.    .   . 45  .   .   . 55  .   . 65  . 67  .   .   .   . 78  .

.    .   . 44 45 46  .   .   .   .   . 67  .   .   .   .   . 82  .
.    .   . 44 45 46  .   .   . 58  .   . 64  . 67  .   . 78  .
.    .   .   .   .   .   .   .   . 67  .   .   .   .   .

.    .   .   . 53  . 55 56  .   . 65  . 67 68  .   .   . 78  .
.    . 43  .   .   .   .   .   . 67 68  .   .   .   .
.    . 43  .   . 50  . 53  . 55  .   . 64  . 67 68  .   .   .

.    . 42 43  . 45 46  .   . 50  . 52  .   . 56  .   .   .   .
.    . 42 43  . 46  .   .   .   .   .   .   .   .

.    .   .   . 53  .   .   .   . 64 65 66 67 68 69  .   .   . 78  .
.    . 42 43  . 45 46  .   .   . 53  . 55  .   . 65  . 67  .   . 78  .
```

continued

Table 9.1 continued

Adherend	Adhesive T...
Polyvinyl chloride (film)	. . . . 22 . . 25 . . . 29 30 . 32 33 34 . . .
Polyvinyl fluoride	. . . . 22 . . 25 26 . . . . . 30 . . . . . .
Polyvinylidene chloride	. . . . 22 . . 25 . . . . . 30 . . . . . .
Styrene acrylonitrile	. . . . 22 . . . .
Urea formaldehyde	
FOAMS	
Epoxy	
Latex	. . . . 16 . 19 . 22 . 25 .
Phenol formaldehyde	. . . . 19 20 . . . 25 . . 28 29 . . . 34 .
Polyethylene	. . . . 21 22 .
Polyethylene—cellulose acetate	. .
Polyphenylene oxide	. . . . 20 . 22 36
Polystyrene	. . . . 22 . 24 . . . 28 29 . . . 34 . .
Polyurethane	. . . . 20 . 22 . 25 . . . .
Polyvinyl chloride	. . . . 22 . 25 26 . 28 29 . 31 . 34 .
Silicone	. . . . 27 . .
Urea formaldehyde	
RUBBERS	
Butyl	. . . . 20 . . 25 .
Butadiene styrene	. . . . 22 . 24 25 .
Butadiene nitrile	. . . . 22 . 25 .
Chlorosulphonated polythene	. . . . 22 .
Ethylene propylene	. . . .
Fluorosilicone	. . . . 27 .
Fluorocarbon	. . . . 27 .
Polyacrylic	
Polybutadiene	. . . . 24 .
Polychloroprene (neoprene)	. . . . 18 . 20 . 22 . 24 25 26 . 28 .
Polyisoprene (Natural)	. . . . . . 16 . 18 19 . 22 . 24 25 . 27 .
Polysulphide	. . . . 26 .
Polyurethane	. . . . 20 21 22 . . 25 . 28 .
Silicone	. . . . 27 .
WOOD, ALLIED MATERIALS	
Cork	1 2 3 4 5 6 . . 9 10 . 12 . . 16 . 18 19 20 . 22 . 24 25 26 27 28 29 30 31 . 33 34 . 3
Hardboard, Chipboard	. . . . 24 .
Wood	1 2 3 . 5 6 . . 9 10 . 12 . . 16 . 18 19 20 21 22 . 24 25 26 27 28 29 . 31 . 33 34 . 3
Wood (laminates)	1 2 3 . 5 6 . . 9 10 . 12 . . 16 . 18 . 20 21 22 . 24 25 26 27 28 29 . 31 . 33 34 . 3
FIBRE PRODUCTS	
Cardboard	1 2 3 4 5 6 . . 9 10 . 12 . . 16 . 18 19 20 21 22 . 24 25 26 . 28 29 30 31 . 33 34 . 3
Cotton	1 2 3 4 5 6 . . 9 10 . 12 . 14 15 16 . 18 19 . 22 . 24 25 26 27 28 29 . 31 32 33 34 . 3
Felt	1 2 3 4 5 6 . . 9 10 . 12 . 14 15 16 . 18 19 . 22 . 24 25 26 27 28 29 . 31 32 33 34 . 3
Jute	1 . 3 4 5 6 . . 9 10 . . 13 14 15 16 . 18 19 . 22 . 24 25 26 27 28 29 . 31 . 33 34 . 3
Leather	1 2 3 4 5 6 . . 9 10 . 12 . . 16 . 18 19 20 21 22 . 24 25 26 27 . 28 . 31 . 33 34 . 3
Paper (bookbinding)	1 2 3 4 5 6 . . 9 10 . 12 . . 16 . 18 19 20 21 . 24 25 26 . 28 29 . 31 . 34 . 3
Paper (labels)	1 2 3 . 5 6 . . 9 10 . 12 13 14 15 16 . 18 19 . 21 22 . 24 25 26 27 28 29 30 31 . 33 34 . 3
Paper (packaging)	1 2 3 4 5 6 . . 9 10 . 12 13 14 15 16 . 18 19 20 21 22 . 24 25 26 . 28 29 30 31 . 33 34 . 3
Wool	1 2 3 4 5 6 . . 9 10 . 12 . 14 15 16 . 18 19 20 . 22 . 24 25 26 27 28 29 . 31 32 33 34 . 3
Rayon	1 . . . 5 6 . . 9 10 . 12 . . 16 . 18 19 20 21 22 . 24 25 26 27 28 29 . 31 . . 34 . 3
Silk	1 2 3 4 5 6 . . 9 10 . 12 . 14 15 16 . 18 19 20 . 22 . 24 25 26 27 28 29 . 31 32 33 34 . 3
INORGANIC MATERIALS	
Asbestos	. . . . 9 . . 12 . . . . 18 19 20 21 22 . 24 25 26 27 . 29 . 31 . . 34 . 3
Carbon	
Carborundum	
Ceramics (porcelain, vitreous)	. . . . 9 10 . 12 . . . . 18 19 20 21 22 . 24 25 26 27 28 29 . 31 . 34 . 37 3
Concrete, stone, granite	
Ferrite	
Glass	. 2 . . 5 . . 8 9 10 11 12 . . . . 18 . 20 . 22 . 24 25 . 27 . . . 30 . . . 34 . 37 3
Magnesium fluoride	
Mica	
Quartz	
Sodium chloride	
Tungsten carbide	

```
.  42 43  .  .  .  .  .  .  .  53  .  55  .  .  .  .  .  .  .  64  .  .  .  .  .  .  .  .  .  78  .  .  .  .  .  .  .
.  .  .  .  .  .  .  .  .  .  .  .  55 56  .  .  .  .  .  .  67  .  .  .  .  .  .  .  .  78  .  .  .  .  .  .  .
.  .  43  .  .  .  .  .  50  .  .  .  55  .  .  .  .  .  .  .  .  .  .  .  74  .  .  .  .  .  .  .  .  .  .
.  .  .  .  .  .  .  .  .  .  .  54  .  .  .  .  .  .  .  .  .  .  .  .  .  .  .  .  .  .  .  .  .  .
.  .  .  .  .  .  .  .  .  .  .  .  57  .  .  .  .  .  .  .  .  .  .  .  .  .  .  .  .  .  .  .  .

.  .  .  .  .  .  .  .  .  .  .  .  .  .  .  .  .  65  .  67  .  .  .  .  .  .  .  .  .  .  .  .  .  .
.  .  .  .  .  46  .  .  .  .  .  .  55  .  .  .  .  .  65  .  67 68  .  .  .  .  .  .  .  .  .  .  .  .
.  .  .  .  .  .  .  .  .  .  .  55  .  58  .  .  .  .  65  .  67 68  .  70  .  .  .  .  .  79  .  81  .  .  .
.  .  .  .  .  .  .  .  .  .  .  .  .  .  .  .  .  .  .  67 68  .  .  .  .  .  .  .  .  .  .  .  .  .
.  .  .  .  .  .  .  .  .  .  .  .  .  .  .  .  .  65  .  .  .  .  .  .  .  .  .  .  .  .  .  .  .
.  .  .  .  .  .  .  .  .  .  .  55  .  .  .  .  .  65  .  67 68  .  .  .  .  .  .  .  .  .  .  .  .

.  .  .  .  .  .  .  .  .  .  .  55  .  .  .  .  .  65  .  67 68  .  70  .  .  .  .  .  .  .  .  .  .
.  42  .  .  .  .  .  .  .  .  .  55  .  58  .  .  .  64 65  .  67 68  .  .  .  .  .  .  .  .  .  .  .
.  .  .  .  .  .  .  .  .  .  .  55 56  .  .  .  .  .  .  67 68  .  .  .  .  .  .  .  .  .  .  .  .

.  .  .  .  .  .  .  .  .  .  57  .  .  61  .  .  .  .  .  .  .  .  .  .  .  .  .  .  .  .  .

.  .  .  .  46  .  .  .  .  .  .  .  .  .  .  .  64  .  .  .  .  .  .  .  .  .  .  .  .  .  .
.  .  .  .  46  .  .  .  .  .  .  55  .  .  .  .  64 65 66  .  68  .  .  .  .  .  .  78  .  81  .  .  .
.  .  .  .  46  .  .  .  .  .  .  55  .  .  .  .  61  .  64 65  .  67 68  .  .  .  .  .  78  .  81  .  .  .

.  .  .  .  46  .  .  .  .  .  .  .  .  .  .  .  .  64 65 66 67 68 69  .  .  .  .  .  .  .  .  .
42 43 44  .  .  .  .  .  .  .  .  .  .  .  .  .  .  .  .  .  .  .  .  .  .  .  .  .  .  .  .

.  .  .  .  46  .  .  .  .  .  55  .  .  .  .  .  62  .  64  .  66 67 68  .  .  .  .  75  .  78 79  .  81  .  .
.  .  .  .  46  .  .  .  .  .  55  .  .  .  .  61  .  64 65 66 67 68  .  .  .  .  76  .  78  .  81  .  .

.  .  .  .  46  .  .  .  .  .  55  .  .  .  .  .  64  .  .  .  .  .  .  .  75  .  .  .  .  .  .
.  .  .  .  .  .  .  .  .  .  .  .  .  .  .  .  64  .  .  .  .  .  .  .  .  .  .  .  .

.  .  .  .  46  .  .  50  .  53  .  55 56 57 58 59 60 61 62  .  65  .  67 68  .  .  .  78 79  .  .  .  85  .  .
.  .  .  .  .  .  .  .  .  .  .  57  .  .  61  .  .  .  .  .  .  .  .  .  .  .  .  .  .
.  .  43  .  45  .  .  50  .  53  .  55 56 57 58 59 60 61 62  .  65  .  67 68  .  70  .  .  .  78  .  .  .  85 86  .  .
.  .  .  .  .  .  .  .  .  .  55 56 57 58 59 60 61 62  .  65  .  67 68  .  .  .  .  .  78  .  .  .  85 86  .  .

.  42 43  .  46  .  49 50  .  53  .  .  .  .  61 62 63  .  65  .  67 68  .  70  .  .  .  .  .  .  .
.  42 43  .  .  .  50  .  54 55 56 57 58  .  61  .  64 65  .  67  .  .  75  .  78 79  .  .  .
.  42 43  .  .  .  50  .  54  .  56 57 58  .  61  .  64 65  .  67  .  .  75  .  78 79  .  .

.  42  .  .  .  50  .  54 55 56 57 58  .  61  .  65  .  67  .  .  75  .  78 79  .  .
.  42 43  .  46  .  50  .  53  .  55 56  .  58  .  65  .  67  .  74 75  .  78 79  .  .
.  43 44  .  .  50  .  52  .  .  .  .  .  .  .  .  .  .  .  .  .  .  .

.  42 43  .  .  50  .  52  .  55  .  .  .  .  .  .  74  .  .  .  .  85 86 87  .  .
.  42 43 44  .  49 50  .  .  .  55  .  .  .  .  .  74  .  .  .  .  85 86 87  .  .
.  42 43  .  .  50  .  54 55 56 57 58  .  61  .  64 65  .  67  .  75  .  78 79  .  .

.  42 43  .  .  .  53 54 55 56  .  .  61  .  64 65  .  67  .  75  .  78  .
.  42 43  .  .  50  .  53 54 55 56 57 58  .  61  .  64 65  .  67  .  75  .  78 79  .  81  .  .

.  .  .  .  53  .  55 56  .  58  .  61 62 63 64 65 66 67 68 69 70  .  .  .  78  .  81  .  .
.  .  .  .  .  .  .  .  .  .  65  .  67  .  .  .  76  .  .  .
.  .  .  .  .  .  .  .  65  .  67  .  .  .  .  .

.  .  .  46  .  .  53  .  55 56 57  .  61 62 63 64 65 66 67 68 69 70  .  .  .  .  .  .  85 86  .  88
.  .  .  46  .  .  53  .  55 56 57  .  61 62  .  64  .  67 68  .  70  .  78 79  .  .  86  .  .
.  .  .  .  .  .  .  .  65  .  67 68  .  .  .

.  43  .  46  .  .  53  .  55 56  .  58  .  63 64 65 66 67 68 69  .  74 75  .  78  .
.  .  .  .  .  .  .  .  67  .  .  .  .  .
.  .  .  .  .  .  64  .  .  .  78  .  81  .  .

.  .  .  .  .  .  65  .  67  .  .  86  .  .
.  .  .  .  .  .  .  67  .  .  .
.  .  .  .  .  65  .  67  .  .  .
```

Table 9.2. ADHESIVE MATERIALS AND PRODUCTS INDEX

Code Nos.	Adhesive Type	Basic Information Page Nos.	Examples of Trade Products Page Nos.	Code Nos.	Adhesive Type	Basic Information Page Nos.	Examples of Trade Products Page Nos.
1.	Animal/Fish glue	35, 48	86, 88	46.	Cyanoacrylate	41	92
2.	Casein	38	88, 92	47.	Acrylamide	33	—
3.	Blood Albumen	36		48.	Ionomer resins	52	Trade So 99
4.	Soy(a) bean glue	75	—	49.	Polyamide (nylon based)	55	*
5.	Dextrine	78	102	50.	Polyamide (versamid based)	62	*
6.	Starch	75	202	51.	Phenolic polyamide	56, 59	*
7.	Gum arabic	—	144	52.	Polystyrene	67	206
8.	Canada balsam	39	*	53.	Polyhydroxy ether	61	170
9.	Cellulose acetate	38	90	54.	Polyesters (linear)	35, 66	146, 184,
10.	Cellulose acetate—butyrate	39	*	55.	Polyester + isocyanate	52, 53	*
11.	Cellulose caprate	39	*	56.	Polyester + monomer	66	*
12.	Cellulose nitrate	39	90	57.	Urea formaldehyde	77	174, 17
13.	Methyl cellulose	40	90	58.	Melamine formaldehyde	53	172, 17
14.	Hydroxy ethyl cellulose	40	90	59.	Urea—melamine formaldehydes	77	*
15.	Ethyl cellulose	39	90	60.	Phenol formaldehyde	57	172
16.	Natural rubber (latex)	54	156				
17.	Rubber hydrochloride	73	*	61.	Resorcinol formaldehyde	72	172, 17
18.	Chlorinated rubber	72	160, 182	62.	Phenolic—resorcinol formaldehydes	72	*
19.	Reclaim rubber	71	60	63.	Phenolic—epoxy	56	172, 21
20.	Butyl rubber	37	88				
21.	Polyisobutylene rubber	37	*	64.	Silicone resins	73	198
				65.	Epoxy (+polyamine)	43	112, 116, 12
22.	Nitrile rubber	55	162, 166	66.	Epoxy (+polyanhydride)	—	201
23.	Polyisoprene rubber	*	*				
24.	Butadiene styrene rubber	36	158	67.	Epoxy (+polyamide)	43, 45	132, 134, 13
				68.	Epoxy—alkyl ester	—	*
25.	Polyurethane rubber	68	186, 188, 192	69.	Epoxy (cycloaliphatic)	43	*
26.	Polysulphide rubber	67	186, 196				
27.	Silicone rubber	74	198, 200	70.	Epoxy—bitumen	—	138
				71.	Epoxy—nylon	45	170, 21
28.	Polychloroprene (neoprene) rubber	65	178–184	72.	Epoxy—polysulphide	46	134, 14
29.	Polyvinyl acetate	69	190, 192, 194	73.	Epoxy—polyurethane	47	Trade Sour
30.	Vinyl acetate—acrylic acid	49	*	74.	Polyethylene imine	—	—
				75.	Polyisocyanate	52	*
31.	Vinyl acetate—ethylene	48	144, 148	76.	Furane resins	48	*
32.	Polyvinyl chloride (PVC)	71	*	77.	Phenolic isocyanate	53	—
33.	Chlorinated PVC	*	*	78.	Phenolic—nitrile	59	162, 164,
34.	Vinyl chloride—vinyl acetate	71	*				
35.	Vinyl chloride—vinylidene	71	*	79.	Phenolic—neoprene	58	*
36.	Polyvinylidene chloride	71, 78	*	80.	Phenolic—polyvinyl butyral	60	134
				81.	Phenolic—polyvinyl formal	60	214, 21
37.	Polyvinyl formal	69	*				
38.	Polyvinyl butyral	69	*	82.	Polyimide	63	170
39.	Polyvinyl alcohol	70	*	83.	Polybenzothiazole	64	*
				84.	Polybenzimidazole	64	178
40.	Polyvinyl alkyl ether	70, 78	*				
41.	Polyvinyl pyridine	—	*	85.	Bitumen (asphalt)	76	*
42.	Polyacrylate	34, 78	*	86.	Silicates (soluble)	50	152
				87.	Mineral waxes	78	*
43.	Polyacrylate (carboxylic)	34	*				
44.	Polyacrylic esters	34	99	88.	Ceramics	40	152
45.	Polymethylmethacrylate	33	82–86	89.	Inorganic	50, 51	150

*Consult manufacturers for information (pp. 320–22)

Table 9.3 BONDING OF PLASTICS WITH SOLVENT CEMENTS

is table indicates the solvents (and mixtures) which may be used to join thermoplastic materials by solvent-welding. The surfaces to be ned are softened, by the application of a suitable solvent and then pressed together to effect a bond. Gap-filling properties of the ment are improved by bodying the solvent with some of the thermoplastic; this reduces shrinkage at the joint which would otherwise sult in stress formation.

Solvent Cements

Plastics	Acetic acid (glacial)	Acetone	Acetone:ethyl acetate:cellulose acetate butyrate (40:40:20)	Acetone:ethyl lactate (90:10)	Acetone:methoxyethyl acetate (80:20)	Acetone:methyl acetate (70:30)	Butyl acetate:acetone:methyl acetate (50:30:20)	Butyl acetate:methyl methacrylate monomer (40:60)	Ethyl acetate	Ethyl acetate:ethyl alcohol (80:20)	Ethylene dichloride	Ethylene dichloride:methylene chloride (50:50)	Glycerine:water (15:85)	Methyl acetate	Methylene chloride	Methylene chloride:methyl methacrylate monomer (60:40)	Methylene chloride:methyl methacrylate monomer (50:50)	Methylethyl ketone	Methyl isobutyl ketone	Methyl methacrylate monomer	Tetrachloroethylene	Tetrachloroethane	Tetrahydrofuran:cyclohexanone (80:20)	Toluene	Toluene:ethyl alcohol (90:10)	Toluene:methylethyl ketone (50:50)	(1,1,2) Trichloroethane	Trichloroethylene	Xylene	Xylene:methyl isobutyl ketone (25:75)
Acrylonitrile butadiene styrene																		x	x							x				x
Cellulose acetate film		x		x	x	x	x		x					x																
Cellulose acetate butyrate		x	x	x	x	x	x		x					x																
Cellulose propionate						x																								
Cellulose nitrate		x							x					x																
Ethyl cellulose											x														x					
Polyamide (nylon)	x																													
Polymethyl methacrylate											x					x	x			x										
Polycarbonate											x	x			x							x					x			
Polystyrene									x						x			x			x			x				x		
Polyvinyl chloride and copolymers (acetate)															x	x							x					x		
Styrene acrylonitrile							x	x										x												
Styrene butadiene		x																x	x											
Polyvinyl alcohol													x																	
Polyphenylene oxide																														x

Table 9.4 PHYSICAL FORMS

Key Nos.	Adhesive Type	Solid	Film	Paste	Liquid	Emulsion/Dispersion
1.	Animal/Fish glue	x	x		x	
2.	Casein	x	x		x	
3.	Blood Albumen	x	x			
4.	Soy(a) bean glue	x				
5.	Dextrine	x			x	
6.	Starch	x				x
7.	Gum arabic				x	
8.	Canada balsam	x			x	
9.	Cellulose acetate	x			x	
10.	Cellulose acetate—butyrate	x			x	
11.	Cellulose caprate	x			x	
12.	Cellulose nitrate	x			x	
13.	Methyl cellulose	x				x
14.	Hydroxy ethyl cellulose	x				x
15.	Ethyl cellulose	x				x
16.	Natural rubber (latex)	x		x		x
17.	Rubber hydrochloride	x		x		x
18.	Chlorinated rubber	x				
19.	Reclaim rubber	x		x		x
20.	Butyl rubber	x	x			
21.	Polyisobutylene rubber	x				
22.	Nitrile rubber	x				x
23.	Polyisoprene rubber	x		x		x
24.	Butadiene styrene rubber	x				x
25.	Polyurethane rubber	x			x	
26.	Polysulphide rubber	x		x		
27.	Silicone rubber	x			x	
28.	Polychloroprene (neoprene) rubber	x				x
29.	Polyvinyl acetate	x				x
30.	Vinyl acetate—acrylic acid					x
31.	Vinyl acetate—ethylene	x	x			x
32.	Polyvinyl chloride (PVC)	x				x
33.	Chlorinated PVC	x				
34.	Vinyl chloride—vinyl acetate	x				x
35.	Vinyl chloride—vinylidene	x				
36.	Polyvinylidene chloride	x				x
37.	Polyvinyl formal	x				
38.	Polyvinyl butyral	x				
39.	Polyvinyl alcohol	x		x		
40.	Polyvinyl alkyl ether	x				x
41.	Polyvinyl pyridine	x				x
42.	Polyacrylate	x				x
43.	Polyacrylate (carboxylic)	x				x
44.	Polyacrylic esters	x				
45.	Polymethylmethacrylate				x	

Key Nos.	Adhesive Type	Solid	Film	Paste	Liquid
46.	Cyanoacrylate				x
47.	Acrylamide				x
48.	Ionomer resins	x	x		
49.	Polyamide (nylon based)	x	x		
50.	Polyamide (versamid based)	x	x		
51.	Phenolic polyamide	x			
52.	Polystyrene	x	x		
53.	Polyhydroxy ether	x	x		
54.	Polyesters (linear)	x			
55.	Polyester + isocyanate			x	x
56.	Polyester + monomer				x
57.	Urea formaldehyde	x	x		
58.	Melamine formaldehyde	x			
59.	Urea—melamine formaldehydes	x	x		
60.	Phenol formaldehyde	x			
61.	Resorcinol formaldehyde	x			
62.	Phenolic—resorcinol formaldehydes		x		
63.	Phenolic—epoxy	x			x
64.	Silicone resins		x		x
65.	Epoxy (+polyamine)				x
66.	Epoxy (+polyanhydride)				x
67.	Epoxy (+polyamide)				x
68.	Epoxy—alkyl ester				x
69.	Epoxy (cycloaliphatic)				x
70.	Epoxy—bitumen	x			
71.	Epoxy—nylon	x	x		
72.	Epoxy—polysulphide				x
73.	Epoxy—polyurethane	x			x
74.	Polyethylene imine	x	x		x
75.	Polyisocyanate				x
76.	Furane resins				x
77.	Phenolic isocyanate				
78.	Phenolic—nitrile	x	x		
79.	Phenolic—neoprene	x	x		x
80.	Phenolic—polyvinyl butyral	x	x		x
81.	Phenolic—polyvinyl formal	x	x		x
82.	Polyimide	x			
83.	Polybenzothiazole	x			
84.	Polybenzimidazole	x	x		
85.	Bitumen (asphalt)	x		x	x
86.	Silicates (soluble)	x		x	
87.	Mineral waxes	x	x		
88.	Ceramics	x			
89.	Inorganic	x		x	x

Table 9.5 PROCESSING   281

Adhesive Type	Solvent release (from emulsion or solution)	Fusion on heating	Chemical reaction	Cured at room temperature	Cured at 65–100°C	Cured at 120–160°C	Cured at +200°C	Vulcanised	Bonding pressure necessary	Bonding pressure unnecessary
Animal/Fish glue	x									x
Casein	x		x							x
Blood Albumen	x									x
Soy(a) bean glue	x									x
Dextrine	x									x
Starch	x									x
Gum arabic	x									x
Canada balsam	x									x
Cellulose acetate	x	x								x
Cellulose acetate—butyrate	x	x								x
Cellulose caprate	x	(x)								x
Cellulose nitrate	x	(x)								x
Methyl cellulose	x									x
Hydroxy ethyl cellulose	x	x								x
Ethyl cellulose	x	x								x
Natural rubber (latex)	x							x		x
Rubber hydrochloride	x									x
Chlorinated rubber	x									x
Reclaim rubber	x							x		x
Butyl rubber	x							x		x
Polyisobutylene rubber	x									x
Nitrile rubber	x	x						x		x
Polyisoprene rubber	x									x
Butadiene styrene rubber	x							x		x
Polyurethane rubber	x	x	x							x
Polysulphide rubber	x		x					x		x
Silicone rubber			x					x		
Polychloroprene (neoprene) rubber	x							x		x
Polyvinyl acetate	x	x								x
Vinyl acetate—acrylic acid	x	(x)								x
Vinyl acetate—ethylene		(x)								x
Polyvinyl chloride (PVC)	x	x								
Chlorinated PVC										x
Vinyl chloride—vinyl acetate	x	x								x
Vinyl chloride—vinylidene	x	x								x
Polyvinylidene chloride										x
Polyvinyl formal	x	(x)								x
Polyvinyl butyral	x	(x)								x
Polyvinyl alcohol	x									x
Polyvinyl alkyl ether	x	x								x
Polyvinyl pyridine	x									x
Polyacrylate	x									x
Polyacrylate (carboxylic)	x									x
Polyacrylic esters	x									x
Polymethylmethacrylate			x							x

Key Nos.	Adhesive Type	Solvent release (from emulsion or solution)	Fusion on heating	Chemical reaction	Cured at room temperature	Cured at 65–100°C	Cured at 120–160°C	Cured at +200°C	Vulcanised	Bonding pressure necessary	Bonding pressure unnecessary
46.	Cyanoacrylate			x							x
47.	Acrylamide										
48.	Ionomer resins		x								x
49.	Polyamide (nylon based)		x								
50.	Polyamide (versamid based)		(x)								x
51.	Phenolic polyamide		x			x					x
52.	Polystyrene	x	x								x
53.	Polyhydroxy ether		(x)			x					x
54.	Polyesters (linear)				x						x
55.	Polyester + isocyanate				x	x				x	
56.	Polyester + monomer				x	x					x
57.	Urea formaldehyde				x	x	x				x
58.	Melamine formaldehyde					x					x
59.	Urea—melamine formaldehydes					x					x
60.	Phenol formaldehyde				x	x	x				x
61.	Resorcinol formaldehyde				x	x					x
62.	Phenolic—resorcinol formaldehydes				x	x					x
63.	Phenolic—epoxy				x	x					x
64.	Silicone resins					x	x				x
65.	Epoxy (+polyamine)				x	x	x				x
66.	Epoxy (+polyanhydride)				x	x	x				x
67.	Epoxy (+polyamide)				x	x	x				x
68.	Epoxy—alkyl ester				x	x					x
69.	Epoxy (cycloaliphatic)				x	x					x
70.	Epoxy—bitumen				x	x	x				x
71.	Epoxy—nylon				x	x				x	
72.	Epoxy—polysulphide				x	x	x				x
73.	Epoxy—polyurethane					x				x	x
74.	Polyethylene imine				x	x					x
75.	Polyisocyanate				x	x					x
76.	Furane resins				x	x	x	x			x
77.	Phenolic isocyanate				x	x					x
78.	Phenolic—nitrile	x			x	x	x	x			x
79.	Phenolic—neoprene	x			x	x	x	x			x
80.	Phenolic—polyvinyl butyral					x	x				x
81.	Phenolic—polyvinyl formal					x	x				x
82.	Polyimide								x		x
83.	Polybenzothiazole								x		x
84.	Polybenzimidazole								x		x
85.	Bitumen (asphalt)	x	(x)			x					x
86.	Silicates (soluble)	x	(x)				x				x
87.	Mineral waxes		(x)								x
88.	Ceramics								x		x
89.	Inorganic					x	x	x			x

also may be fused by ultrasonics *see* HAUSER, R. L., 'Ultra-adhesives for ultrasonic bonding', *Adhes. Age,* **12** (3), 27 (1969).

Table 9.6 RESISTANCE RATINGS

KEY: Excellent (e);
Good (g);
Moderate (m);
Fair (f);
Poor (p)

Key Nos.	Adhesive Type	Heat	Cold	Water (hot)	Water (cold)	Alcohols	Hydrocarbons (aliphatic)	Hydrocarbons (aromatic)	Hydrocarbons (chlorinated)	Mineral oils/greases	Ketones	Esters	Acids	Alkalies	Biodeterioration	Peel stress	Shear stress
1.	Animal/Fish glue	m	g	p	p	p	g	g	g	g					f		g
2.	Casein	m	m	p	p	m	g	g	g	g					f	f	
3.	Blood Albumen	f	f	p	m	m	m	m	m	m			p	p	f	f	
4.	Soy(a) bean glue	f	p	p	p	p	m	m	m	m				p		f	
5.	Dextrine	f	f	p	p	p	m	m	m	m						f	
6.	Starch	f	f	p	p	p	m	m	m	m						p	
7.	Gum arabic	f	f	p	p		m	m	m	m						p	g
8.	Canada balsam	f	f														m
9.	Cellulose acetate	f	f	p	m	p	p	p	p	p	p	p	p	m	e	p	g
10.	Cellulose acetate—butyrate	f	f	m	g	p	g	g	p	m	p	p	m	g	e	f	g
11.	Cellulose caprate	f	f	m	g	p	g	g	p	m	p	p	f	g	e	f	g
12.	Cellulose nitrate	p	f	f	f	p	m	p	p	m	p	f	f	g	e	p	g
13.	Methyl cellulose	g	g	p	p	p	g	p	p	g	p	p	p	p	e	p	
14.	Hydroxy ethyl cellulose	g	g	p	p	f	g	g	g	g	g	g	g	f	e	p	
15.	Ethyl cellulose	m	e	p	g	p	g	p	g	g	p	p	f	g	e	p	
16.	Natural rubber (latex)	f					p	p	p	f	f	f			g	f	g
17.	Rubber hydrochloride	g	g	g	g	m	e	g	f	e	f	f	g	g	e		
18.	Chlorinated rubber	p	p	g	p	p	g	p	p	g	p	p	p	g		f	g
19.	Reclaim rubber	m	m	g	g	g	p	p	p	p	p	p	f	f	e	f	g
20.	Butyl rubber	p	p	p	g	g	p	p	p	p	p	p	f	f	e	e	f
21.	Polyisobutylene rubber	p	f	p	g	g	p	p	p	g	g	g	g	g	e	p	p
22.	Nitrile rubber	m	f	p	g	f	g	f	p	g	p	p	p	p	e	f	g
23.	Polyisoprene rubber	f	m	g	g	g	p	p	p	p	p	p	f	f	e		
24.	Butadiene styrene rubber	f	f	m	g	g	p	p	p	m	p	p	f	g	e	f	g
25.	Polyurethane rubber	p	f	p	g	g	p	p	p	g	g	g	g	g	e	g	m
26.	Polysulphide rubber	p	g	p	e	g	g	g	p	g	p	p	g	g	e	g	g
27.	Silicone rubber	e	e	g	g	f	f	f	f	g	f	f	m	f	e	p	f
28.	Polychloroprene (neoprene) rubber	m	f	g	g	f	g	f	p	g	p	p	g	g	e	f	g
29.	Polyvinyl acetate	p	p	p	f	f	g	p	p	g	p	p	f	f	e	p	g
30.	Vinyl acetate—acrylic acid	p	g	m	g	f	p	p	p	m	p	p	f	p	e	p	p
31.	Vinyl acetate—ethylene	p	g	m	g	f	p	p	g	g	g	p	m		m	f	m
32.	Polyvinyl chloride (PVC)	p	p	p	g	g	g	f	p	g	g	p	g	g	g	f	g
33.	Chlorinated PVC	p		p	g	g	g	p	p	g	p	p	g	g	e	f	
34.	Vinyl chloride—vinyl acetate	p		f	g	g	g	p	p	g	p	p	g	g	g	p	g
35.	Vinyl chloride—vinylidene	p		f	g	g	g	p	p	g	p	p	g	g	g	p	p
36.	Polyvinylidene chloride	p		f	g	g	g	p	p	g	p	p	g	g	g	p	
37.	Polyvinyl formal	p	g	m	g	g	g	g	g	g	g	p	p	p		p	g
38.	Polyvinyl butyral	p	g	m	g	p	g	p	g	g	p	p	p	p	m	p	g
39.	Polyvinyl alcohol	f	f	p	p	f	e	e	e	g	e	e	p	p		g	
40.	Polyvinyl alkyl ether	p	p	p	p	p	p	p	p	p	p	p	f	g		f	m
41.	Polyvinyl pyridine																
42.	Polyacrylate	p	p	p	g	p	e	f	p	p	p	p	f	g	e	p	p
43.	Polyacrylate (carboxylic)	p	p	p	g	p	e	p	p	p	p	p	f	g	e	e	m
44.	Polyacrylic esters	p	f	p	m	m	f	p	p	p	p	p	m	g	e	p	f
45.	Polymethylmethacrylate	p	g		m	f	f	p	p	p	m	m	f		e		g

cont

Table **9.6** continued

Key Nos.	Adhesive Type	Heat	Cold	Water (hot)	Water (cold)	Alcohols	Hydrocarbons (aliphatic)	Hydrocarbons (aromatic)	Hydrocarbons (chlorinated)	Mineral oils/greases	Ketones	Esters	Acids	Alkalies	Biodeterioration	Peel stress	Shear stress
46.	Cyanoacrylate	p	p	p	p	p	m	p	p	m	p	p	p	p	e		g
47.	Acrylamide	p	f	p	m	m	p	p	p	p	p	p	m	g	e	p	f
48.	Ionomer resins	m	g	f	m	p	g	g	g	g	g	g	g	p	e	g	m
49.	Polyamide (nylon based)	p	g	p	p	p	g	g	p	g	g	g	p	g	e	g	g
50.	Polyamide (versamid based)	m	g				m	m	p							f	g
51.	Phenolic polyamide	p	g	p	g	g	g	p	p	p	p	p	g	g		g	g
52.	Polystyrene	f	g	g	g	f	p	p	p	f	g	g	f	g		p	g
53.	Polyhydroxy ether	f	g	m	g	f	g	p	g	g	g	g	f	g		p	g
54.	Polyesters (linear)	f	g	m	g	f	g	p	g	g	g	g	f	g		m	g
55.	Polyester + isocyanate	p	f	p	f	g	g	p	p	g	p	p	m	p		p	g
56.	Polyester + monomer	f	f	p	g	g	g	g	g	g	g	g	g	g		p	g
57.	Urea formaldehyde	g	g	p	g	g	g	g	g	g	g	g	g	g			g
58.	Melamine formaldehyde	g	g	g	g	g	g	g	g	g	g	g	m	m			g
59.	Urea—melamine formaldehydes	g	g	g	g	g	g	g	g	g	g	g	m				g
60.	Phenol formaldehyde	g	g	g	g	g	g	g	g	g	g	g	g	g			g
61.	Resorcinol formaldehyde	g	g	g	g	g	g	g	g	g	g	g	g	g			g
62.	Phenolic—resorcinol formaldehydes	e	f	g	g	g	m	g		m	p	p	g	g	e		g
63.	Phenolic—epoxy	e	g		g	g	g	m	p	g	p	p	g	g	e	p	e
64.	Silicone resins	m	p	f	m	m	m	m	p	g	p	p	g	g	e		g
65.	Epoxy (+ polyamine)	e	p	g	g	g	g	g	g	p	p	p	g	g	e	p	g
66.	Epoxy (+ polyanhydride)	p	g	p	g	m	g	f	m	g	p	p	f	p	e	p	g
67.	Epoxy (+ polyamide)	m	p	g	g	f	g	f	p	g	p	p	g	p	e	g	g
68.	Epoxy—alkyl ester	e														p	g
69.	Epoxy (cycloaliphatic)	f	f	f	g		p	p	p		p	p			e	g	m
70.	Epoxy—bitumen	p	g	p	g	f	g	p	p		p	p				g	g
71.	Epoxy—nylon	g	e	g	g			p	p		p	p				e	g
72.	Epoxy—polysulphide	g	g													g	g
73.	Epoxy—polyurethane	g	e	g	g	g	g	g	p	g	p	p	p	p		g	g
74.	Polyethylene imine	g															
75.	Polyisocyanate			g	g	g	g	g		g	g		m	g			
76.	Furane resins	g															
77.	Phenolic isocyanate	g	m	g	g	g	g	p	p	g	p	p	g	g			
78.	Phenolic—nitrile	f	g		g	f	g	p	p	g	p	p	f	g		g	g
79.	Phenolic—neoprene	f	f	f	g	p	g	p	p	g	p	p	p	g		f	g
80.	Phenolic—polyvinyl butyral	p	p	p	g	p	g	p	p	g	p	p	p	g		f	g
81.	Phenolic—polyvinyl formal	e	e	p	g	p	g	p	p	g	p	p	p	m		f	g
82.	Polyimide	e	e	e	e	e	e	e	e	e	e	e	e	p		m	g
83.	Polybenzothiazole	e	e											p			
84.	Polybenzimidazole	e	e	g	g	g	g	g	g	g	g	g	e			m	g
85.	Bitumen (asphalt)	g	g	m	g	p	g	g	g	g	p	g	g	p	e	m	
86.	Silicates (soluble)	p	g	m	g	p	p	p	p	p	p	p	g	g		m	
87.	Mineral waxes	m	g	m	m	m	p	p	p	p	p	m	m		m	m	
88.	Ceramics	e		g	g	g	g	g	g	g	g	g				p	g
89.	Inorganic															p	g

Table 9.7 INDUSTRIAL USAGE
(see also review articles in the Bibliography p. 346)

Key Nos.	Adhesive Type	Aircraft	Automobile	Bookbinding	Building	Civil Engineering	Footwear	Foundry work	Furniture	Packaging	Scientific Instrument: Optical	Scientific Instrument: Electronic	Shipbuilding	Textiles and Clothing	Woodworking
1.	Animal/Fish glue			x					x	x					x
2.	Casein		x						x	x					x
3.	Blood Albumen									x					x
4.	Soy(a) bean glue														x
5.	Dextrine									x					
6.	Starch							x		x					
7.	Gum arabic			x						x					
8.	Canada balsam										x				
9.	Cellulose acetate									x					
10.	Cellulose acetate—butyrate										x				
11.	Cellulose caprate										x				
12.	Cellulose nitrate					x	x			x					
13.	Methyl cellulose									x					
14.	Hydroxy ethyl cellulose									x					
15.	Ethyl cellulose									x					
16.	Natural rubber (latex)	x	x	x		x	x							x	
17.	Rubber hydrochloride														
18.	Chlorinated rubber														
19.	Reclaim rubber	x	x		x	x	x								
20.	Butyl rubber	x	x			x									
21.	Polyisobutylene rubber														
22.	Nitrile rubber	x	x			x	x					x	x		
23.	Polyisoprene rubber														
24.	Butadiene styrene rubber	x													
25.	Polyurethane rubber	x	x			x	x							x	
26.	Polysulphide rubber	x	x		x	x							x		
27.	Silicone rubber	x									x	x			
28.	Polychloroprene (neoprene) rubber	x	x		x	x	x			x				x	
29.	Polyvinyl acetate				x	x			x	x	x				
30.	Vinyl acetate—acrylic acid														
31.	Vinyl acetate—ethylene		x	x		x			x	x	x				
32.	Polyvinyl chloride (PVC)														
33.	Chlorinated PVC														
34.	Vinyl chloride—vinyl acetate														
35.	Vinyl chloride—vinylidene														
36.	Polyvinylidene chloride														
37.	Polyvinyl formal	x													
38.	Polyvinyl butyral										x				
39.	Polyvinyl alcohol									x					
40.	Polyvinyl alkyl ether									x					
41.	Polyvinyl pyridine									x					
42.	Polyacrylate														
43.	Polyacrylate (carboxylic)														
44.	Polyacrylic esters	x				x					x				
45.	Polymethylmethacrylate	x				x					x				

con

Table 9.7 continued

Key Nos.	Adhesive Type	Aircraft	Automobile	Bookbinding	Building	Civil Engineering	Footwear	Foundry work	Furniture	Packaging	Scientific Instrument: Optical	Scientific Instrument: Electronic	Shipbuilding	Textiles and Clothing	Woodworking
46.	Cyanoacrylate	x	x			x						x			
47.	Acrylamide														
48.	Ionomer resins										x				
49.	Polyamide (nylon based)		x			x			x						
50.	Polyamide (versamid based)		x			x						x			
51.	Phenolic polyamide														
52.	Polystyrene														
53.	Polyhydroxy ether														
54.	Polyesters (linear)														
55.	Polyester + isocyanate														
56.	Polyester + monomer														
57.	Urea formaldehyde					x			x						x
58.	Melamine formaldehyde					x									x
59.	Urea—melamine formaldehydes					x									x
60.	Phenol formaldehyde					x			x						x
61.	Resorcinol formaldehyde					x			x						x
62.	Phenolic—resorcinol formaldehydes														x
63.	Phenolic—epoxy	x	x			x									
64.	Silicone resins	x				x					x	x			
65.	Epoxy (+polyamine)	x	x		x	x					x	x	x		x
66.	Epoxy (+polyanhydride)														
67.	Epoxy (+polyamide)	x	x		x	x					x	x	x		x
68.	Epoxy—alkyl ester				x	x									
69.	Epoxy (cycloaliphatic)														
70.	Epoxy—bitumen				x	x									
71.	Epoxy—nylon	x													
72.	Epoxy—polysulphide	x				x							x		
73.	Epoxy—polyurethane														
74.	Polyethylene imine														
75.	Polyisocyanate														
76.	Furane resins				x	x		x							
77.	Phenolic isocyanate														
78.	Phenolic—nitrile	x	x			x									
79.	Phenolic—neoprene	x	x			x									
80.	Phenolic—polyvinyl butyral	x				x							x		
81.	Phenolic—polyvinyl formal	x				x									
82.	Polyimide														
83.	Polybenzothiazole														
84.	Polybenzimidazole														
85.	Bitumen (asphalt)		x		x	x							x		
86.	Silicates (soluble)					x		x		x					x
87.	Mineral waxes										x				
88.	Ceramics										x				
89.	Inorganic										x				

Table 9.8.

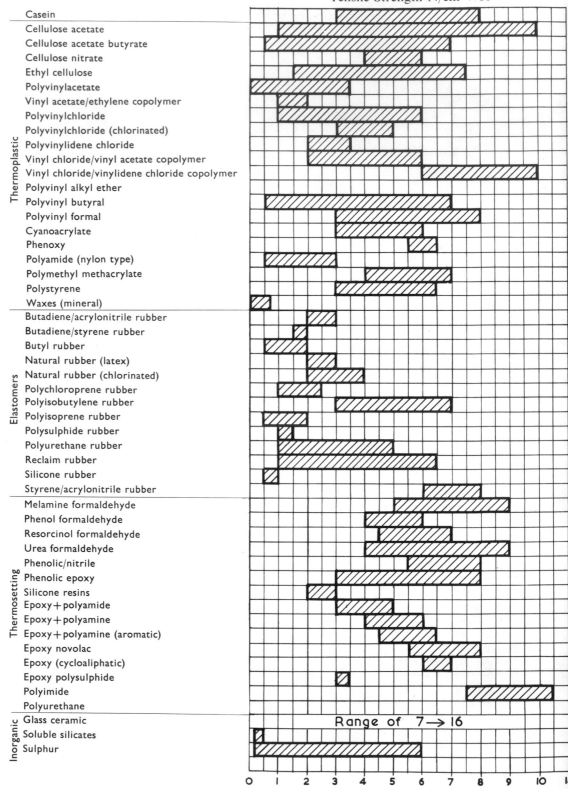

Tensile Strength N/cm² × 10³

Table 9.9.

Modulus of Elasticity N/cm$^2 \times 10^{-5}$

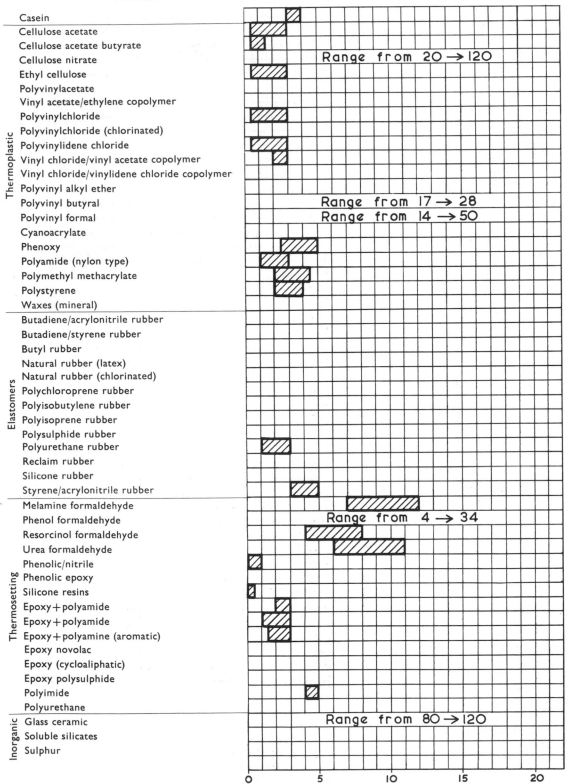

Thermoplastic	Casein
	Cellulose acetate
	Cellulose acetate butyrate
	Cellulose nitrate
	Ethyl cellulose
	Polyvinylacetate
	Vinyl acetate/ethylene copolymer
	Polyvinylchloride
	Polyvinylchloride (chlorinated)
	Polyvinylidene chloride
	Vinyl chloride/vinyl acetate copolymer
	Vinyl chloride/vinylidene chloride copolymer
	Polyvinyl alkyl ether
	Polyvinyl butyral
	Polyvinyl formal
	Cyanoacrylate
	Phenoxy
	Polyamide (nylon type)
	Polymethyl methacrylate
	Polystyrene
	Waxes (mineral)

Range from 20 → 120

Range from 17 → 28

Range from 14 → 50

Elastomers
Butadiene/acrylonitrile rubber
Butadiene/styrene rubber
Butyl rubber
Natural rubber (latex)
Natural rubber (chlorinated)
Polychloroprene rubber
Polyisobutylene rubber
Polyisoprene rubber
Polysulphide rubber
Polyurethane rubber
Reclaim rubber
Silicone rubber
Styrene/acrylonitrile rubber

Thermosetting
Melamine formaldehyde
Phenol formaldehyde
Resorcinol formaldehyde
Urea formaldehyde
Phenolic/nitrile
Phenolic epoxy
Silicone resins
Epoxy+polyamide
Epoxy+polyamide
Epoxy+polyamine (aromatic)
Epoxy novolac
Epoxy (cycloaliphatic)
Epoxy polysulphide
Polyimide
Polyurethane

Range from 4 → 34

Inorganic
Glass ceramic
Soluble silicates
Sulphur

Range from 80 → 120

0   5   10   15   20

288

Table 9.10.

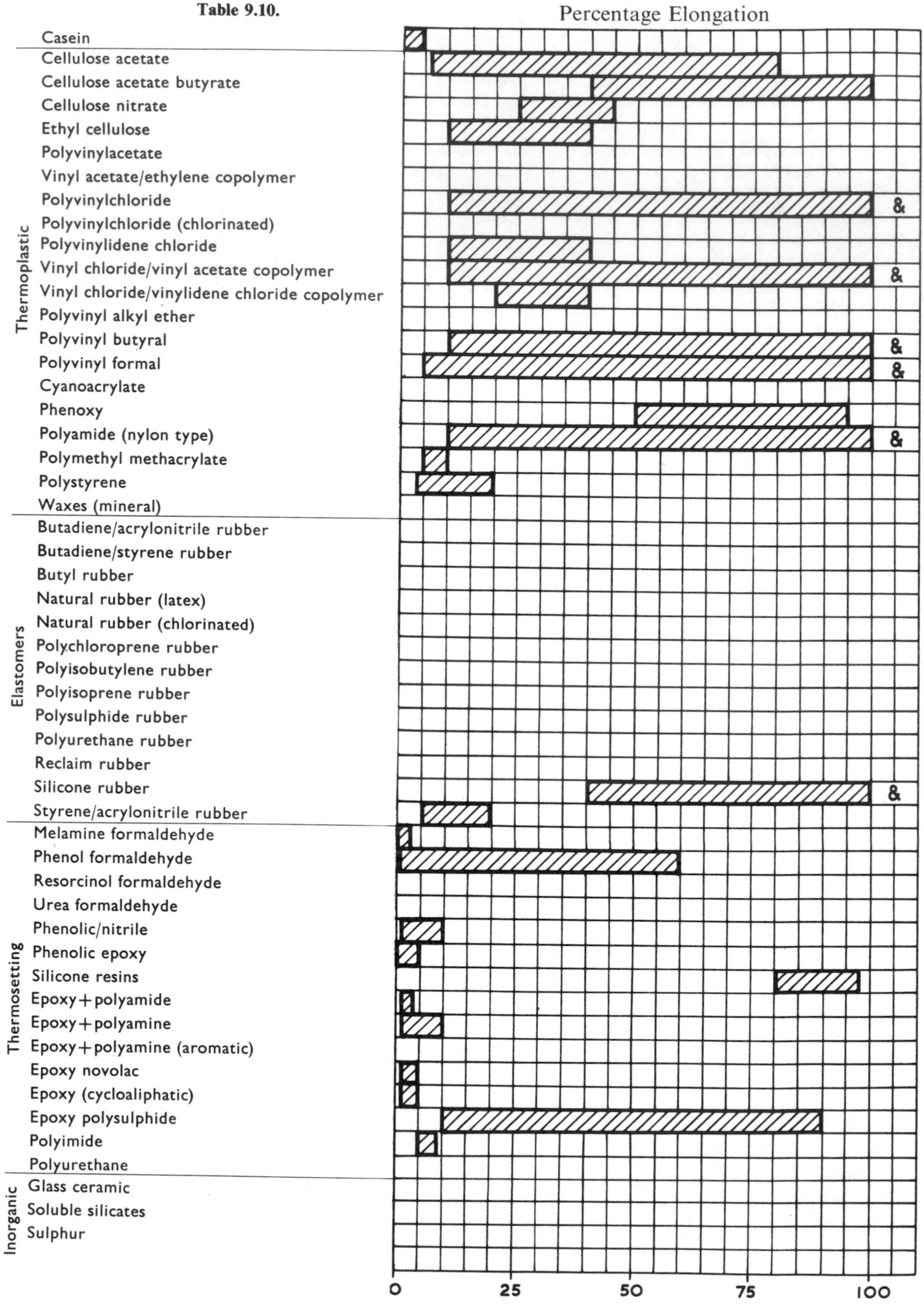

Percentage Elongation

	Material	Percentage Elongation (approx. bar range)
Thermoplastic	Casein	0–3
	Cellulose acetate	8–82
	Cellulose acetate butyrate	40–100
	Cellulose nitrate	26–42
	Ethyl cellulose	10–40
	Polyvinylacetate	
	Vinyl acetate/ethylene copolymer	
	Polyvinylchloride	14–100 &
	Polyvinylchloride (chlorinated)	
	Polyvinylidene chloride	14–40
	Vinyl chloride/vinyl acetate copolymer	14–100 &
	Vinyl chloride/vinylidene chloride copolymer	25–42
	Polyvinyl alkyl ether	
	Polyvinyl butyral	10–100 &
	Polyvinyl formal	5–100 &
	Cyanoacrylate	
	Phenoxy	50–82
	Polyamide (nylon type)	8–100 &
	Polymethyl methacrylate	0–5
	Polystyrene	5–14
	Waxes (mineral)	
Elastomers	Butadiene/acrylonitrile rubber	
	Butadiene/styrene rubber	
	Butyl rubber	
	Natural rubber (latex)	
	Natural rubber (chlorinated)	
	Polychloroprene rubber	
	Polyisobutylene rubber	
	Polyisoprene rubber	
	Polysulphide rubber	
	Polyurethane rubber	
	Reclaim rubber	
	Silicone rubber	52–100 &
	Styrene/acrylonitrile rubber	8–14
Thermosetting	Melamine formaldehyde	0
	Phenol formaldehyde	2–60
	Resorcinol formaldehyde	
	Urea formaldehyde	
	Phenolic/nitrile	0–10
	Phenolic epoxy	0–8
	Silicone resins	80–90
	Epoxy+polyamide	0–5
	Epoxy+polyamine	2–10
	Epoxy+polyamine (aromatic)	
	Epoxy novolac	0–3
	Epoxy (cycloaliphatic)	0–3
	Epoxy polysulphide	8–60
	Polyimide	2–5
	Polyurethane	
Inorganic	Glass ceramic	
	Soluble silicates	
	Sulphur	

0   25   50   75   100

Percentage Elongation

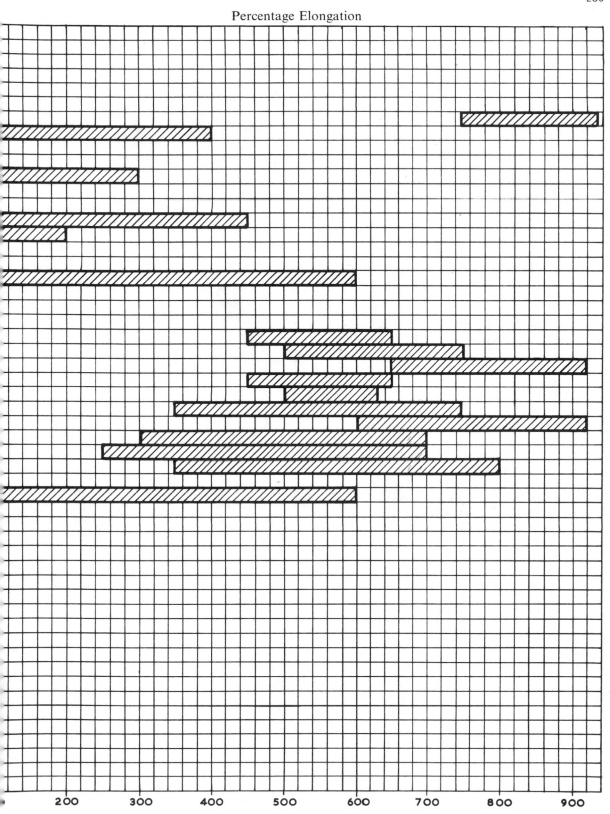

Table 9.11.

Compressive Strength $N/cm^2 \times 10^3$

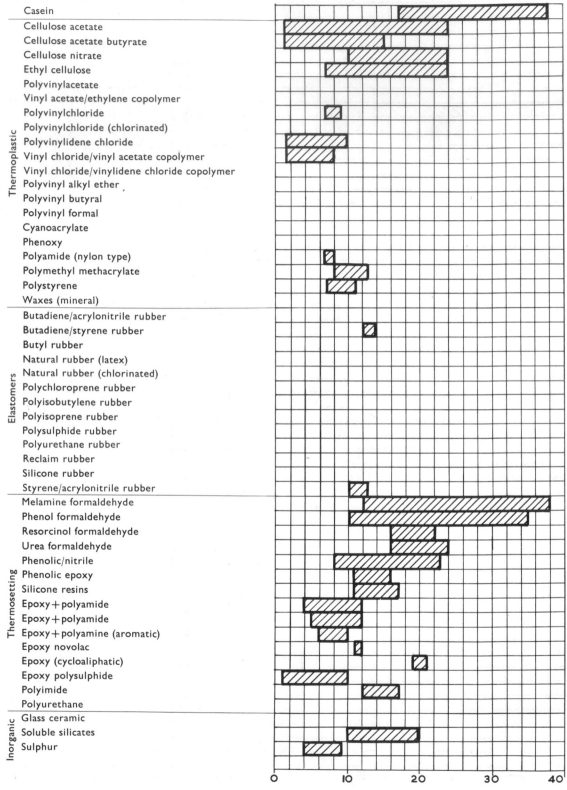

Thermoplastic
- Casein
- Cellulose acetate
- Cellulose acetate butyrate
- Cellulose nitrate
- Ethyl cellulose
- Polyvinylacetate
- Vinyl acetate/ethylene copolymer
- Polyvinylchloride
- Polyvinylchloride (chlorinated)
- Polyvinylidene chloride
- Vinyl chloride/vinyl acetate copolymer
- Vinyl chloride/vinylidene chloride copolymer
- Polyvinyl alkyl ether
- Polyvinyl butyral
- Polyvinyl formal
- Cyanoacrylate
- Phenoxy
- Polyamide (nylon type)
- Polymethyl methacrylate
- Polystyrene
- Waxes (mineral)

Elastomers
- Butadiene/acrylonitrile rubber
- Butadiene/styrene rubber
- Butyl rubber
- Natural rubber (latex)
- Natural rubber (chlorinated)
- Polychloroprene rubber
- Polyisobutylene rubber
- Polyisoprene rubber
- Polysulphide rubber
- Polyurethane rubber
- Reclaim rubber
- Silicone rubber
- Styrene/acrylonitrile rubber

Thermosetting
- Melamine formaldehyde
- Phenol formaldehyde
- Resorcinol formaldehyde
- Urea formaldehyde
- Phenolic/nitrile
- Phenolic epoxy
- Silicone resins
- Epoxy+polyamide
- Epoxy+polyamide
- Epoxy+polyamine (aromatic)
- Epoxy novolac
- Epoxy (cycloaliphatic)
- Epoxy polysulphide
- Polyimide
- Polyurethane

Inorganic
- Glass ceramic
- Soluble silicates
- Sulphur

0   10   20   30   40

Table 9.12.

Flexural Strength N/cm$^2 \times 10^3$

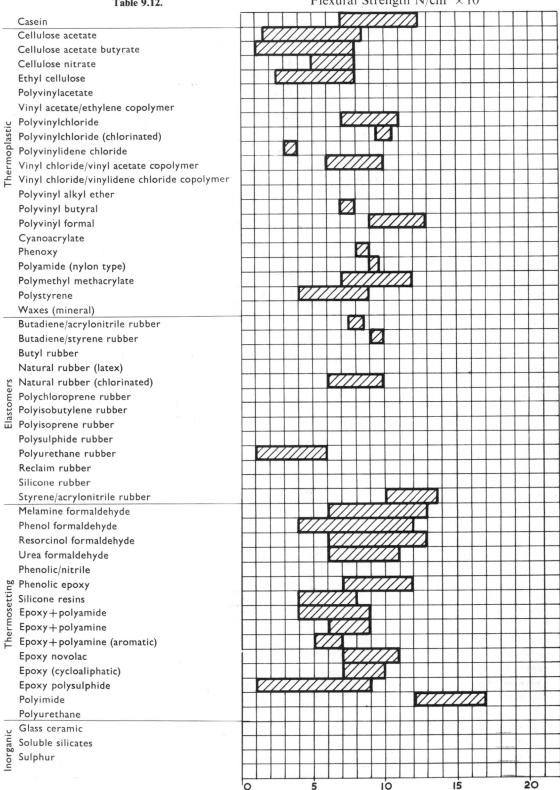

Table 9.13.

Izod Impact Strength J/2·54 cm

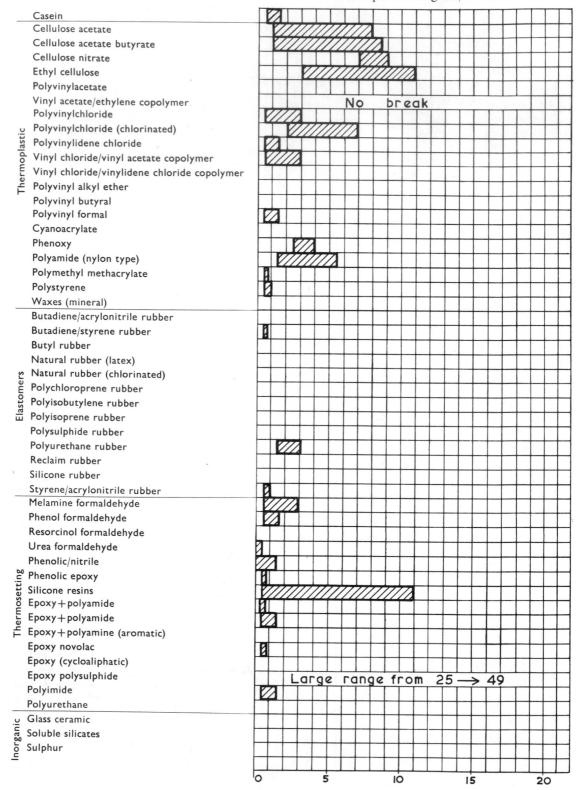

Table 9.14.   Tensile Impact Strength (Charpy) J/2·54 cm

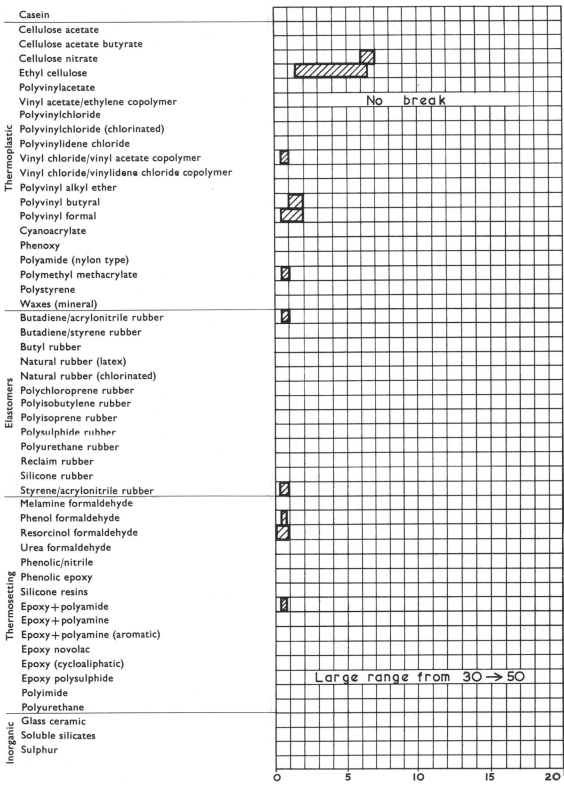

Table 9.15

Percentage Water Absorption in 24 h

Thermoplastic	Casein
	Cellulose acetate
	Cellulose acetate butyrate
	Cellulose nitrate
	Ethyl cellulose
	Polyvinylacetate
	Vinyl acetate/ethylene copolymer
	Polyvinylchloride
	Polyvinylchloride (chlorinated)
	Polyvinylidene chloride
	Vinyl chloride/vinyl acetate copolymer
	Vinyl chloride/vinylidene chloride copolymer
	Polyvinyl alkyl ether
	Polyvinyl butyral
	Polyvinyl formal
	Cyanoacrylate
	Phenoxy
	Polyamide (nylon type)
	Polymethyl methacrylate
	Polystyrene
	Waxes (mineral)
Elastomers	Butadiene/acrylonitrile rubber
	Butadiene/styrene rubber
	Butyl rubber
	Natural rubber (latex)
	Natural rubber (chlorinated)
	Polychloroprene rubber
	Polyisobutylene rubber
	Polyisoprene rubber
	Polysulphide rubber
	Polyurethane rubber
	Reclaim rubber
	Silicone rubber
	Styrene/acrylonitrile rubber
Thermosetting	Melamine formaldehyde
	Phenol formaldehyde
	Resorcinol formaldehyde
	Urea formaldehyde
	Phenolic/nitrile
	Phenolic epoxy
	Silicone resins
	Epoxy+polyamide
	Epoxy+polyamide
	Epoxy+polyamine (aromatic)
	Epoxy novolac
	Epoxy (cycloaliphatic)
	Epoxy polysulphide
	Polyimide
	Polyurethane
Inorganic	Glass ceramic
	Soluble silicates
	Sulphur

0 1 2 3 4 5 6 7 8 9 10 11

Table 9.16.　　　　Continuous Heat Resistance Temperature Range °C

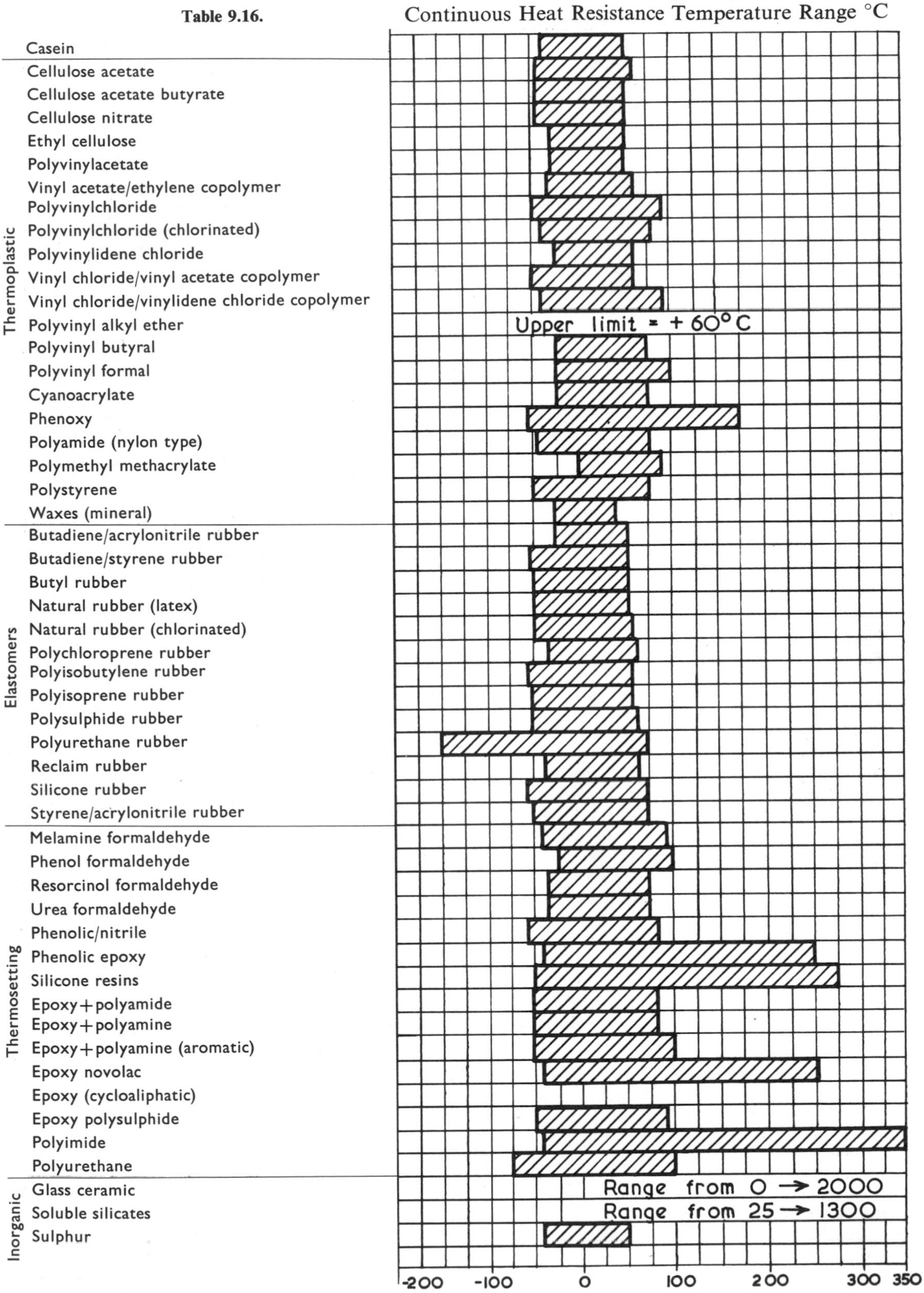

Thermoplastic

- Casein
- Cellulose acetate
- Cellulose acetate butyrate
- Cellulose nitrate
- Ethyl cellulose
- Polyvinylacetate
- Vinyl acetate/ethylene copolymer
- Polyvinylchloride
- Polyvinylchloride (chlorinated)
- Polyvinylidene chloride
- Vinyl chloride/vinyl acetate copolymer
- Vinyl chloride/vinylidene chloride copolymer
- Polyvinyl alkyl ether
- Polyvinyl butyral
- Polyvinyl formal
- Cyanoacrylate
- Phenoxy
- Polyamide (nylon type)
- Polymethyl methacrylate
- Polystyrene
- Waxes (mineral)

Upper limit = + 60°C

Elastomers

- Butadiene/acrylonitrile rubber
- Butadiene/styrene rubber
- Butyl rubber
- Natural rubber (latex)
- Natural rubber (chlorinated)
- Polychloroprene rubber
- Polyisobutylene rubber
- Polyisoprene rubber
- Polysulphide rubber
- Polyurethane rubber
- Reclaim rubber
- Silicone rubber
- Styrene/acrylonitrile rubber

Thermosetting

- Melamine formaldehyde
- Phenol formaldehyde
- Resorcinol formaldehyde
- Urea formaldehyde
- Phenolic/nitrile
- Phenolic epoxy
- Silicone resins
- Epoxy+polyamide
- Epoxy+polyamine
- Epoxy+polyamine (aromatic)
- Epoxy novolac
- Epoxy (cycloaliphatic)
- Epoxy polysulphide
- Polyimide
- Polyurethane

Inorganic

- Glass ceramic
- Soluble silicates
- Sulphur

Range from 0 → 2000
Range from 25 → 1300

-200　-100　0　100　200　300　350

296

Table 9.17.   Heat Distortion Temperature at 66 lbf/in² (45·5 N/cm²) °C

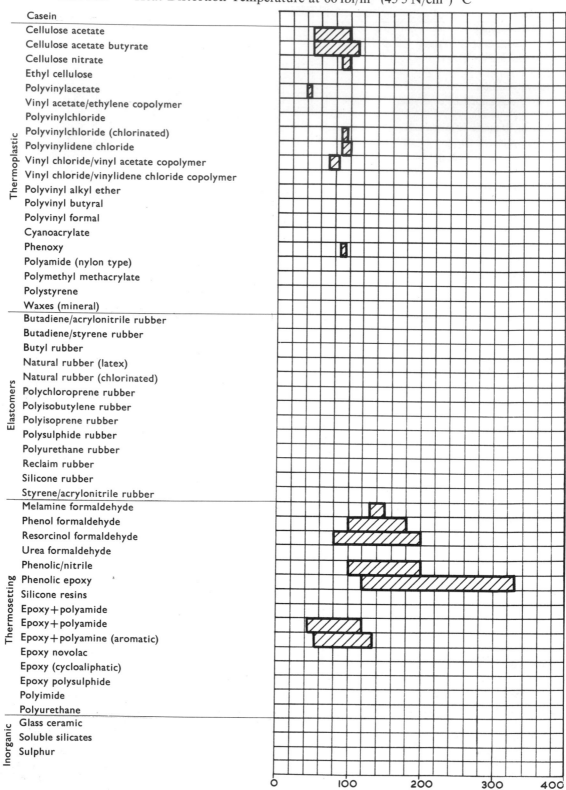

297

Table 9.18.   Heat Distortion Temperature at 264 lbf/in² (182 N/cm²) °C

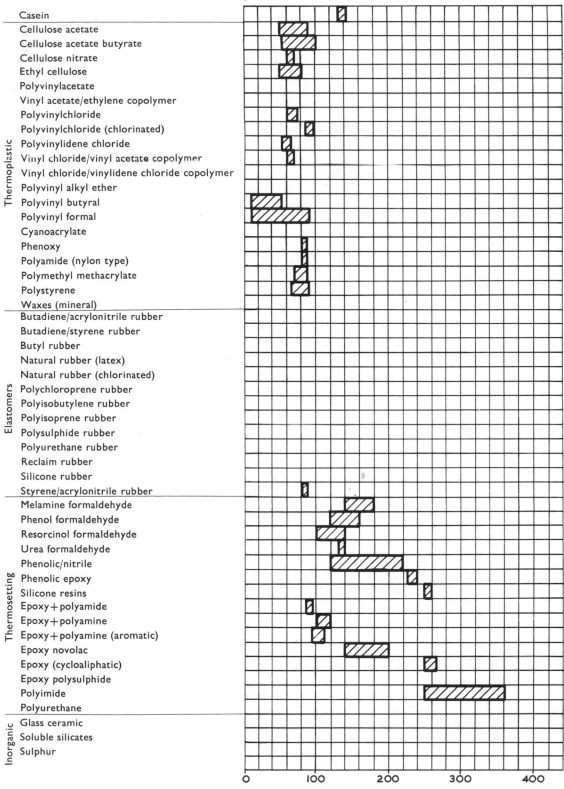

Table 9.19.   Thermal Conductivity $J.m^{-2}.s^{-1}.°C^{-1} \times 10^{-1}$

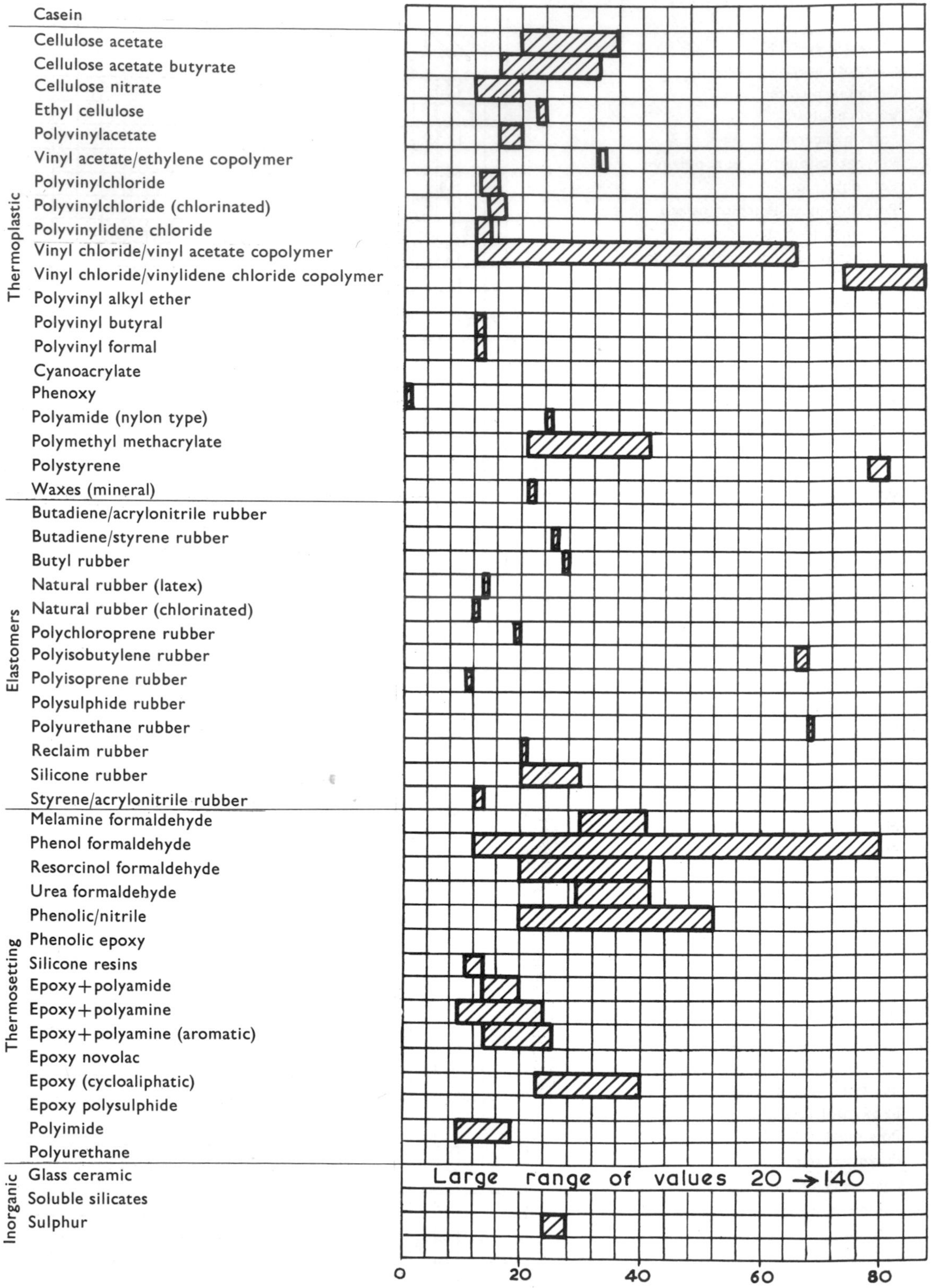

Thermoplastic

- Casein
- Cellulose acetate
- Cellulose acetate butyrate
- Cellulose nitrate
- Ethyl cellulose
- Polyvinylacetate
- Vinyl acetate/ethylene copolymer
- Polyvinylchloride
- Polyvinylchloride (chlorinated)
- Polyvinylidene chloride
- Vinyl chloride/vinyl acetate copolymer
- Vinyl chloride/vinylidene chloride copolymer
- Polyvinyl alkyl ether
- Polyvinyl butyral
- Polyvinyl formal
- Cyanoacrylate
- Phenoxy
- Polyamide (nylon type)
- Polymethyl methacrylate
- Polystyrene
- Waxes (mineral)

Elastomers

- Butadiene/acrylonitrile rubber
- Butadiene/styrene rubber
- Butyl rubber
- Natural rubber (latex)
- Natural rubber (chlorinated)
- Polychloroprene rubber
- Polyisobutylene rubber
- Polyisoprene rubber
- Polysulphide rubber
- Polyurethane rubber
- Reclaim rubber
- Silicone rubber
- Styrene/acrylonitrile rubber

Thermosetting

- Melamine formaldehyde
- Phenol formaldehyde
- Resorcinol formaldehyde
- Urea formaldehyde
- Phenolic/nitrile
- Phenolic epoxy
- Silicone resins
- Epoxy+polyamide
- Epoxy+polyamine
- Epoxy+polyamine (aromatic)
- Epoxy novolac
- Epoxy (cycloaliphatic)
- Epoxy polysulphide
- Polyimide
- Polyurethane

Inorganic

- Glass ceramic
- Soluble silicates
- Sulphur

Large range of values 20 → 140

0   20   40   60   80

Table 9.20.

Linear Coefficient of Expansion $\frac{\Delta l}{l} \times 10^{-5}\,°C$

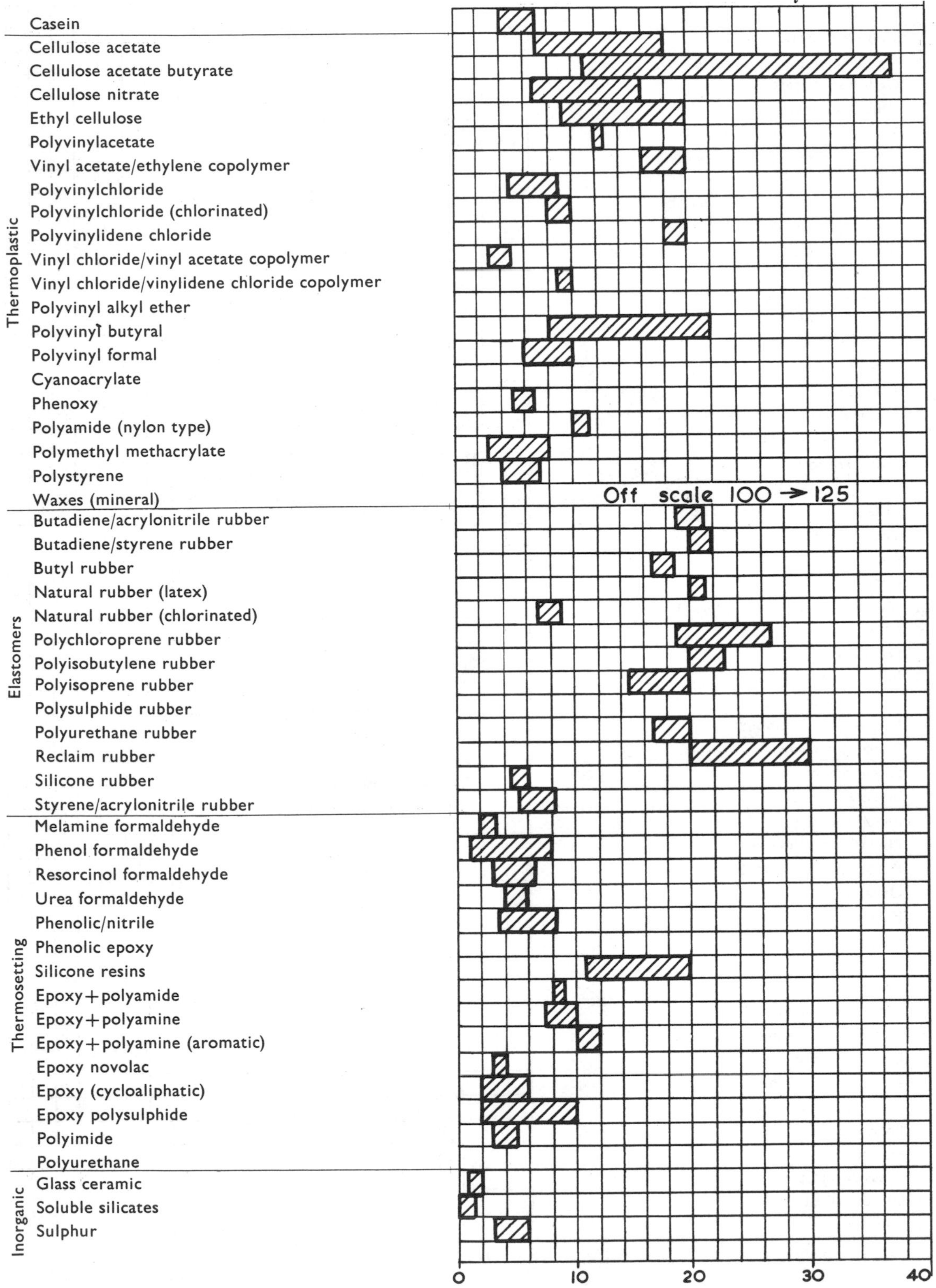

Thermoplastic

- Casein
- Cellulose acetate
- Cellulose acetate butyrate
- Cellulose nitrate
- Ethyl cellulose
- Polyvinylacetate
- Vinyl acetate/ethylene copolymer
- Polyvinylchloride
- Polyvinylchloride (chlorinated)
- Polyvinylidene chloride
- Vinyl chloride/vinyl acetate copolymer
- Vinyl chloride/vinylidene chloride copolymer
- Polyvinyl alkyl ether
- Polyvinyl butyral
- Polyvinyl formal
- Cyanoacrylate
- Phenoxy
- Polyamide (nylon type)
- Polymethyl methacrylate
- Polystyrene
- Waxes (mineral)

Off scale 100 → 125

Elastomers

- Butadiene/acrylonitrile rubber
- Butadiene/styrene rubber
- Butyl rubber
- Natural rubber (latex)
- Natural rubber (chlorinated)
- Polychloroprene rubber
- Polyisobutylene rubber
- Polyisoprene rubber
- Polysulphide rubber
- Polyurethane rubber
- Reclaim rubber
- Silicone rubber
- Styrene/acrylonitrile rubber

Thermosetting

- Melamine formaldehyde
- Phenol formaldehyde
- Resorcinol formaldehyde
- Urea formaldehyde
- Phenolic/nitrile
- Phenolic epoxy
- Silicone resins
- Epoxy+polyamide
- Epoxy+polyamine
- Epoxy+polyamine (aromatic)
- Epoxy novolac
- Epoxy (cycloaliphatic)
- Epoxy polysulphide
- Polyimide
- Polyurethane

Inorganic

- Glass ceramic
- Soluble silicates
- Sulphur

0   10   20   30   40

300

Table 9.21.

Dielectric Constant at 60 Hz

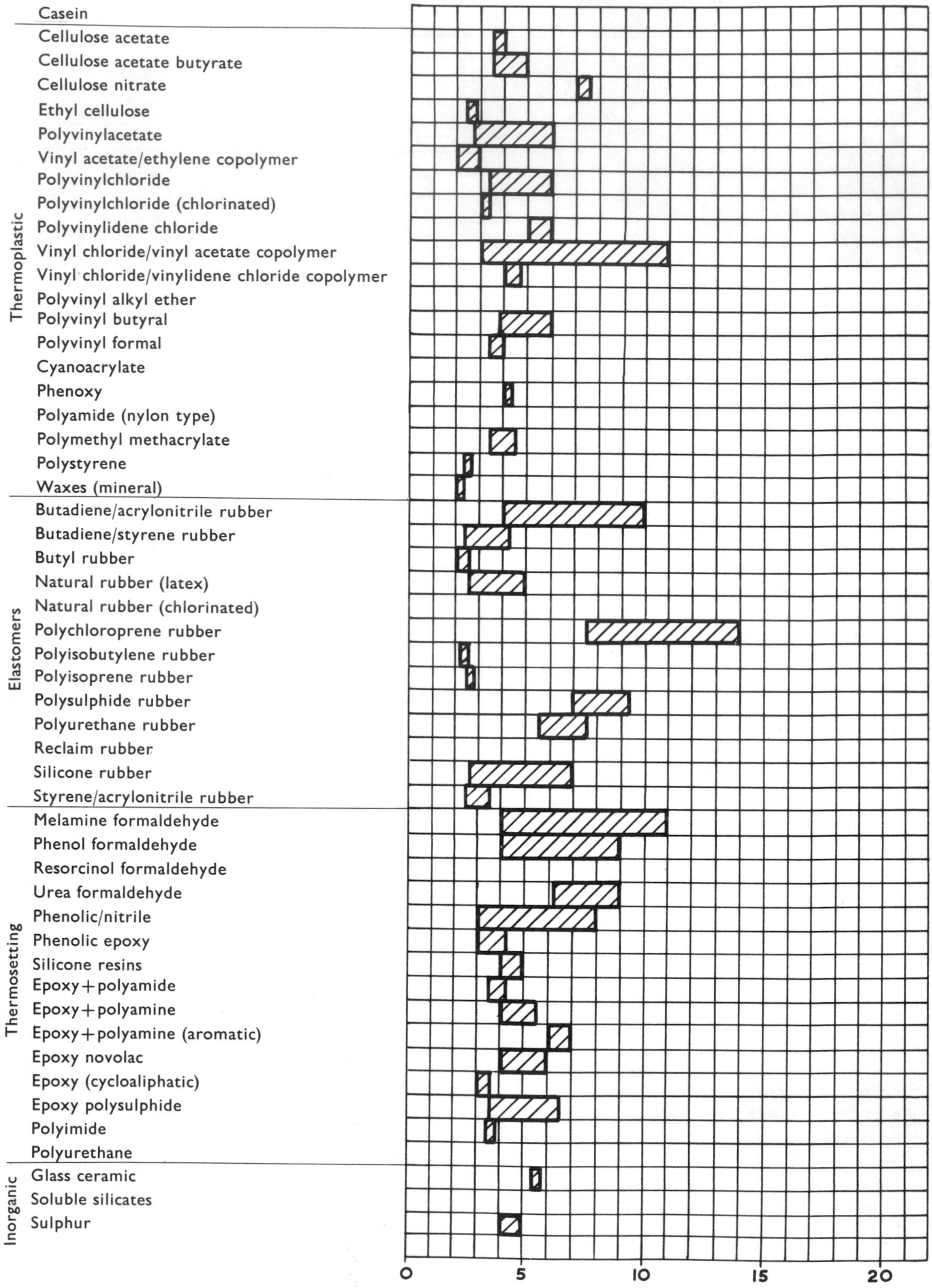

Thermoplastic
- Casein
- Cellulose acetate
- Cellulose acetate butyrate
- Cellulose nitrate
- Ethyl cellulose
- Polyvinylacetate
- Vinyl acetate/ethylene copolymer
- Polyvinylchloride
- Polyvinylchloride (chlorinated)
- Polyvinylidene chloride
- Vinyl chloride/vinyl acetate copolymer
- Vinyl chloride/vinylidene chloride copolymer
- Polyvinyl alkyl ether
- Polyvinyl butyral
- Polyvinyl formal
- Cyanoacrylate
- Phenoxy
- Polyamide (nylon type)
- Polymethyl methacrylate
- Polystyrene
- Waxes (mineral)

Elastomers
- Butadiene/acrylonitrile rubber
- Butadiene/styrene rubber
- Butyl rubber
- Natural rubber (latex)
- Natural rubber (chlorinated)
- Polychloroprene rubber
- Polyisobutylene rubber
- Polyisoprene rubber
- Polysulphide rubber
- Polyurethane rubber
- Reclaim rubber
- Silicone rubber
- Styrene/acrylonitrile rubber

Thermosetting
- Melamine formaldehyde
- Phenol formaldehyde
- Resorcinol formaldehyde
- Urea formaldehyde
- Phenolic/nitrile
- Phenolic epoxy
- Silicone resins
- Epoxy+polyamide
- Epoxy+polyamine
- Epoxy+polyamine (aromatic)
- Epoxy novolac
- Epoxy (cycloaliphatic)
- Epoxy polysulphide
- Polyimide
- Polyurethane

Inorganic
- Glass ceramic
- Soluble silicates
- Sulphur

0   5   10   15   20

Table 9.22.

Dielectric Constant at 1 MHz

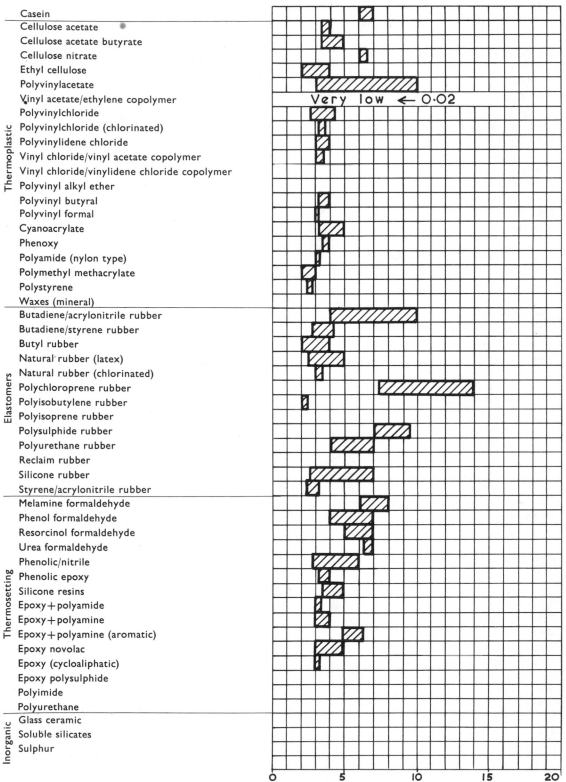

Very low ← 0·02

Thermoplastic
- Casein
- Cellulose acetate
- Cellulose acetate butyrate
- Cellulose nitrate
- Ethyl cellulose
- Polyvinylacetate
- Vinyl acetate/ethylene copolymer
- Polyvinylchloride
- Polyvinylchloride (chlorinated)
- Polyvinylidene chloride
- Vinyl chloride/vinyl acetate copolymer
- Vinyl chloride/vinylidene chloride copolymer
- Polyvinyl alkyl ether
- Polyvinyl butyral
- Polyvinyl formal
- Cyanoacrylate
- Phenoxy
- Polyamide (nylon type)
- Polymethyl methacrylate
- Polystyrene
- Waxes (mineral)

Elastomers
- Butadiene/acrylonitrile rubber
- Butadiene/styrene rubber
- Butyl rubber
- Natural rubber (latex)
- Natural rubber (chlorinated)
- Polychloroprene rubber
- Polyisobutylene rubber
- Polyisoprene rubber
- Polysulphide rubber
- Polyurethane rubber
- Reclaim rubber
- Silicone rubber
- Styrene/acrylonitrile rubber

Thermosetting
- Melamine formaldehyde
- Phenol formaldehyde
- Resorcinol formaldehyde
- Urea formaldehyde
- Phenolic/nitrile
- Phenolic epoxy
- Silicone resins
- Epoxy+polyamide
- Epoxy+polyamine
- Epoxy+polyamine (aromatic)
- Epoxy novolac
- Epoxy (cycloaliphatic)
- Epoxy polysulphide
- Polyimide
- Polyurethane

Inorganic
- Glass ceramic
- Soluble silicates
- Sulphur

0   5   10   15   20

Table 9.23. Power Factor ($\times 10^{-3}$) at 60 Hz

Thermoplastic	
Casein	
Cellulose acetate	
Cellulose acetate butyrate	
Cellulose nitrate	
Ethyl cellulose	
Polyvinylacetate	
Vinyl acetate/ethylene copolymer	
Polyvinylchloride	
Polyvinylchloride (chlorinated)	
Polyvinylidene chloride	
Vinyl chloride/vinyl acetate copolymer	
Vinyl chloride/vinylidene chloride copolymer	
Polyvinyl alkyl ether	
Polyvinyl butyral	
Polyvinyl formal	
Cyanoacrylate	
Phenoxy	
Polyamide (nylon type)	
Polymethyl methacrylate	
Polystyrene	
Waxes (mineral)	

Elastomers	
Butadiene/acrylonitrile rubber	
Butadiene/styrene rubber	
Butyl rubber	
Natural rubber (latex)	
Natural rubber (chlorinated)	
Polychloroprene rubber	
Polyisobutylene rubber	
Polyisoprene rubber	
Polysulphide rubber	
Polyurethane rubber	
Reclaim rubber	
Silicone rubber	
Styrene/acrylonitrile rubber	

Thermosetting	
Melamine formaldehyde	
Phenol formaldehyde	
Resorcinol formaldehyde	
Urea formaldehyde	
Phenolic/nitrile	
Phenolic epoxy	
Silicone resins	
Epoxy+polyamide	
Epoxy+polyamine	
Epoxy+polyamine (aromatic)	
Epoxy novolac	
Epoxy (cycloaliphatic)	
Epoxy polysulphide	
Polyimide	
Polyurethane	

Inorganic	
Glass ceramic	
Soluble silicates	
Sulphur	

Axis: 0   50   100   150   200

Table 9.24.   Power Factor ($\times 10^{-3}$) at 1 MHz

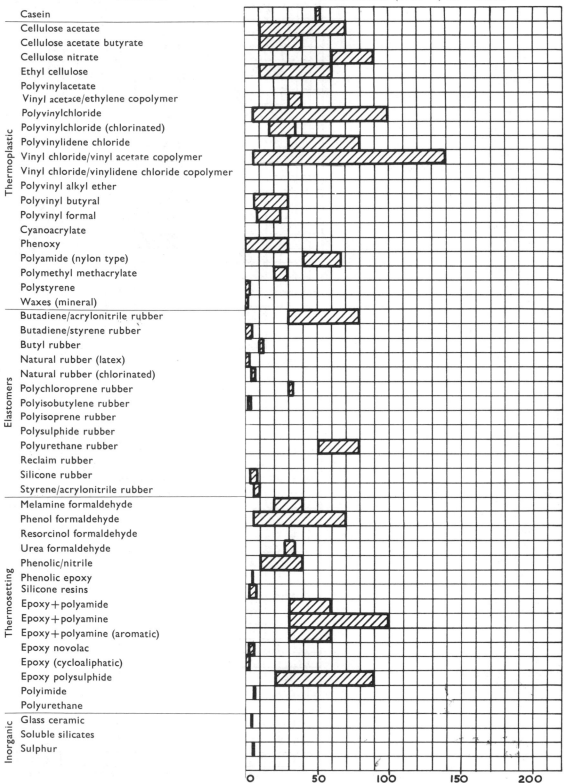

Thermoplastic
- Casein
- Cellulose acetate
- Cellulose acetate butyrate
- Cellulose nitrate
- Ethyl cellulose
- Polyvinylacetate
- Vinyl acetate/ethylene copolymer
- Polyvinylchloride
- Polyvinylchloride (chlorinated)
- Polyvinylidene chloride
- Vinyl chloride/vinyl acetate copolymer
- Vinyl chloride/vinylidene chloride copolymer
- Polyvinyl alkyl ether
- Polyvinyl butyral
- Polyvinyl formal
- Cyanoacrylate
- Phenoxy
- Polyamide (nylon type)
- Polymethyl methacrylate
- Polystyrene
- Waxes (mineral)

Elastomers
- Butadiene/acrylonitrile rubber
- Butadiene/styrene rubber
- Butyl rubber
- Natural rubber (latex)
- Natural rubber (chlorinated)
- Polychloroprene rubber
- Polyisobutylene rubber
- Polyisoprene rubber
- Polysulphide rubber
- Polyurethane rubber
- Reclaim rubber
- Silicone rubber
- Styrene/acrylonitrile rubber

Thermosetting
- Melamine formaldehyde
- Phenol formaldehyde
- Resorcinol formaldehyde
- Urea formaldehyde
- Phenolic/nitrile
- Phenolic epoxy
- Silicone resins
- Epoxy+polyamide
- Epoxy+polyamine
- Epoxy+polyamine (aromatic)
- Epoxy novolac
- Epoxy (cycloaliphatic)
- Epoxy polysulphide
- Polyimide
- Polyurethane

Inorganic
- Glass ceramic
- Soluble silicates
- Sulphur

0   50   100   150   200

Table 9.25.

Dielectric Strength kV/cm

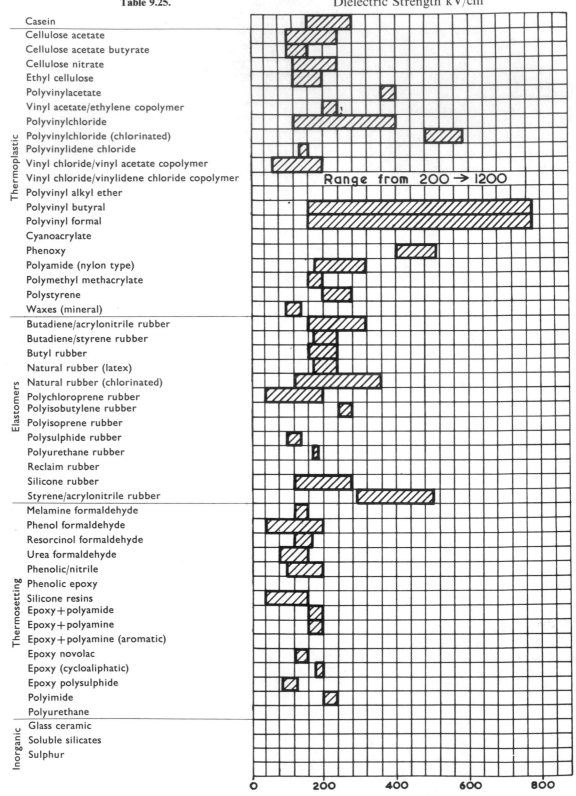

Range from 200 → 1200

Table 9.26. Volume Resistivity Ω/cm

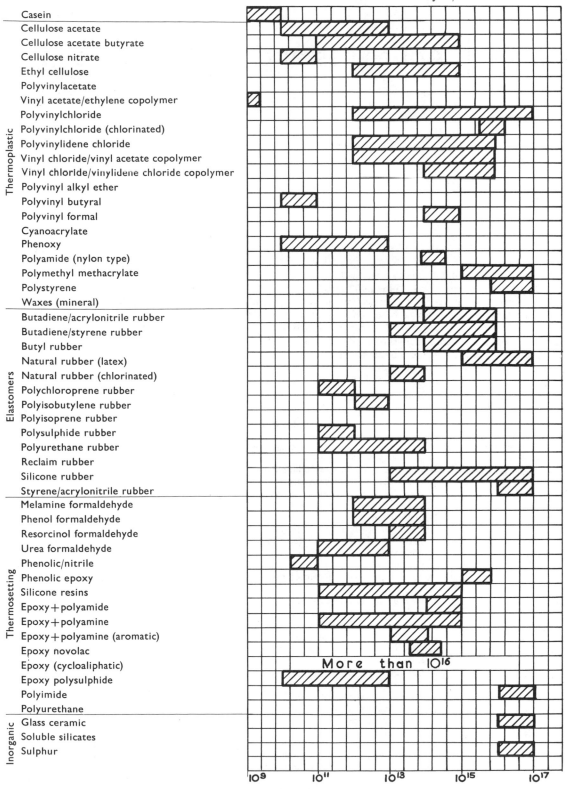

| # Adhesives and Trade Sources

10.1. TRADE NAMES LIST

Adhesives are known and sold commercially under a variety of trade names which, to the uninitiated, give little information as to their chemical nature and application. The table below is based on an analysis of current trade literature and other published material and lists the trade names of adhesives which are presently available in the United Kingdom.

The trade names, arranged alphabetically, are followed in turn by descriptions which designate the basic chemical composition and/or give some

indication of the form and use of the adhesive. The list includes only those materials which are actual adhesive products and, with few exceptions, no representation has been given to the considerable number of raw materials which are known to form the basis of adhesive systems.

A list of manufacturers against code numbers is given on p. 322 (Table 10.3).

Additional sources of information on adhesive materials will be found in the references at the end of Table 10.1.

Table 10.1.

Trade Names	Type, Description or Uses	Code No.
Aatex	Natural and synthetic latex adhesives for footwear manufacture	13
Acalor	Range of chemically resistant cements for tiles	1
Acebrand	Animal glues for paper and paint manufacture	2
Acrifix	Range of adhesives based on acrylic resins	81
Acrofix	Dental cement	93
Acronal	Acrylic ester polymers and co-polymer emulsions, solids, and solutions	30
Acrulite	Cold curing acrylic resins used as adhesives for perspex	237
Adamastic	Polishing wheel liquid cement	2
Adam's	Liquid skin glue	2
Adcol	Range of starch based adhesives for paper bag manufacture, labelling and bookbinding	3
Adcote	Extensive range of adhesives for packaging, plastic films, laminates and metals, etc	197
Adeer	Cements for floor coverings and fibre glass insulation	257
Adpep	Adhesive for polystyrene tile bonding	
Adsol	Solvent based rubber adhesives	5
Aero	Animal glue for woodworking trades	88
Aerocol	Polyvinyl acetate adhesives	73
Aerodux	Resorcinol formaldehyde wood glues	73
Aerolastic	Rubber-pitch joint sealant	109
Aerolite	Urea formaldehyde wood glues	73
Aerophen	Phenol formaldehyde resins	73
Aerosol, 873B	Synthetic rubber/resin adhesive in aerosol form	92
Aerospace	One part silicone rubber adhesive and sealant	96
AF	Epoxy resin	193
Agf	Urea formaldehyde gap filling adhesive for porous materials	278
Agomet	Epoxy resin based adhesives for metals, ceramics, plastics, etc	59

continued

Table 10.1 continued

Trade Names	Type, Description or Uses	Code No.
Agrippa	Range of adhesives for civil engineering and building applications	178
Albertol	Phenolic resins with resin modifier	136
Albumenoid	General purpose fish glue	8
Alcotex	Polyvinyl alcohol products	230
Alcyn	Cyanoacrylate based adhesives	107
Alfadite	Range of casein glues	2
Alfatalat	Alkyd resin products	230
Alfrax	Aloxite cements	63
Alganol	Dental cement	93
Alpha-Ace	Cyanoacrylate adhesive	260
Amco	Adhesive for bonding expanded polystyrene to metal	46
Amset	Range of fast setting animal glue adhesives for packaging industry	264
Amtex	Range of adhesives based on latex and polyvinyl acetate	46
Ancamide	Polyamide curing agents for epoxy resins	11
Ancamine	Amine based curing agents for epoxy resins	11
Anchor-weld	Range of adhesives based on polyvinyl acetate, nitrile, neoprene and other synthetic rubbers	11
Anderson-Mastic	Adhesive for bituminous substrates	12
Anglo	Natural and synthetic rubber solvent based adhesives for the footwear industry	13
AP Adhesive	Nitrocellulose/synthetic rubber adhesive	128
Aquapaint	Range of epoxy resin adhesives and coatings	226
Aquaseal	Bitumen/rubber based adhesives and sealants for flooring applications	34
Aquastik	Polyvinyl acetate based adhesives	34
Arabol	Trade name for self-adhesive tapes	46
Arafix	General purpose impact adhesive	46
Araldite	Range of epoxy resin adhesives and casting resins	73
Araprene	Range of adhesives based on neoprene rubber	46
Arasite	Adhesive for bonding insulation materials to metal	46
Arastik	Adhesive for bonding insulation materials to metals	46
Arastix	Rubber-based adhesive	46
Ardux	Urea adhesive for bonding rigid panels to vertical or uneven wall surfaces	73
Armourglaze	Range of epoxy resin coatings and adhesives	164
Aro Mark III	Hot-melt adhesive system	15
As	Synthetic resin pastes and emulsions	5
Atlas	Two-part synthetic resin adhesive	18
Atlas-Ago	Synthetic resin adhesives	59
Atmosel	Silica based cement	126
Autobond	General purpose adhesive	235
Autoplax	Range of adhesives for polythene and polyvinyl chloride films based on epoxy and other synthetic resins	20
Autostic	Sodium silicate based cement	64
Avdelbond	Range of cyanoacrylate adhesives	21
Bakelite	Range of synthetic resin based adhesives	50
Bakelite cement	Stoving cement based on phenolic resins	23
Bal-epoxy	Epoxy resin adhesive	58
Bal-fix	Lightweight ceramic tile adhesive	58
Bal-flex	Two-part rubber-latex cement based ceramic tile adhesive	58
Bal-floor	Bitumen based adhesive for ceramic and mosaic floors	58
Bal-mix	Cement based wall tile fixative for interior or exterior use	58
Bal-proof	Synthetic rubber/resin based adhesive for ceramic materials and tiles	58
Balsa	General purpose adhesive for balsa wood models	128, 262
Bal-tad	Polyvinyl acetate based adhesive for ceramic tiles	58
Bateman's	Non-inflammable rubber based adhesives	31
Beckopox	Epoxy resins and hardeners	136
Beetle	Range of adhesives based on urea or melamine formaldehyde in liquid or powder form for furniture assembly	38
Belzona	Adhesive/sealant with metallic powder base	
Betaseal	Adhesive/sealants based on polysulphide, butyl and acrylic materials	35
Betastay	Range of rubber based adhesives	35
Betol	Range of adhesives, pastes and gums	36

continued

Table 10.1 continued

Trade Names	Type, Description or Uses	Code No.
Betta Bonda	Polyvinyl acetate adhesive	16
Bevaloid	Vinyl acetate copolymer based adhesives for packaging industry	135
Bexol	Range of synthetic/rubber based adhesives and solutions	23
Bilaseal	Epoxy resin based putty	37
Bimul	Range of adhesives based on bitumen emulsions	46
Bodex	Range of dextrine gums for bottle labelling	46
Bondaglass	Glass fibre and synthetic polyester resin repair kit for metals	42
Bondapaste	Two-part synthetic polyester resin adhesive and filler for metal repairs	42
Bond-a-snak	Range of flexible adhesives for packaging	200
Bondcrete	Polyvinyl acetate bonding and sealing agent for concrete and woodwork	33
Bondfast	Polyvinyl acetate based adhesive	262
Bond GT	Thermosetting resin adhesive for film lamination	243
Bonding Agent 007	Epoxy resin based adhesive	212
Bondmaster	Range of adhesives based on synthetic resins/rubbers	242
Boscoprene	Two-part synthetic resin/rubber cements for sealing and coating	47
Boscotex	Aqueous synthetic polymer dispersions for PVC	47
Bostik	Range of synthetic resin rubber adhesives, sealing and coating compounds	47
Boston	Range of adhesives for the footwear industry	47
BPL	Range of adhesives and elastomeric compounds	54
Breon	Range of nitrile rubbers and latices	49
Brimor	High temperature strain gauge cements based on aluminium phosphate and silica	234
Britfix	Epoxy resin and polystyrene based adhesives for general purpose use	140
Brown Plastic Cement	Adhesive for floor coverings and fibre glass insulations	257
BSL	Range of film adhesives and surface protection solutions	73
Bulldog	Synthetic rubber cement for bonding polyurethane foam; gummed tapes for packaging	47
Butterfly	Range of gummed, self adhesive and heat-seal papers and tapes	160
Butvar	Polyvinyl butyral adhesive resins	250
Cabufix	Range of adhesives for cellulose acetate butyrate bonding	185
Calbar	Liquid animal glue	26
Carbofrax	Silicon carbide refractory cement	63
Cariflex	Range of synthetic rubber formulations based on styrene butadiene, polybutadiene, polyisoprene	252
Carpetex	Rubber latex adhesive for backing carpets and rugs	166
Casco	Casein glues in powder form for timber and plywood	46
Cascodex	Vegetable glues for bookbinding and packaging	46
Cascogel	Animal glues for packaging and bookbinding	46
Cascomelt	Range of hot-melt adhesives for packaging and bookbinding	46
Cascomite	Range of urea formaldehyde glues for wood	45
Casco-ML	Anaerobic polyester adhesive/sealants for locking metal screws	45
Cascophen	Range of phenol/resorcinol/cresylic resin based adhesives	45
Casco-resin	Range of urea formaldehyde based adhesives for wood	45
Casco-rez	Range of polyvinyl acetate adhesives for woodworking	45
Cascosel	Casein, latex, polyvinyl acetate, rubber/resins, based adhesives for packaging	46
Cascoset	Emulsion adhesives for floor tiles	45
Cascotape	Self adhesive tapes	46
Cascote	Furane resins	45
Catabond	Phenolic and resorcinol formaldehyde adhesives and binders for wood, metals	67
Catabrase	Phenolic formaldehyde adhesives for the abrasives industry	67
Catacol	Phenolic resin based adhesives for bonding wood, metal foils to phenolic plastics	67
Cebond	Polyvinyl acetate based adhesives, cements, and concrete additives	70
Ceemar	Range of adhesive cements for the building industry	263
Cement II	Adhesive for bonding acrylic plastics	171
Cellobond	Polyester resins, phenolic and cresylic resins, urea and modified phenolic resins	55
Cellodex	Vegetable adhesives	26
Cellofas	Carboxy methyl cellulose adhesive for wallpaper applications	144
Cellu-gum	Adhesive for stationery applications	200
Cerafix	Polyvinyl acetate paste for ceramic tiles	278
Certite	Polyester resin based materials	241
Certofix	Fish glue for general purposes	68
Chain	Pearl and powder animal glue products	232

continued

Table 10.1 continued

Trade Names	Type, Description or Uses	Code No.
Chelatex	Rubber latex–water based products	66
Chelsea	Range of adhesives for shoe trade, based on neoprene and natural rubbers	66
Chemex	Adhesive for repairing rubber tyres	235
Chemgrip	Epoxy resin based adhesives for bonding fluoropolymers	184
Chemlok	Range of epoxy resin based adhesives, and rubber to metal bonding primers	100
Chemtite	Bottle labelling adhesive resistant to water/ice	200
Chinafix	Neoprene based adhesive for repairing porcelain and pottery	158
Chromix	High temperature, mineral based cement	188
Chuk'ka	Cement for repairing china, glass, porcelain, pottery, etc	79
Clam	Range of adhesives based on synthetic rubbers and resins for packaging, fabrics, tiles, etc	177
Claysil	Sodium silicate based material with clay filler	89
Cleanfix	Polyvinyl acetate adhesive for wood and porous materials	262
Clear Adhesive	General purpose adhesive for domestic use	262
Clipfas	Liquid adhesive for bonding nylon	18
CMW	Range of dental cements and carpet bonding adhesives	75
Coat-rez	Solvent-resin coating formulations for textiles and food packaging materials	61
Cobond	Polyvinyl acetate emulsions for woodworking and packaging applications	85
Coflex	Polyvinyl acetate emulsions for bookbinding, packaging, stationery, etc	85
Colegrip	Industrial adhesives	77
Collapress 95	Adhesive for woodwork	260
Colma-fix	Epoxy resin based adhesive	254
Colma Grout	Filled epoxy resin levelling compound for building applications	254
Colset	Range of adhesives based on bitumen	39
Comet	Liquid skin glue	2
Comet Cold	Cold liquid skin glue	2
Conductive adhesive	Synthetic polymer adhesive with metal powder filler	8
Copper	Dental cement	83
Copybind	Carpet binding paste	79
Copydex	General purpose latex adhesive for textiles, carpet binding, paper, bookbinding	79
Coragum	Range of starch, dextrine, polyvinyl acetate, and hot-melt adhesives	82
Corkbond	Synthetic solvent based adhesives	215
Corro-proof	Epoxy and polyester resin cements for ceramic tiles and concrete repair	83
Cow	Range of adhesives based on synthetic rubbers and resins	84
Cox's	Range of animal hide glues	85
Creteform	Epoxy based adhesive with filler, for the building trades	133
Crispin	General purpose adhesives	87
Croid	Range of industrial adhesives based on natural and synthetic polymers and rubbers	88
Crystal Clean 513	Methyl cellulose based adhesive	260
CT	Polyvinyl acetate adhesive for woodworking	83
CTF	Ceramic tile fix plasticised cement for floor and wall tiles	58
CT-S	Synthetic rubber blend adhesive	98
Cubix	Animal glues for woodworking and bookbinding, in solid form	251
Culminal	Methyl cellulose based adhesive for wallpaper application	260
CXL	Range of epoxy resin adhesives and coatings	77
Cyanolit	Ethyl or methyl cyanoacrylate based adhesives	149
Cybond	Modified polyurethane based adhesive	91
3D	Range of adhesives for industrial applications	4
4D	Polyvinyl acetate adhesive	128
Daltoflex	Polyurethane based formulations for fabric, plastics, textiles, leather bonding	144
Davol	Latex based adhesive for surgical appliances	119
D-B	Range of epoxy resin based, gap-filling adhesives	133
DCMC	General purpose contact adhesive, of solvent-rubber type, in aerosol form	92
DEK	Epoxy resin based adhesives for repairing masonry, china, ceramics, glass, etc	16
Delseal	Adhesive coated paper for labelling	255
Delta-bond	Epoxy resin adhesive with filler	157
Denex	Epoxy resin adhesives for strain gauges	271
Desmocoll	Polyurethane and polyester based resins for adhesives	32
Desmodur	Range of isocyanates for adhesive manufacture	32
Desogrip	Neoprene resin adhesives in solvent base, for bonding natural and synthetic rubbers	215

continued

Table 10.1 continued

Trade Names	Type, Description or Uses	Code No.
Desoplas	Polyurethane based adhesives	215
Devcon	Range of polyurethane based adhesives (Flexanes) for industrial applications; epoxy resins	94
Dex	Dextrine based adhesive	16
DFL	Range of dental cements	93
Domolac	Cellulose based adhesives for glass fibre and electrical insulants	95
Double-bond	Range of epoxy resin based adhesives with fillers	133
Dow	Range of epoxy resins and adhesives for industrial applications	97
Dow-Corning	Range of adhesives and sealants based on silicone rubber formulations	97
Drystick	One-part screen adhesives for labels, nameplates and transfers	249
Dufix	Polyvinyl acetate emulsion based adhesive for general use	144
Dulite	Adhesive for wallpaper application	143
Dunlop	Range of rubber based adhesives and sealants	98
Dupoxy-resin	Epoxy resin adhesives for flooring applications	101
Durofast	Synthetic rubber based contact adhesive	224
Durofix	Heat resistant cellulose acetate cement for repairing pottery, porcelain, ceramics, etc	224
Duroflex	Flexible packaging adhesives	200
Duroglue	Animal glue for bonding leather, wood, paper and porous materials	224
Duro-lok	Thermosetting emulsions for woodworking applications	200
Duro-tac	Solvent based adhesive for pressure sensitive tapes, and film lamination applications	200
Duroxyn	Adhesives based on epoxy resins	136
Dustfas	Rubber based adhesive	18
Dynaglue	Adhesive for bonding flooring materials based on wood	281
Dyox	Cellulose based adhesive for wallpaper	260
Easiclean	Range of emulsion adhesives for packaging	200
Eastman 910	Cyanoacrylate based adhesive	103, 73
Eastobond	Hot-melt adhesives for the packaging industry	
EC	Range of synthetic rubber adhesives	193
Ecco amp	Electrically conductive adhesives and coatings	105
Eccobond	Adhesives, cements and sealants, based on epoxy resins, synthetic polymers and rubbers	105
Eccoceram	Inorganic adhesives	105
Eccosil	Silicone polymer based adhesive	105
EE	Polyvinyl acetate emulsions for the packaging industry	16
EL-Bond	Thermoplastic resin for film lamination	243
Elecolit	Electrically conductive epoxy/silver adhesives and coatings	
Elvacite	Acrylic resins for adhesives, lacquers and inks	99
Elvanol	Polyvinyl alcohol adhesive for tube winding and lamination	99
Elvax	Ethylene vinyl acetate based copolymer resins for hot-melt adhesives	99
Embit	Polyvinyl acetate based adhesives for envelope flaps	200
Enva-dex	Natural and synthetic polymer blends for envelope adhesives	200
Envalok	Synthetic resin adhesive for envelope flaps	46
EP	Polyester adhesive lacquers	61
EPA	Filled polyvinyl acetate adhesive for polystyrene and wood fibre ceiling tiles	278
Epic	Range of epoxy and synthetic resin adhesives	106
Epikote	Epoxy resins	252
Epikure	Range of curing agents for epoxy resin based adhesives	252
Epok	Surface coating resins and resin emulsion adhesives	49
Epophen	Epoxy resins in powder or liquid form	45
Epoxi-patch	Range of epoxy resin based adhesives and sealants	109, 117, 142
Epoxy-pitch	Epoxy-pitch based adhesive for bonding epoxy coatings	164
EPS	Expanded polystyrene tile adhesive	218
Eskabond	Neoprene based adhesive for the footwear industry	269
Evatane	Hot-melt adhesive resins	110
Evode-insul	Range of adhesives and accessory materials for thermal insulation	107
Evo-stik	Range of adhesives based on natural and synthetic rubbers and polymers	107, 283
Evo-tex	Non-inflammable emulsions for wallpanel applications	107
Excel	Range of adhesives based on natural rubbers	108
Expandabond	Polyvinyl acetate emulsion	109

continued

Table 10.1 continued

Trade Names	Type, Description or Uses	Code No.
Expandite	Range of epoxy resin based adhesives	109
Fabrex	Rubber/latex carpet adhesive	88
Fastbond	Contact cement for wallboards, wood veneers, laminates, etc	193
Fiborclad	Inorganic adhesive cement for wall and sealing boards and tiles, etc	114
Filmgrip	Laminating adhesive for bag manufacture	189
Fimodyn	Cyanoacrylate based adhesive	
Fimofix	Cyanoacrylate based adhesive	
Firefly	Asbestos tape with polyvinyl ether adhesive, for oven door insulation	275
Firefrax	Alumina/silicate based cements	63
'Five minute' epoxy	Fast setting epoxy resin adhesive	94
Fixol	Adhesives based on natural products, for packaging and bookbinding	16
Fixuma	Polymer emulsion based plastic cement for wall tiles	56
Flamingo	Polystyrene tile adhesive	218
Flash Seal	Adhesive gums for envelope manufacture	189
Flat Rez	Adhesives for stationery manufacture	189
Flatsheet	Dextrine coated sheet reactivated by moistening	255
Flexane	Polyurethane based adhesives	94
Flexfas	Filled latex cement adhesive for thermal insulation applications	18
Flexhide	Flexible animal hide glues for bookbinding	2
Flexibond	Range of polyvinyl acetate emulsions, hot melts and other types of adhesive for bookbinding and joinery	291
Flexobond	Adhesive-mastic	254
Flintkote	Range of adhesives for building and flooring industries, based on bitumen, polyvinyl acetate, synthetic rubber-resin emulsions	112
Floortex	Compounds based on rubber for cement flooring compositions	238
Foamgrip	Acrylic emulsion	
Fobel Thriftile	Adhesive for ceramic tiles	170
Fobel warm	Adhesive for wall veneering	170
Formvar	Polyvinyl formal resins	250
Fortafix	High temperature adhesive cements based on sodium silicate or inorganic minerals for thermal, dielectric and mechanical applications	114
Foss	Range of rubber based adhesives	118
Foster	Coatings, sealants and adhesives for thermal insulation applications	18
FPT/NA	Range of nitrile rubber based adhesives	115
FPT/NEA	Range of neoprene rubber based adhesives	115
Gasil	Silica gels for use in adhesive formulations	89
Gelva	Polyvinyl acetate resins for hot-melt and pressure sensitive adhesive formulations	250
Gelvatol	Polyvinyl alcohol resins	250
Genclor S	Chlorinated PVC adhesive resin	110
Girder	Contact adhesives in film form	47
Gloy	Range of polyvinyl acetate, starch, dextrine, and gum arabic based adhesives for general use	16
Gluak	High viscosity dextrine based adhesives	16
Glucine	Starch based glue for paper and cardboard	175
Glumaster	Animal glue adhesives for packaging industry	264
Glutina	Liquid animal glue	251
GMS	Range of pressure sensitive adhesives	195A
Goldengrain	Range of bone and hide glues, available as powders, jellies and liquids	2
Gripfix	Dextrine based adhesive	262
Gripmount	Supported, pressure sensitive adhesive film for general purpose	47
Gripso	Range of adhesives	25
Gripsobond	Range of adhesives	25
Gripsotex	Range of adhesives	25
Gripsotite	Neoprene and polyurethane based adhesive cements for the footwear industry	25
Guarcol C10	Vegetable colloid glue for bonding leather to glass	9
Handy-set	Epoxy resin based adhesives	164
Harcosuper	Silicate based adhesive	130
Heatherbond	Epoxy resin based adhesives	110
Heatherbond Plasteel	Epoxy resin based adhesive with steel filler	110
Heat Set	Heat activated liquid adhesive which sets by gelation	200

continued

Table 10.1 continued

Trade Names	Type, Description or Uses	Code No.
Held	Range of animal glue products	131
Heldite	Jointing compound	131
Henkel	Range of industrial adhesives and resins	260
Herberts EP	Adhesive lacquers based on saturated polyesters	61
Hermatite	Range of jointing and sealing compounds	133
Hi-bond	Range of adhesives for the building industry	10
Hidux	High temperature structural adhesives based on modified phenolic resins	73
Hifix	Adhesive for paper-hanging and bill posting	144
HMG	Range of adhesives based on polyvinyl acetate, polystyrene, cellulosic and other polymers	128
Hobar	Flexible jelly glue	26
Hodgson's	Range of animal glues for paper and packaging	135
Holdenite	Range of adhesives for lamination	137
Hold-Tac	Pressure sensitive adhesives	200
Holdtite	Range of adhesives based on natural and synthetic rubbers, for packaging trades	16
Howbond	Epoxy and resorcinol formaldehyde based adhesive cements for flooring applications	139
Howfix	Synthetic rubber/resin adhesives for general purpose bonding	139
How Stik	Inflammable solvent/rubber based adhesive	139
Howtex	Non-inflammable water based adhesives for indoor building and repair applications	139
HT	Phenolic epoxy structural adhesive for metals	91
Hysol	Range of epoxy resin adhesives for industrial and aerospace use	142
Ice-tight	Casein based adhesive for packaging	160
Idenden	Polyvinyl acetate copolymer emulsion adhesives for packaging and bookbinding	145
Igas	Range of adhesive-mastic jointing materials	254
Imfix	Range of adhesives for paper-hanging and bill posting	144
Impact	Range of synthetic rubber/resin contact adhesives	107
Imseal	Synthetic resin coated heat sealing paper	255
Indasol	Range of solvent based adhesives	148
Indatex	Water based adhesive	148
Index	Range of adhesives	146
Insol	Casein cements	88
Instabond	Neoprene based adhesive for plastic laminate bonding	113
Instanter	Liquid glue	199
Instantex	Latex based adhesive for automobile industry	269
Instantflex	Range of hot-melt adhesives for bookbinding	200
Instant-Lok	Range of hot-melt adhesives for packaging	200
Instaweld	Hot-melt adhesives for woodworking and industrial assembly	200
Insul	Range of adhesives for bonding thermal insulation materials	107
Intutherm	Fire-proof coating for steel	95
Invisibond	Two-part polyester adhesive with gap filling properties	79
IS-products	Rapid setting cyanoacrylate adhesives	161
Isobond	Polyvinyl acetate emulsion adhesive and primer	253
Isoclad	Two-part polyol based adhesive	190
Isoseal	Oleoresinous sealant for wood and cork floors	214
Jelly gum	Bottle labelling adhesive products	200
Jet-patch	Dextrine and polyvinyl acetate based adhesives for the stationery trades	200
Jet-seam	Dextrine and vegetable based adhesives for envelope manufacture	200
Jewellers cement	Adhesive for fixing gem stones	128
Jiffy bind	Carpet binding adhesive paste	79
Jiffytex	Range of latex adhesives for floor coverings and carpets	158
Jiffyweld	Two-part adhesives	158
Joy	Polyvinyl acetate, cellulose acetate, cements for workshop and domestic use	274
Joy-peardrop	Cellulose nitrate based adhesive for glass, pottery, cardboard, etc	274
Joy-plane	Polystyrene cement for domestic use	274
Joy-stixin	Cellulose nitrate based adhesive for china, pottery, etc	274
Kaybond	Natural rubber based adhesive	269

continued

Table 10.1 continued

Trade Names	Type, Description or Uses	Code No.
Kaytex	Latex based adhesive	269
Kelbond	Neoprene based adhesive sealant	162
Kelseal	Epoxy resin based adhesive for bonding metal and concrete; range of adhesive/sealants	162
Kenafix	Ceramic tile adhesive	10
Kendall	Range of adhesives	265
Kervit	Tile adhesive	217
Keystone	Starch, dextrine, polyvinyl acetate emulsion adhesives for packaging industry	16
Kingsnorth	Bitumen based flooring adhesives for wood blocks and asphalt tiles	34
Kleiberit	Acrylic adhesive	205
Kollerbond	Epoxy-polyamide resin adhesives	266
Kollercast	Epoxy resins for moulding	266
Korafix	Neoprene/resin based adhesives for general purpose contact bonding	215
Kor-lok	Thermosetting emulsions for the woodwork trades	200
Kuroplast	Hot-melt adhesive materials	30
Kwikfill	Polyester filler material for repairing metal	133
Kwik-Klamp	Emulsion adhesives for wood lamination	200
Kwik-Seal	Synthetic adhesives for envelope flaps	200
Lacol	Rubber-bitumen flooring adhesive	98
Lagfas	Mineral filled adhesive coating for canvas insulating materials	18
Laitzo XXX	Casein glue	65
Lamacrest	Range of adhesives based on epoxy resins and polyurethanes for the building industry	164
Laminac	Polyester adhesives for laminating	91
Lamodex	Water soluble dextrine adhesives for laminating	85A
Lamwax	Laminating waxes	61
Lankrothane	Polyurethane adhesive and coating products (Lankro Chemicals Ltd)	
Lapcell	Cellulose based adhesive for wallpaper	217
Lap-Lok	Synthetic resin adhesives	200
Laycock's	Animal glues for joinery and bookbinding	168
Lay-flat	Dextrine based adhesives for envelope manufacture	200
Lay-Tex	General purpose rubber-latex adhesive	120
Lepage	Range of glues and gums for domestic use	262
Linoglue	Natural resin based adhesive for linoleum and floor tiles	281
Lino Paste	Linoleum tile adhesive	253
Lion	Starch, dextrine and gum arabic glues for paper and cardboard, etc	175
Liquapruf	Bitumen/rubber adhesive and waterproofing emulsion	214
Load-lok	Vegetable/resin based adhesive for paper, sacks and cardboard	200
Loctite	Range of cyanoacrylate and anaerobic adhesives and sealants	161
Lok-Master	Emulsion and latex based packaging adhesives	200
Lutofan	Vinyl chloride copolymer solutions and dispersions	30
Lutonal	Range of polyvinyl ether materials for adhesives	30
3M	Range of adhesives based on epoxy resins and other synthetic polymers and rubbers	193
Macnair	Film laminating adhesives	88
Marb-l-cote Mastic	Adhesive sealant	182
Marbond II	Adhesive cement for ABS plastics	11
Marley	Range of adhesives based on bitumen, polyvinyl acetate, epoxy and other synthetic resins, and resin oil mastics	183
Mecufix	Laminating adhesives for metallised polyester films to vinyl based substrates	185
Megabond	Powder adhesive for wallpaper	206
Melomain	Range of adhesives for standard and post formed plastic laminates	73
Mendex	Epoxy resin based adhesive	155
Mendit	Range of adhesives based on starch, dextrine, polyvinyl acetate, and animal glues	187
Meno	Resin adhesive film for veneering and plywood	191
Mereco	Epoxy resin based adhesives and encapsulants	81
Mermend	Rubber latex based adhesive for textile binding	166
Metabond	Epoxy resin based adhesives	81
Metacast	Epoxy resin materials for moulding	81
Methofas	Methyl cellulose adhesives for general decorating materials	144

continued

Table 10.1 continued

Trade Names	Type, Description or Uses	Code No.
Metrotex	Rubber latex based adhesive	215
Metylan	Cellulose adhesive	260
Mhobond	Electrically conductive epoxy resin adhesive	109, 117, 142
Microlak	Adhesive lacquer	17
Micromelt	Range of resin and copolymer based hot-melt adhesives	17
Micromul	Polyvinyl acetate based emulsions	17
Milvex	Thermoplastic polyamide adhesive	86
Min-u-sil	Crystalline silica adhesive filler material in uniform micron sizes	59
Molit	Synthetic resin adhesive for rigid or flexible PVC laminating to wooden surfaces other than timber	260
Monatex	Aqueous latex or emulsion adhesives	195
Monsol	Solvent based contact adhesives	195
Moto-stik	General purpose adhesive	106, 176, 138
Mouldrite	Urea formaldehyde resins for use as adhesives in the woodworking trades	144
Multibond	Synthetic rubber adhesive for the footwear industry	283
Multiglue	Polyvinyl acetate emulsion adhesive	16
Multigrip	Range of adhesives based on synthetic rubber/resin materials	198
Multi-lok	Range of rubber contact cements and structural adhesives based on neoprene phenolics and other polymers	200
Mycroseal	Two-part adhesive and solvent solution	17
NABL	Range of adhesives	203
National	Range of adhesives for the packaging, woodworking and other industries	200
Necol	Nitrocellulose based adhesive	110, 160, 144
Nestor	Range of polyvinyl acetate emulsion and acrylic resin based adhesives for paper and packaging	111
Nevaset	Water resistant plumbers' mastic	201
Nic-o-bond	General purpose and tile fixing adhesives and cements	201
Nic-o-grout	Waterproof adhesive grouting material	201
Nitobond	Epoxy resin adhesives for concrete repair	70
Nitofix	General purpose epoxy adhesive for the building industry	70
Nitoflor	Epoxy resin based adhesive for concrete	70
Nitotile	Epoxy resin adhesive for bonding linoleum and vinyl tiles to damp screeds	70
Nitowall	Epoxy resin based surface coating material	70
Nylaweld	Adhesive for bonding nylon to nylon	219
007	Epoxy resin adhesive with filler	212
Octopus	Adhesive for wallpaper	16
ONX	Range of silica based refractory adhesive cements	90
Oradhesive	Adhesive bondage for surgical use, based on cellulose-polyisobutylene-gelatin formulation	
PA	Range of adhesives for bonding polyvinyl chloride and expanded polystyrene	219
Pac	Synthetic rubber based adhesive	79
PAFRA	Range of adhesives based on synthetic resins and rubbers for the packaging industry	207
Palerapier	Dextrine based adhesive for paper and board laminating	2
Panel	Adhesive for flooring application	98
Panoglue	Neoprene based adhesives for hardboard, etc	281
Parex	Polyvinyl acetate based adhesive for wood, wallboards and laminates	114
Pastex	Bookbinding adhesive products	200
Patex	Range of impact adhesives	260
Penloc	Range of anaerobic adhesives for sealing and fastening screws, shafts, bearings, etc	149
Perbunan	Range of synthetic rubber emulsions	32
Perflex	Non-inflammable rubber based adhesive for insulation applications	46
Permabond	Ethyl and methyl cyanoacrylate based adhesives for rapid assembly	208A, 259
Permaloc	Range of anaerobic adhesives for rapid bonding	208A, 259
Pertec	Range of contact adhesives and polyvinyl acetate emulsions, and hot-melt adhesives for bookbinding	209

continued

Table 10.1 continued

Trade Names	Type, Description or Uses	Code No.
Pevafix	Adhesive for bonding polyvinyl alcohol films	185
Phillips	Latex, neoprene and polyurethane based adhesives for the footwear industry	210
Phillisol	Natural rubber based adhesive for leather and rubber bonding	210
Pick-up	Adhesive for single- and two-ply paper tissues	264
Pitabond	Adhesive emulsion for building applications	107
Planatol	Range of adhesives for bookbinding	211
Plasbronze	Epoxy resin based adhesive with bronze powder filler	110
Plasilica	Epoxy resin based adhesive with silica filler	110
Plasinium	Epoxy resin based adhesive with aluminium filler	110
Plaskitex	Neoprene based adhesive for polyvinyl chloride	158
Plaslead	Epoxy resin based adhesive with lead filler	110
Plasmount	Supported two-way adhesive film	47
Plasteel	Epoxy resin based adhesive with steel filler	110
Plasticel	Synthetic rubber based contact adhesive	57
Plastick	Liquid animal glues for wood venecring and laminating	2
Plastic steel	Epoxy resin based adhesive with steel filler	94
Plastiflex	Polyvinyl acetate emulsion adhesives for carton sealing and packaging	46
Plastikon	Range of rubber/resin adhesives for industrial use	57
Plastilock	Range of structural adhesives	57
Plastisol	Polyvinyl chloride based adhesive	57
Plextol	Acrylic resin based adhesives	82
Pliobond	Nitrile rubber based adhesive cements and neoprene based contact adhesives	123
Pliogrip	Range of adhesive sealants and corking compounds	123
Plusbond	Synthetic rubber resin based adhesives	213
Plusgrip	Range of natural rubber based adhesives	213
Plustex	Water based polyvinyl acetate adhesives for the building and decorating industries	213
Plusweld	Polyvinyl chloride resin based adhesives for metal bonding	213
Plycol	Range of adhesives, sealants and coatings, for flooring applications	214
Plycol Isobond	Polyvinyl acetate emulsion adhesive	214
Plycolay	Cement type screed for levelling floors	214
Ply-Lok	Cross linking emulsion adhesive for wood lamination	200
Plymaster	Dry film adhesives (which may be coated on metal foils) for general industrial applications	242
Pochin	Range of adhesives based on neoprene and other synthetic rubbers	215
Polidene	Vinylidene chloride copolymer emulsions	244
Polybond	Polyvinyl acetate copolymer adhesive and concrete additive	216
Polybond	Range of adhesives for laminating packaging materials based on plastics and foils	76
Polycell	Water soluble cellulose ether based adhesive for wallpaper	217
Polyester T series	Hot-melt adhesives based on terephthalic acid esters	71
Polyfix	Cement for ceramic tiles	217
Poly-lok	Latex based adhesive for laminating polythene to paper	200
Polyox	Water soluble resins	280
Polypenco	Adhesive for bonding nylon to nylon	219
Polystik	Polyvinyl acetate emulsion for bonding wood and concrete, etc	88
Polystyrene	Cement for bonding polystyrene film based on polystyrene	262
Polytak	Polyurethane based adhesives	26
Polyvale	Polyvinyl acetate based adhesives	26
Ponal	Synthetic resin based adhesives for woodworking and joinery	260
Pontex	Adhesive paste for stationery use	78
PRC	Range of synthetic elastomers	55
Propiofan	Polyvinyl propionate homo- and copolymer dispersions for adhesive formulations	30
Pyramid	Sodium silicate based materials	89
Pyruma	Heat resistant putty and filling cement for firebrick materials	56
Quentbond	Epoxy resin based structural adhesives for metals, ceramics and plastics	222
Quentcourse	Epoxy resin adhesives for masonry	222
Quentflex	Polyurethane adhesives	222
Quentgrip	Epoxy-pitch adhesives	222
Quentoglaze	Polyurethane adhesives which cure by moisture activation	222
Quickstick	Synthetic rubber/resin adhesive for hardboard, wall laminates, etc	214

continued

Table 10.1 continued

Trade Names	Type, Description or Uses	Code No.
Rabex	Range of synthetic rubber type adhesives	223
Racin	Dental cement	93
Rapier	Dextrine based cold glues for packaging and carton sealing, etc	2
Rawlbond	Urea formaldehyde based adhesive for wall and tile fixing, etc	224
Rayobond	Range of adhesive products for the building industry	4
Rayofix	Range of adhesives based on synthetic and natural rubbers, acrylic resins, and polyvinyl acetate emulsions	4
Rayweir	Building industry adhesives	4
Redfern	General purpose contact adhesive	225
Redi-bond	Contact adhesive based on polyvinyl acetate	169
Redux	Polyvinyl formal-phenolic structural adhesive for metal bonding	73
Resemul	Synthetic emulsion adhesive	160
Revell	Polystyrene based cement	229
Reverfix	Range of rubber based adhesives	230
Revertex	Natural rubber latex	230
Rex	Starch based wallpaper paste	231
Rexcell	Cellulose based wallpaper adhesive	231
Rhealit	Dental cement	93
Rhodopas	Polyvinyl acetate resin	125
Rhodoviol	Polyvinyl alcohol esters	125
Ribbonseal	Flexible joint sealing compound based on butyl rubber	193
Richafix	Range of rubber based adhesives for ceramic tile bonding	58
Rito	Range of jointing and waterproofing mastics	178
Ritolastic	Bituminous and non-bituminous weatherproofing adhesive coatings	178
Romac	General purpose adhesive for bonding polyvinyl chloride and leather to metal, etc	235
Romanite	Polyvinyl acetate based adhesive and cement additive	178
Ross Bond	Contact adhesive for polystyrene foam	236
RTV	Room temperature curing silicone rubber based adhesives	153
Rubber-resin	General purpose contact adhesive	47
Sactex	Rubber based adhesive compound for repairing sacks and bags	238
Safetee Ductfas	Fire resistant adhesive for thermal insulation materials	18
Sairset	Gap filling, high temperature adhesive cement	126
Samson	Range of adhesives based on synthetic and natural polymers for packaging applications	160
Satellit	Dental cement	93
Saxit	Epoxy resin based adhesive for tiles	260
Schjel-bond	Polyester resin based adhesives on insulating film substrates	243
Scotch	Permanent and pressure sensitive aerosol adhesive products	193
Scotchweld	Range of epoxy resin based structural adhesives	193
Sealobond	Polyvinyl acetate based adhesive	246
Sealocrete	Epoxy resin adhesive for concrete and building applications	246
Season	Hot-melt adhesive for wood	199
Seccotine	General purpose animal glue	188
Security	Rubber based adhesive	121
Seel-m	Putty grade epoxide adhesive filler for general purpose application	133
Self-Seal	Rubber-latex adhesives for 'self-seal' cartons and envelopes	200
Sellobond	Polyvinyl acetate based adhesives	248
Serifix	Two-part adhesive for Screen Mounting (silk, nylon, terylene or metal fabrics)	249
Servofix	Electrically conductive adhesive	188
Shawnad	Range of laminating adhesives for plastic films, paper, foils, etc	197
Sheppy	Dextrin and animal glues for joinery and bookbinding	251
Siccollin	Water soluble starch powder adhesives for laminating and paper sack manufacture	85A
Sichel	Range of adhesives based on starch, dextrine and casein glues	82
Sika	Range of waterproofing adhesive materials	254
Sikafix	Adhesive additive	254
Sikalatex	Adhesive additive based on latex	254
Silastic RTV	Silicone rubber adhesives and sealants	96
Silastoseal	Cold curing silicone rubber adhesive	97
Silcoset	RTV silicone rubber adhesives for ceramics, glass, and selected metals and plastics	144

continued

Table 10.1 continued

Trade Names	Type, Description or Uses	Code No.
Silikophen	Phenolic adhesive modified with silicones	81
Silikopon	Epoxy resin modified with silicone resin	81
Silverlock	Range of adhesives based on synthetic polymers and rubbers for industrial applications	57
Sirawax	Adhesive wax for attachment to glass, metals, and other substrates	
Solbit	Bitumen based adhesives	39
Sondaflex	Urea formaldehyde resin glues	258
Sondal	Range of adhesives based on animal glue, casein, polyvinyl acetate, and formaldehyde, for joinery, etc	258
Special H	Adhesive for bonding rubber and polyvinyl chloride tiles and sheets	281
Spectrabond	General purpose contact adhesive	259
Spencer	Acrylic resin adhesive for bonding polystyrene: epoxy resin adhesive for electrical applications	8
Spencobond	Polyvinyl acetate emulsion adhesive for wood	8
Spray-Lok	Sprayable synthetic adhesives in the form of emulsions and solutions	200
Spray 77	General purpose contact adhesive in aerosol form	193
Spynflex	Synthetic resin adhesive for bookbinding	200
Stabilit	Epoxy resin based adhesive	260
Stanfast	Natural rubber based adhesive in solvent	215
Stanmelt	Industrial hot-melt adhesives	247
Stanox	Range of synthetic adhesives for industrial use	260
Statoset	Inorganic refractory cement	126
Stefix	Adhesive for wood, pottery and fabrics	262
Stera-Stic	Range of self-adhesive films, foils and papers	262A
Stera-Tape	Double sided self-adhesive tape products	262A
Stikall	Latex based materials for bonding fabrics and rubber	215
Stopgap	Range of adhesives based on latex and acrylic resins, for flooring applications	25
Stronghold	Adhesive cement for tiles	289
Structobond	Waterproofing epoxy resin for flooring applications	263
Structoplast	Range of adhesives based on epoxy resins and polyvinyl acetate emulsions for concrete and flooring applications	263
Stuk	Range of adhesives for footwear manufacture	165
Stukko	Starch, dextrine, polyvinyl acetate, and animal glues for domestic use	187
Stycast	Rigid and flexible moulding materials based on epoxy resins and other materials	105
Stycco	Range of latex and emulsion adhesives	25
Styccobond	Range of adhesives based on rubbers, emulsions and epoxy resins	25
Styccoprene	Range of sythetic rubber based adhesives	25
Styccoscreed	Binding emulsion and floor levelling powder	25
Stypol	Adhesive for polyester glass fibre laminates, etc	117
Styroglue	Mastic adhesive for polystyrene and ceramic tiles	281
Supastik	Range of adhesives based on polyvinyl acetate, synthetic rubbers/resins, and oleo resins	109
Superbond	Polyvinyl acetate based bonding liquid for the building industry	62
Superfix	Polyvinyl acetate based adhesive	128
Supergrip	Adhesive for footwear manufacture	269
Super Lay-Flat	Synthetic and natural adhesives for envelope applications	200
Suprapal	Synthetic resin derived from styrene and maleates	30
Suprasec	Isocyanates for all purposes	144
Surebond	Polychloroprene based adhesive	87
Surestik	Range of adhesives based on synthetic rubbers and polyvinyl acetate and acrylic emulsions	13
Surgical simplex	Adhesive paste for surgical application	204
Swift's	Range of adhesives based on dextrines, rubbers, hot melts, for the packaging industry	264
Symplast	Range of adhesives based on thermosetting polymers	26
Syncap	Adhesive cements for lamp capping	40
Synglu	Adhesive for foundry applications	40
Synphorm	General purpose adhesive	40
Tack-coat	Adhesive primer for bonding epoxy coatings to porous surfaces	164
Tackmaster	Pressure sensitive adhesives (supported and unsupported) in dry film form	242
Takstrip	Self-adhesive paper tape	160

continued

Table 10.1 continued

Trade Names	Type, Description or Uses	Code No.
Talon	Range of adhesives based on animal glue, casein, dextrine and hot-melt materials	102
Tapwata	Range of adhesives based on starch and cellulose for wallpaper and cardboard	16
Technimelts	Hot-melt adhesives	61
Tego	Adhesive in dry film form	191
Tekna	Thermosetting resin adhesive with metal filler	267
Tenace	Vegetable based adhesive sealing tape	189
Tenaxis	Range of synthetic adhesive products	200
Tensol	Acrylic resin based cements	144
Tensulac	Bitumen based adhesive for roofing felt	34
Texicote	Vinyl acetate and copolymer adhesive emulsions	244
Texicryl	Acrylic resin emulsions	164
Texigel	Polyacrylic thickening agents for adhesive materials	244
Texilac	Vinyl acetate and acrylic polymers and copolymers for pressure sensitive, heat-seal and general purpose adhesives	244
Thermaflo	Range of hot-melt adhesives based on ethylene vinyl acetate for the footwear industry and packaging	107
Thermogrip	Synthetic hot-melt adhesives	47
Thermseal	Resin coatings for textiles and food packaging materials	189
Thistle	Adhesive for linoleum tiling	214
Thistlebond	Epoxy resin based adhesive	110
Thixofix	Thixotropic contact adhesive for general use	98
Thomsit	Synthetic resin dispersion for linoleum and polyvinyl chloride flooring materials	260
Thor	Urea formaldehyde and phenol formaldehyde foundry resins	45
Three dimensional	Range of adhesives for industrial applications	4
Tiluma	Cement for firebrick materials and tiles	56
Timbabond	Adhesive for wood	273
Titazel	Adhesives based on epoxy resins	239
Titebond	Range of impact adhesives on nitrile rubbers	16
Tivocoll	Synthetic latex	269
Tivodur	Neoprene rubber formulation for the footwear industry	269
Tivogum	Synthetic rubber adhesive for the automobile industry	269
Tivoli	Range of adhesives for the footwear industry	269
Tivolit	Synthetic rubber/resin adhesives for general industry	269
Tivolux	Neoprene based adhesives for the footwear industry	269
Tivopal	Synthetic rubber/resin adhesives for flooring applications	269
Tivosan	Nitrocellulose for general industry	269
Tivostar	Neoprene adhesive for footwear industry	269
Tivotex	Natural latex	269
Tivovult	Two-part natural rubber	269
Tixo	Cyanoacrylate based adhesive	266
TML	Range of strain gauge adhesives	104
Top	Animal gums	232
Torqseal	Anaerobic adhesive/sealant	133
Tranco	Polyurethane polymers used as adhesives, binders and coatings	81
Transmount	Unsupported transfer adhesive films	47
Tremco	Range of adhesive sealants, mastics and caulking compounds based on polysulphides and acrylic resins	272
Tretobond	Range of adhesives for building industry	273
Tri-mor	Aluminous cements	196
Trinasco	Epoxy resin, polyvinyl acetate, and urea formaldehyde adhesives for the building industry	173
Trojan	Dextrine powder adhesives for box manufacture	85A
Trufix	Starch based paste for paper and cardboard	262
Tuf-Bond	Polyvinyl acetate adhesives	276
Tufskin	Polyvinyl acetate emulsion adhesives for joinery, veneering, etc	2
Tuf-Stick	Synthetic rubber/resin contact adhesives	276
Tuf-tak	Polyvinyl acetate bonding cements and contact adhesives for the building industry	276
Tug	Liquid gelatin glue	199
Twinbond	Epoxy resin based adhesive	138

continued

Table 10.1 continued

Trade Names	Type, Description or Uses	Code No.
Twinstik	Double sided pressure sensitive transfer tape for packaging, labelling and civil engineering applications	107
Two-part 81–80	Polyurethane based adhesive	19
Ty-Ply	Range of adhesives and primers for rubber to metal bonding	11
Uhu	Range of adhesives based on polyvinyl acetate, neoprene, epoxy resins, for domestic applications	172
Unbrako	Anaerobic adhesive/sealants	277
Unibond	Polyvinyl acetate emulsion	278
Unicol	Casein emulsion adhesives	281
Uni-filla	Water-soluble-powder adhesive sealant for vinyl materials	278
Unifix	Natural product liquid and jelly adhesives	281
Uni-last	Neoprene adhesive	278
Unilite Arbrisso	Range of synthetic adhesives for industrial use; wood flooring, packaging, plastics, building, etc	281
Unilite	Range of adhesives for joinery, flooring, building and textiles	281
Unilok	Range of adhesives for wood, rubbers, fabrics, plastics, metals, and paper	282
Unimatic	Bonding compound for engineering applications	279
Union	Animal and fish glue products	281
Unipaste	Starch based wallpaper adhesive	278
Uniprene	Synthetic resin/rubber based adhesive for tile and sheet floorings	281
Unirod 100	Thermoplastic cement for the footwear industry	47
Unistik	General purpose contact adhesive	278
Unitimb	Adhesive for wood	278
Universal	Liquid animal glue	88
Univip	Polyvinyl acetate emulsion for expanded polystyrene	278
U.P.7	Neoprene based contact adhesive	278
Uvefix	Adhesive for ethyl cellulose films	185
Vale	Urea formaldehyde resin	26
Vandike	Polyvinyl acetate or copolymer adhesives	52
Vantac	Aqueous pressure sensitive adhesive products	52
Versalon	Polyamide resin for hot-melt adhesives	81
Versamid	Polyamide resins for hot-melt and structural adhesives	81
Vincit	Range of adhesives based on polyvinyl acetate emulsions, acrylic resins, neoprene rubber, epoxy resins, and hot-melt materials	63
Vinnaplas	Polyvinyl acetate resins and copolymers available as solids or dispersions	59
Vinnol	Vinyl chloride/vinyl acetate copolymers	59
V-mat	Epoxy resin based adhesives for concrete repair	221
Vulcabond	Mixed polyisocyanates in xylene for rubber to textile bonding	144
Vulcabond T	Aqueous dispersion of resorcinol formaldehyde resin and latex for bonding natural and synthetic rubbers to rayon	144
Vulcabond TVPN	Aqueous dispersion of resorcinol formaldehyde resin and rubber latices for bonding natural and synthetic rubbers to rayon	144
Vylok	Formulated dextrine adhesives for stationery	200
Wallcol	Synthetic resin emulsion for polyvinyl chloride and textiles	281
Wallmount	Adhesive film	47
WBSP	Magnesite cement	114
Weatherban	Polysulphide rubber based sealing compounds	193
Weldtite	Polystyrene cements for model construction	22
Whitecraft	Polyvinyl acetate based adhesive	128
Widespread	Thixotropic contact adhesive for the building industry	98
Wilson	Range of adhesives for the packaging trade	289
Witcobond	Range of solvent adhesives based on neoprene, nitrile, natural, polyurethane rubbers, and various copolymers	290
Witcodex	Starch and dextrine based adhesives	290
Witcogrip	Range of adhesives based on polyvinyl acetate, acrylic resins, latex, neoprene, reclaim rubber, and other polymers in aqueous solution	290
Wood-lok	Range of copolymer emulsions for woodworking	200
Woodworker	Adhesive for woodworking and building trade	98
Wresilac	Amino based adhesive for woodworking	228

continued

Table 10.1 continued

Trade Names	Type, Description or Uses	Code No.
Wresimul	Emulsion adhesive for refractory materials	228
Wresinoid	Range of adhesives for foundry applications and abrasive wheels	228
Wresinol	Abrasive paper adhesive	228
Wresitex	Adhesive for non-woven fabrics and textiles	228
XL	Range of adhesives based on natural rubber	108
YDC	Range of starch, dextrine, polyurethane, urea formaldehyde and other resin based adhesives for joinery and plastics	291
Zet	Adhesive cement for china and porcelain repairs	208
Zipgrip 10	Cyanoacrylate adhesive	94
ZZ	Epoxy polyamide adhesives for structural bonding of rigid materials	16

REFERENCES

HURD, J., *Adhesives Guide,* SIRA, (1959).
Adhesives Directory (1974–75), A. S. O'Connor & Co. Ltd., Richmond, Surrey (annual publication).
New Trade Names in the Rubber and Plastics Industries, RAPRA, (1974) (annual publication).

European Plastics Year Book, IPC Scientific and Technical Press, Guildford, (1975) (annual publication).
European Rubber Directory, Maclaren & Sons Ltd., (1974).
Surface Coating Resin Index, The British Resin Manufacturers' Association, London, (1974) (annual publication).

10.2. BASIC ADHESIVE TYPES AND TRADE SOURCES

Manufacturers of adhesives products based on the raw materials listed below are specified by the code number system given on page 322 (Table 10.3).

Table 10.2. BASIC ADHESIVE TYPES AND TRADE SOURCES

Basic Material	Trade Source
Acrylic	2, 4, 6, 13, 21, 25, 33, 45, 52, 74, 87, 88, 91, 98, 107, 109, 118, 144, 145, 148, 160, 161, 165, 183, 193, 195, 200, 207, 213, 217, 223, 253, 258, 260, 264, 269, 278, 281, 289, 290, 291.
Alkyds	82, 146, 269.
Alkyd/phenolic	82, 146, 269.
Animal glues	2, 26, 36, 88, 140, 160, 200, 213, 231, 232, 251, 258, 260, 262, 264, 281, 285, 289.
Bitumen and asphalts	18, 25, 34, 47, 58, 74, 107, 109, 144, 145, 155, 178, 183, 193, 213, 214, 233, 239, 246, 253, 289.
Bitumen/latex	34, 46, 83, 97, 107, 109, 112, 145, 193, 213, 238, 239, 290.
Butyl rubber	4, 47, 84, 88, 98, 109, 115, 148, 152, 162, 193, 195, 200, 213, 269, 290.
Casein	4, 5, 16, 17, 25, 26, 36, 45, 46, 47, 82, 85A, 87, 88, 102, 148, 160, 177, 200, 203, 215, 232, 258, 264, 281, 285, 289.

Basic Material	Trade Source
Casein/latex	5, 16, 26, 45, 46, 47, 66, 88, 98, 102, 107, 145, 146, 148, 158, 160, 165, 177, 195, 200, 203, 214, 215, 223, 238, 264, 269, 283, 289.
Cellulose acetate	4, 103, 224, 262, 269, 274.
Cellulose acetate-butyrate	200.
Cellulose nitrate	4, 140, 144, 155, 172, 189.
Ceramic	114, 105, 221, 241.
Chlorinated rubber	4, 25, 47, 66, 84, 87, 98, 109, 115, 118, 139, 146, 152, 160, 165, 200, 213, 216, 223, 264, 269.
Coumarone-indene	16, 47, 82, 147, 223, 269, 281.
Cresylic resins	45, 144, 221, 281.
Cyanoacrylate	4, 21, 47, 73, 103, 105, 149, 161, 208A, 213, 259A, 260.
Dextrin	2, 12, 45, 85A, 88, 118, 140, 160, 189, 195, 200, 203, 231, 232, 260, 262, 264, 288, 289, 291.
Epoxy	5, 16, 25, 45, 57, 63, 67, 73, 81, 84, 91, 94, 98, 106, 107, 109, 118, 133, 139, 146, 148, 155, 165, 183, 189, 193, 197, 201, 242, 211, 213, 214, 221, 224, 239, 246, 258, 260, 263, 266, 269, 281, 285.
Epoxy/alkyd ester	98, 269.
Epoxy/phenolic	40, 91, 98, 146, 148, 193, 221, 269.
Epoxy/polyamide	16, 45, 47, 73, 91, 98, 109, 118, 133, 146, 155, 167, 193, 197, 213, 221, 246, 266, 269.
Epoxy/polyester	98, 146, 148, 197, 221, 269.

continued

Table 10.2 continued

Basic Material	Trade Source	Basic Material	Trade Source
Epoxy/polysulphide	45, 47, 73, 98, 155, 221, 269.	Polyisobutylene	4, 13, 47, 66, 87, 98, 107, 109, 118, 146, 152, 160, 193, 195, 200, 213, 216, 264, 269.
Ethyl cellulose	4.		
Ethylene/vinyl acetate copolymer	2, 4, 13, 45, 51, 66, 87, 88, 107, 148, 160, 165, 195, 195A, 200, 207, 209, 223, 264, 289, 291.	Polyphenylene resins	195A.
		Polystyrene	4, 6, 13, 88, 98, 118, 140, 148, 152, 172, 195A, 200, 213, 223, 262, 264.
Fish glues	88, 195, 232, 260, 262, 281, 289.	Polysulphide rubber	47, 53, 84, 98, 109, 242.
Furane resins	40, 83.	Polyurethane	4, 13, 18, 25, 47, 53, 66, 84, 87, 88, 91, 94, 98, 107, 115, 118, 144, 148, 152, 155, 165, 193, 195, 197, 200, 203A, 209, 213, 216, 222, 223, 245, 264, 269, 281, 288, 290, 291.
Gums (natural)	2, 88, 195, 232, 260, 262, 281, 289.		
Hot melt	16, 26, 36, 45, 46, 47, 63, 66, 81, 82, 88, 98, 102, 103, 107, 108, 112, 133, 139, 144, 146, 155, 160, 165, 195, 200, 201, 209, 211, 213, 242, 258, 260, 264, 269, 270, 281, 289, 291.	Polyvinyl acetate	2, 4, 5, 6, 13, 17, 18, 23, 25, 46, 47, 58, 66, 73, 74, 83, 84, 85A, 88, 98, 107, 109, 114, 118, 139, 145, 148, 160, 165, 172, 178, 183, 189, 195, 200, 203, 206, 207, 209, 213, 216, 223, 233, 241, 245, 246, 251, 253, 258, 260, 262, 264, 269, 276, 278, 281, 285, 289, 290, 291.
Hydroxyl propyl methyl cellulose	142, 216, 260, 289.		
Inorganic	4, 18, 25, 58, 63, 83, 114, 176, 183, 217, 241, 252, 278.	Polyvinyl acetate/ polyvinyl chloride copolymer	4, 6, 13, 23, 46, 47, 87, 98, 118, 140, 148, 160, 165, 195, 200, 203A, 209, 213, 216, 223, 269, 288.
Isocyanate	23, 25, 26, 45, 46, 47, 81, 88, 136, 144, 146, 148, 155, 165, 200, 213, 215, 223, 269, 281, 290.	Polyvinyl alcohol	2, 4, 33, 66, 85A, 88, 118, 148, 160, 165, 176, 195, 200, 207, 216, 217, 223, 264, 269, 281, 289, 291.
Melamine/ formaldehyde	38, 67, 73, 85A, 144, 211, 260, 291.	Polyvinyl butyral	4, 23, 47, 146, 148, 216, 242, 250.
Methyl cellulose	16, 26, 46, 102, 139, 144, 183, 201, 223, 260, 264, 269, 289.	Polyvinyl chloride	4, 6, 13, 47, 66, 79, 84, 88, 98, 118, 146, 152, 162, 195, 200, 213, 223, 260, 264, 269.
Neoprene/phenolic	5, 13, 16, 26, 47, 66, 84, 88, 98, 107, 109, 123, 136, 146, 147, 148, 165, 193, 198, 200, 210, 213, 223, 242, 264, 269, 281, 283.	Polyvinyl ether	4, 6, 38, 47, 87, 88, 98, 107, 118, 146, 152, 195, 213, 223, 269, 281.
		Polyvinyl formal	250.
Neoprene rubber (polychloroprene)	2, 4, 13, 18, 25, 45, 46, 47, 58, 66, 74, 84, 87, 88, 98, 107, 109, 115, 118, 123, 139, 148, 152, 162, 165, 172, 183, 189, 193, 195, 200, 203A, 209, 213, 223, 233, 237, 260, 262, 264, 269, 276, 278, 281, 285, 288, 290.	Polyvinyl/phenolic	23, 67, 73, 91, 193, 213, 242.
		Resorcinol formaldehyde	23, 40, 45, 67, 73, 85A, 87, 88, 258, 285, 291.
		Resorcinol/phenolic	23, 73, 258.
		Rubber (cyclised)	4, 98, 148, 216.
Nitrile/phenolic	5, 13, 23, 26, 47, 66, 84, 91, 98, 107, 123, 136, 146, 147, 148, 165, 193, 200, 209, 213, 221, 242, 264, 269, 283.	Rubber (natural latex)	4, 13, 25, 46, 47, 58, 66, 74, 79, 88, 94, 98, 107, 109, 118, 140, 145, 148, 160, 165, 166, 195, 200, 203, 203A, 209, 213, 216, 223, 233, 237, 245, 264, 269, 281, 285, 288, 289, 290, 291.
Nitrile rubber (acrylonitrile butadiene)	2, 4, 5, 13, 25, 45, 46, 47, 66, 84, 88, 91, 98, 107, 115, 118, 123, 139, 145, 146, 152, 183, 193, 195, 200, 213, 223, 233, 237, 260, 264, 269, 288, 290.		
		Rubber (reclaim)	2, 47, 87, 94, 98, 107, 109, 118, 123, 145, 148, 193, 200, 203A, 213, 288.
		Rubber (synthetic latex)	5, 11, 16, 25, 46, 47, 66, 88, 98, 107, 109, 148, 160, 165, 195, 200, 213, 223, 264, 269, 289, 290.
Oleoresins	18, 109, 193, 288.	Rubber (vulcanised)	98, 123, 148, 165, 193, 200, 203A, 264, 269, 288.
Phenol formaldehyde	23, 40, 45, 73, 85A, 109, 258, 281, 285, 291.		
Phenolic/polyvinyl	23, 67, 73, 146, 213, 242.	Shellac	133, 155.
Phenolic/polyamide	73, 146.	Silicate	2, 4, 16, 83, 89, 114, 152, 160, 165, 200, 288, 289.
Polyamides	4, 11, 12, 47, 66, 91, 107, 165, 209, 216, 264, 269, 281, 291.	Silicones	4, 47, 73, 74, 88, 98, 109, 144, 148, 153, 191.
Polybenzimidazole	287.	Sodium carboxy methyl cellulose	139, 144, 176, 216, 217, 231, 260.
Polyester (unsaturated)	4, 13, 23, 25, 26, 47, 50, 66, 81, 83, 98, 103, 107, 118, 133, 136, 148, 165, 189, 197, 200, 213, 221, 222, 233, 241, 242, 252, 260, 269, 281, 291.	Starch	2, 3, 12, 45, 74, 85A, 88, 160, 176, 195, 200, 217, 231, 262, 264, 281, 288, 289.
Polyimide	91, 99, 144, 185, 195A.		

continued

Table 10.2 continued

Basic Material	Trade Source
Styrene-butadiene	2, 4, 13, 18, 25, 47, 66, 83, 87, 88, 98, 107, 109, 118, 139, 148, 152, 160, 165, 183, 193, 195, 200, 203A, 213, 216, 223, 233, 237, 246, 264, 269, 281, 288.
Urea formaldehyde	38, 40, 45, 67, 73, 87, 88, 144, 165, 258, 276, 285, 291.
Vegetable protein (soya, etc)	2, 45, 47, 88, 160, 177, 200, 232, 291.
Vinyl chloride/ vinylidene copolymer	13, 98, 136, 146, 147, 160, 207, 211, 213, 269, 281.
Waxes	17, 61, 288.

10.3. ADHESIVES MANUFACTURERS

The following table comprises the names and addresses of firms and companies manufacturing adhesives products available to the U.K. The list includes some overseas trade sources where these are known to have a marketing agent or representative company, in this country, able to supply foreign adhesives products.

The code numbers provide the key to the Trade Sources for the Trade Names given in Table 10.1.

Table 10.3. ADHESIVES MANUFACTURERS

Code No.	Manufacturer
1	ACALOR LTD Kelvin Way, Crawley, Sussex. 0293 23271
2	ADAMS, ALFRED & CO LTD Reliance Works, Church Lane, West Bromwich, Staffs. 021–553 0263
3	ADCOL LTD Cavenham House, By-pass Road, Colnbrook, Bucks. 964 4411
4	ADHESIVE & ALLIED PRODUCTS LTD 2 Highbridge Road, Barking, Essex. 01–594 4066
5	ADHESIVE SOLUTIONS LTD Woodcock Hill Industrial Estate, Harefield Road, Rickmansworth, Herts. 87 79949
6	ADHESIVE SPECIALITIES LTD 10 Dryden Chambers, 119 Oxford Street, London W1. 01–437 5338/3063/8391
7	AIRBORNE INDUSTRIES LTD Airborne Works, Arterial Road, Leigh-on-Sea, Essex. 0702 525265

Code No.	Manufacturer
8	ALBUMENOID PRODUCTS CO LTD, THE PO Box 76, Albert Quay, Aberdeen, Scotland. 0224 20202/4
9	ALLIANCE DYE & CHEMICAL CO LTD Grecian Mill, Lever Street, Bolton, Lancs. 0204 21971/3
10	ALLIED BUILDING COMMODITIES LTD Queens Mill Road, Huddersfield, Yorks.
11	ANCHOR CHEMICAL CO LTD Clayton, Manchester M11 4SR. 061–223 2461
12	ANDERSON (ADHESIVES) LTD, D Newman Lane, Alton, Hants. GU34 2QR. 0420 84849
13	ANGLO CHEMICAL CO (LEICESTER) LTD Lee Circle, Leicester. 0533 26271/2
14	APPLICATIONS TECHNOLOGY (RESINS) LTD Main Road, West Kingsdown, Nr Sevenoaks, Kent. 047485 2852
15	ARO CORPORATION (UK) LTD Caernarvon, North Wales. 0286 3551
16	ASSOCIATED ADHESIVES LTD 8th Avenue, Manor Park, London E12. 01–478 0061
17	ASTOR PETROCHEMICALS LTD 9 Savoy Street, Strand, London WC2. 01–836 5082
18	ATLAS PRODUCTS & SERVICES LTD Fraser Road, Erith, Kent DA8 1PN. 38 32255
19	AUSTEN CHEMICALS LTD Ederby, Leicester LE9 5NH. 0533 729 2282
20	AUTOMOBILE PLASTICS LTD Autoplax House, 7 Henry Road, New Barnet, Herts. 01–449 9147
21	AVDEL ADHESIVES LTD Welwyn Garden City, Herts AL7 1EZ. 96 28161
22	BAGGS LTD, C B Claremont Industrial Estate, Claremont Way, London NW2. 01–458 4448
23	BAKELITE XYLONITE LTD Thermosetting Division, Redfern Road, Tyseley, Birmingham B11 2BJ. 021–706 3322
24	BAKER-PERKINS LTD Wewswood Works, Peterborough.
25	BALL & CO LTD, F 628/640 Garrett Lane, Tooting, London SW17. 01–946 2256
26	BARDENS (BURY) LTD Hollins Vale Works, PO Box 1, Bury, Lancs. Waitefield 2084/5

continued

Table 10.3 continued

Code No.	Manufacturer
27	BARKER LTD, E R Lion Works, Elliott Road, Selly Oak, Birmingham 29. 021–472 1929
28	BARLOW WHITNEY LTD Watling Street, Bletchley, Bucks. 0908 23571
29	BARR & STROUD LTD Caxton Street, Anniesland, Glasgow G13 1HZ. 041–954 9601
30	BASF UNITED KINGDOM LTD PO Box 4, Earl Road, Cheadle Hulme, Cheadle, Cheshire SK8 6QG. 061–485 7181
31	BATEMAN'S ADHESIVES LTD British Grove, London W4. 01–748 3187
32	BAYER CHEMICALS LTD Kingsway House, 18–24 Paradise Road, Richmond, Surrey TW9 1SJ. 01–940 6077
33	B.C. PRODUCTS (BONDCRETE) LTD 29 Eve Road, Woking, Surrey. 04862 61719
34	BERRY WIGGINS & CO LTD Kingsnorth on the Medway, Hoo, Rochester, Kent. 063427 304
35	BETASEAL CO LTD Eastbourne Road, Trading Estate, Slough, Bucks. 75 20262
36	BETOL ADHESIVES LTD Brook Street Works, Hazel Grove, Nr Stockport, Cheshire. 061–483 3049
37	B.I.C.C. LTD Blackwall Lane, London SE10. 01–858 7011
38	B.I.P. CHEMICALS LTD PO Box 6, Warley, Worcs. 021–552 1551
39	BITUMEN INDUSTRIES LTD Ajax Avenue, Slough, Bucks. 75 23274
40	BLACKBURN & OLIVER LTD No 2 Factory, Lamberhead Industrial Estate, Nr Wigan, Lancs. 0942 82252/3
41	BLOORE, G H LTD 480 Honeypot Lane, Stanmore, Middlesex. 01–952 2391
42	BONDAGLASS LTD 158 Ravenscroft Road, Beckenham, Kent. 01–778 0071/3
43	BONDING SYSTEMS (SALES) LTD Vines Cross Road, Horam, Heathfield, Sussex. 04353 2226/7
44	BONE BROS LTD Manor Farm Road, Wembley, Middlesex. 01–997 9555

Code No.	Manufacturer
45	BORDEN CHEMICAL CO (UK) LTD, THE North Baddesley, Southampton, Hants. 0421 232131 also at Hollins Vale Works, Bury, Lancs.
46	BORDEN CHEMICAL CO (UK) LTD, THE (Arabol-Edwardson Adhesives Division), Marsh Lane, Ware, Herts. 0920 2394
47	BOSTIK LTD Ulverscroft Road, Leicester LE4 6BW. 0533 50015
48	B.P. CHEMICALS INTERNATIONAL LTD Devonshire House, Mayfair Place, Piccadilly, London W1X 6AY. 01–493 8131
49	BP CHEMICALS (UK) LTD Hayes Road, Sully, Penarth, Glamorgan. 04462 2321
50	BRITISH INDUSTRIAL PLASTICS LTD (Sheet & Film Division), Brantham, Manningtree, Essex CO11 1NJ. 020–639 2401
51	BRITISH INDUSTRIAL PLASTICS LTD (Chemical Division), PO Box 6, Oldbury, West Midlands B69 4PD 021–552 1551
52	BRITISH OXYGEN CHEMICALS LTD Hammersmith House, London W6. 01–748 2020
53	BRITISH PAINTS & CHEMICALS Elastomers Division, Portland Road, Newcastle-upon-Tyne NE2 1BL. 0632 25151
54	BRITISH PAINTS LTD Portland Road, Newcastle-on-Tyne NE2 1BL 0632 25151
55	BRITISH RESIN PRODUCTS LTD Devonshire House, Mayfair Place, London W1X 6AY. 01–493 8131
56	BRITISH SISALKRAFT LTD Refractories Division, The Hill, Ilford, Essex. 01–478 5341
57	BTR INDUSTRIES LTD Silvertown House, Vincent Square, London SW1. 01–834 3848
58	BUILDING ADHESIVES LTD Federation House, Stoke-on-Trent, Staffs ST4 2SE. 0782 45411
59	BUSH BEACH & SEGNER BAYLEY LTD 175 Tottenham Court Road, London W1P 0BJ. 01–580 8041
60	CAMPBELL & CO LTD, REX 7 Idol Lane, Eastcheap, London EC3. 01–626 8190

continued

Table 10.3 continued

Code No.	Manufacturer
61	CAMPBELL TECHNICAL WAXES LTD Thames Road, Crayford, Kent. 29 24555
62	CANNON & CO LTD, B 11 Mill Road, Wellingborough, Northants. 09333 2621
63	CARBORUNDUM CO LTD, THE Trafford Park, Manchester 17. 061–872 2381
64	CARLTON BROWN & PARTNERS LTD Elford Mill, Elford, Tamworth, Staffs. 082785 355
65	CASEIN INDUSTRIES LTD 20 Linford Street, London SW8. 01–622 9090
66	CASWELL & CO LTD Chelsea Works, St Michael's Road, Kettering, Northants. 0536 2041
67	CATALIN LTD Farm Hill Road, Waltham Abbey, Essex EN9 1NL. 97 23344
68	CERTOFIX LTD St Andrew's Dock, Hull, Yorks. 0482–27361
69	CHALLENGE ADHESIVES LTD Chalfont Works, 19 Coleman Street, Southend-on-Sea, Essex. 0702 66100
70	CHEMICAL BUILDING PRODUCTS LTD Warple Works, Cleveland Road, Hemel Hempstead, Herts. 0442 4901
71	CHEMICALS TRADING CO LTD 25 Berkeley Square, London W1X 6DH. 01–499 1246
72	CHEMIDUS PLASTICS LTD Brunswick Road, Cobbs Wood, Ashford, Kent. 0233 22271
73	CIBA-GEIGY (UK) LTD Plastics Division, Duxford, Cambridge. 0220 32121
74	CLAM-BRUMMER LTD Maxwell Road, Boreham Wood, Elstree, Herts. 01–953 2992
75	CMW LABORATORIES LTD Polymer Division, Preston New Road, Blackpool, Lancs. 0253 61301
76	COATES BROTHERS (INDUSTRIAL FINISHES) LTD Easton Street, London WC1X 0DP. 01–837 2810
77	COLEBRAND LTD 15 Hampden Gurney Street, Marble Arch, London W1M 5AL. 01–262 8948
78	CO-OPERATIVE WHOLESALE LTD Bone Products Department, PO Box 3, 86 South Baileygate, Pontefract, Yorks. 0977 3311/2

Code No.	Manufacturer
79	COPYDEX LTD 1 Torquay Street, Harrow Road, London W2. 01–286 7391
80	CORNBROOK RESIN CO LTD Clough House Mill, Wardle, Rochdale, Lancs. 0706 78616/7
81	CORNELIUS CHEMICAL CO LTD Ibex House, Minories, London EC3. 01–709 0221
82	CORN PRODUCTS (SALES) LTD Claygate House, Esher, Surrey. 78 62181
83	CORROSION TECHNICAL SERVICES LTD Oakcroft Road, Chessington, Surrey. 01–397 3344
84	COW (PROOFINGS) LTD, P B Eastbourne Road, Trading Estate, Slough, Bucks. 75 22274/6
85	COX, J & G LTD Grogie Mills, Edinburgh 11, Scotland. 031–337 6222/4
85A	CPC (UNITED KINGDOM) LTD Adhesives Division, Trafford Park, Manchester M17 1PA. 061–872 2571
86	CRAY VALLEY PRODUCTS LTD St Mary Cray, Orpington, Kent. 66 32545
87	CRISPIN CHEMICAL CO LTD Coleman Road, Leicester LE5 4NQ. 0533 66331/2
88	CRODA GLUES LTD Winthorpe Road, Newark, Notts NG24 2AL 0636 6711
89	CROSFIELD & SONS LTD, JOSEPH PO Box 26, Warrington, Lancs. 0925 31211
90	CURTIS, A L (ONX) LTD Westmoor Laboratory, Chatteris, Cambs. 03543 2561
91	CYANAMID INTERNATIONAL CORPORATION Bloomingdale Dept, 154 Fareham Road, Gosport, Hants. 03292 6131
92	DCMC INDUSTRIAL AEROSOLS LTD 291 Edgware Road, London W2. 01–723 1003
93	DENTAL FILLINGS LTD 49 Grayling Road, London N16. 01–800 7444
94	DEVCON LTD Station Road, Theale, Reading, Berks RG7 4AB. 0734 302304
95	DOMOLAC, DURESCO LTD Abbey Wood, London, SE2. 01–855 2451/8

continued

Table 10.3 continued

Code No.	Manufacturer
96	DOW CHEMICAL CO (UK) LTD 105 Wigmore Street, London W1. 01–935 4441 also at Heath Row House, Bath Road, Hounslow, Middlesex. 01–759 2600
97	DOW-CORNING INTERNATIONAL LTD Reading Bridge House, Reading, Berks. RG1 8PW. 0734 57251
98	DUNLOP CHEMICAL PRODUCTS DIVISION Chester Road, Erdington, Birmingham B35 7AL. 021–373 8101
99	DU PONT (UK) LTD Du Pont House, 18 Bream's Buildings, Fetter Lane, London EC4A 1HT.
100	DURHAM RAW MATERIALS LTD 1–4 Great Tower Street, London EC3. 01–626 4333
101	DUSSEK-BITUMEN & TAROLEUM LTD Loushers Lane, Widerspool, Warrington, Lancs.
102	EAGLE PACKAGING & PRINTING CO LTD Compsall Mill, Compsall, Cheshire.
103	EASTMAN CHEMICAL INTERNATIONAL AG PO Box 66, Kodak House, Station Road, Hemel Hempstead, Herts. 0442 62441
104	ELECTRO MECHANISMS LTD 218–221 Bedford Avenue, Slough, Bucks. 75 27242
105	EMERSON & CUMING (UK) LTD Colville Road, Acton, London W3. 01–992 6692
106	EPIC ADHESIVES CHEMICAL CO LTD Beaconsfield Place, East Street, Epsom, Surrey. 78 27171
107	EVODE LTD Industrial Adhesives Division, Common Road, Stafford. 0785 2241 also at 450/52 Edgware Road, London W2. 01–262 2425
108	THE EXCEL POLISH CO LTD Victoria Street, Northampton. 0604 37985
109	EXPANDITE LTD Chase Road, London NW10. 01–965 4321
110	FERGUSON & TIMPSON LTD Glasgow SW2, Scotland. 041–882 4691
111	FERGUSON & SONS LTD, JAMES Lea Park Works, Prince George's Road Merton Abbey, London SW19. 01–648 2283

Code No.	Manufacturer
112	THE FLINTKOTE CO LTD 1 Fitzroy Square, London W1. 01–387 7224
113	FORMICA LTD PO Box 2, De La Rue House, 84–86 Regent Street, London W1R 6AB. 01–734 8020
114	FORTAFIX LTD First Drove, Fengate, Peterborough. 0733 66136
115	FPT INDUSTRIES LTD The Airport, Portsmouth, Hants PO3 5PE. 0705 62301/4
116	FREEMAN CHEMICALS LTD PO Box 8, Ellesmere Port, Wirral, Cheshire. 051–355 1999
117	FURSE & CO LTD, W J Traffic Street, Nottingham NG2 1NF. 0602 863471
118	FUSSELL'S RUBBER CO LTD (LARS FOSS KEMI LTD) Knightstone Mills, Worle, Weston-Super-Mare, Somerset. 0934 3573
119	GLOVER DENTAL SUPPLIES Lancaster Road, Shrewsbury, Shropshire. 0743 52657/8
120	GLOY 8th Avenue, Manor Park, London E12 5JW. 01–478 0061
121	GOODLIFF MANUFACTURING CO LTD Chaucer Street, Northampton. 0604 34543
122	GOODRICH, B F & CO LTD 44–45 Chancery Lane, London WC2. 01–405 6803
123	GOODYEAR TYRE & RUBBER CO (GREAT BRITAIN) LTD, THE Distribution Centre, Fordhouses, Wolverhampton, Staffs WV2 4BU. 0902 26471
124	GORTON, EDWARD LTD Paddington Chemical Works, Warrington, Lancs. 0925 34631
125	GREEFF & CO LTD, R W Acorn House, Victoria Road, London W3. 01–993 0531
126	GREEN, A P, REFRACTORIES LTD York House, Wembley, Middlesex. 01–903 0455
127	GRIPPE ADHESIVES LTD, 1 Bothwell Lane, Glasgow W2, Scotland. 041–334 3240
128	GUEST LTD, H MARCEL Riverside Works, Collyhurst Road, Manchester 9. 061–205 2644
129	HADFIELDS (MERTON) LTD Mitcham, Surrey. 01–648 3422

continued

Table 10.3 continued

Code No.	Manufacturer
130	HARBOROUGH CONSTRUCTION CO LTD Harbilt Works, Market Harborough, Lincs. 0645 2254/6
131	THE HELD GLUE & COMPOUNDS CO LTD Brent Way, Brentford, Middlesex. 01–560 8429
132	HERCULES POWDER CO LTD 1 Great Cumberland Place, London W1H 8AL. 01–262 7766
133	HERMETITE PRODUCTS LTD Tavistock Road, West Drayton, Middlesex. 933 3731
134	HERTFORD SEALANTS LTD Eastbourne Road, Trading Estate, Slough, Bucks. 75 20262
135	HODGSON & SONS LTD, RICHARD Flemingate, Beverley, Yorks. 0482 881133
136	HOECHST UK LTD Hoechst House, Salisbury Road, Hounslow, Middlesex. 01–570 7712
137	HOLDEN & SONS LTD, ARTHUR Bordesley Green Road, Birmingham 9. 021–772 2761
138	HOLT PRODUCTS LTD Vulcan Way, New Addington, Surrey. 503 3011
139	HOWLETT & SONS LTD, W E Kingsthorpe Road, Northampton. 0604 33826
140	HUMBROL LTD Marfleet, Hull, Yorks. 0482 74121
141	HYDE & CO (LONDON) LTD 17A St Mary's Road, London W5. 01–567 5126
142	HYSOL STERLING LTD Hysol House, Heddon Street, London W1R 8BP. 01–734 9931/2
143	ICI LTD Paints Division, Wexham Road, Slough, Bucks. 75 81151
144	ICI LTD Imperial Chemical House, Millbank, London SW1P 3JF. 01–834 4444
145	IDENDEN ADHESIVES LTD Blackwater Way Industrial Estate, Ash Road, Aldershot, Hants. 0252 26855
146	INDEPENDENT ADHESIVES & CHEMICALS LTD Wennington Road, Rainham, Essex. 76 53344
147	INDIARUBBER & TYRE ADHESIVES CO LTD, THE Bankfield Mill, Stoneclough, Radcliffe, Manchester M26 9EX. 0204 73736/8

Code No.	Manufacturer
148	INDUSTRIAL ADHESIVES LTD Moor Road, Chesham, Bucks. 02405 4444
149	INDUSTRIAL SCIENCE (Proprietors: DENIS LEADER (MACHINERY) LTD) Leader House, Snargate Street, Dover, Kent. 0304 2656
150	INDUSTRIAL WAXES LTD 11 Penrhyn Road, Kingston-upon-Thames, Surrey. 01–549 2211
151	INTERSEM LTD.
152	ITAC LTD Bankfield Mills, Stoneclough, Radcliffe, Manchester M26 9EX. 0204 73736/8
153	JACOBSON VAN DEN BERG & CO (UK) LTD Jacoberg House, 231 The Vale, London W3 7RN. 01–743 9121
154	JACOB, WHITE & CO LTD Westminster Mill, Horton Kirby, Dartford, Kent. 320 3511
155	JEFFERY & CO, ALFRED Marshgate Lane, London E15. 01–514 2457
156	JENNER & GILL LTD Landor Street, Birmingham 8. 021–327 2075
157	JERMYN INDUSTRIES Vestry Estate, Sevenoaks, Kent. 0732 51174
158	JIFFYTEX PRODUCTS LTD Rex Barrow Works, Pewsey, Wilts. 06726 3238
159	JOHNSON & BLOY LTD Metana House, Hind Court, Fleet Street, London EC4. 01–353 1261
160	JONES & CO LTD, SAMUEL Butterfly House, Dingwall Road, Croydon, Surrey CR9 3DA. 01–686 5588 also at St Neots Mill, St Neots, Huntingdon PE19 4EE. 0480 75351
161	KANE GROUP LTD, DOUGLAS Swallowfields, Welwyn Garden City, Herts. 96 21261
162	KELSEAL LTD Vogue House, Hanover Square, London W1. 01–493 1411 also at Kelsey House, Wood Lane, Hemel Hempstead, Herts. 0442 50241
163	KLINGER, RICHARD Klingerit Works, Sidcup, Kent. 01–300 7777

continued

Table 10.3 continued

Code No.	Manufacturer	Code No.	Manufacturer
164	LAMACREST LTD Harrogate, Yorks. 0423 66656/7	182	MARB-L-COTE MFG (CB) LTD 49 Whitepost Lane, London E9. 01–985 9641/2
165	LARKHILL SOLING CO LTD Adhesives & Chemicals Division, Street, Somerset. 04584 2121	183	MARLEY FLOORS LTD Adhesive & Allied Products Division, Bath Road, Beenham, Reading, Berks RG7 5PU. 073521 3456
166	LATEX IMPORT & SACK CO LTD Burnett House, 82/3 Mytongate, Hull, Yorks. 0482 27476	184	MARSHALL-HOWLETT LTD 44 Tower Hill, London EC3. 01–709 1461
167	LATTER, A & CO LTD 43 South End, Croydon, Surrey CR9 1AN. 01–688 9335/8	185	MAY & BAKER LTD Dagenham, Essex RM10 7XS. 01–592 3060
168	LAYCOCK'S (ASHTON-UNDER-LYNE) LTD William Street Works, Ashton-under-Lyne, Lancs. 061–330 1503	186	MCCAW, STEVENSON & ORR LTD Castlereagh Road, Belfast 5.
169	LAYMATT FLOORING CO LTD 30–40 Seaborne Road, Bournemouth, Hants. 0202 44241	187	MEND-IT LTD Orphanage Road, Birmingham 24. 021–373 6951
170	LEBOFF, S (FOBEL) LTD Hyde House, The Hyde, Edgware Road, Colindale, London NW9. 01–205 0044	188	METALS RESEARCH LTD Melbourn, Royston, Herts. Melbourn (Cambs) 611
171	LENNIG CHEMICALS LTD Lennig House, 21 Mason's Avenue, Croydon, Surrey CR9 3NB. 01–668 4181	189	MEYHALL CHEMICAL (UK) LTD 1 Boundary Road, New Ferry, Bebington, Wirral, Cheshire L62 5AL. 051–645 8220/9
172	LIBERTA-IMEX LTD Liberta House, Scotland Hill, Sandhurst, Berks. 02517 829613	190	MG PLASTICS LTD P O Box 7, Preston, Lancs. 0772 54107
173	LIMMER & TRINIDAD CO LTD Trinidad Lake House, 232/242 Vauxhall Bridge Road, London SW1. 01–828 4388	191	MICANITE & INSULATORS CO LTD, THE Westinghouse Road, Trafford Park, Manchester M17 1PR. 061–872 2431
174	LION EMULSIONS LTD Three Mill Lane, Bromley-by-Bow, London E3 3DY. 01–980 6161	192	MILLER, HENRY & CO LTD Church Road, Harold Wood, Romford, Essex. 45 43937
175	LION INK LTD Century Works, Booth Street, Walkden, Manchester. 061–790 2328	193	MINNESOTA MINING & MANUFACTURING CO LTD 3M House, Wigmore Street, London W1. 01–486 5522
176	LIVERPOOL ADHESIVE PASTE CO LTD 9 Roberts Street, Liverpool 3. 051–236 2086	194	MOLECULAR CONSERVATION LTD Clare Road, Harrogate, Yorks. 0423 67641
177	LONDON ADHESIVE CO LTD Maxwell Road, Borehamwood, Elstree, Herts. 01–953 2992	195	MONARCH ADHESIVES LTD Schoolfield Road, West Thurrock, Essex RM16 1HR. 792 6567/8
178	L.T.D. BUILDING PRODUCTS Church Road, Litherland, Liverpool L21 8NX. 051–928 5214	195A	MONSANTO LTD Monsanto House, 10–18 Victoria Street, London SW1H 0NQ. 01–222 5678
179	LUCENT PRODUCTS LTD 202 Cambridge Road, Kingston, Surrey. 01–549 0669	196	MORGAN REFRACTORIES LTD Neston, Wirral, Cheshire. 051–336 1406
180	MACFARLANE & CO LTD, N S Tapes Division, Clansman House, Sutcliffe Road, Glasgow C3, Scotland. 041–959 3396	197	MORTON-WILLIAMS LTD Greville House, Hibernia Road, Hounslow, Middlesex. 01–570 7766
181	MAC-TAC INTERNATIONAL LTD, INC Northampton.	198	MULTIGRIP LTD Airborne Works, Arterial Road, Leigh-on-Sea, Essex. 0702 525265

continued

Table 10.3 continued

Code No.	Manufacturer	Code No.	Manufacturer
199	MURRAY & JONES LTD Meredith Street, London E13. 01–476 5262/4	214	PLYCOL LTD (see code no. 253)
		215	POCHIN, F H & H S LTD Spinney Hill Road, Leicester. 0533 36575
200	NATIONAL ADHESIVES & RESINS LTD Slough, Bucks. 75 29191 also at Braunston, nr Rugby, Warwicks. Braunston 624	216	POLYBOND LTD Warsash, Southampton. 04895 5272
		217	POLYCELL PRODUCTS LTD Polycell House, Broadwater Road, Welwyn Garden City, Herts. 96 28131
201	NICHOLLS & CLARKE LTD Niclar House, 3/10 Shoreditch High Street, London E1. 01–247 5432	218	POLYGLAZE LTD Victoria Way, Burgess Hill, Sussex. 04446 5206
202	NICOBOND (PRODUCTS) LTD 12/13 Blossom Street, London E1. 01–247 7763	219	POLYPENCO LTD Gate House, Welwyn Garden City, Herts. 96 21221
203	NORTH BRITISH ADHESIVES LTD 'Nabl' Works, Dunedin Street, Edinburgh 7, Scotland. 021–556 3535	220	PORTANAL LABORATORIES LTD 4 Greenholm Road, London SE9. 01–303 7956
203A	NORTHERN COUNTIES ADHESIVES 11 Parish Ghyll Road, Ilkley, Yorks. 09433 3282	221	PROTECTIVE MATERIALS LTD Watery Lane, Birmingham 9. 021–772 6259
204	NORTH HILL PLASTICS LTD Manley Court, Stoke Newington High Street, London N16. 01–800 3773	222	QUENTSPLASS LTD Thorpe Arch Trading Estate, Wetherby, Yorks LS23 7BZ. 0937 2126 and 3388
205	OHLENSCHAGER BROS LTD 49 Weston Street, London SE1. 01–407 0888/1318	223	RADBURNE & BENNETT (RUSHDEN) LTD Graveley Street, Rushden, Northants. 09334 2077
206	OMEGA PLASTICS LTD Northwick Road, Canvey Island, Essex SS8 0PT.	224	THE RAWLPLUG CO LTD Rawlplug House, 147 London Road, Kingston, Surrey. 01–546 2191 also at 3 Trading Estate, Humber Road, Cricklewood, London NW2. 01–452 6687
207	PAFRA LTD Bentalls, Basildon, Essex. 0268 3351/2	225	REDFERN POLYMERS LTD Hyde, Cheshire. 061–368 2621
208	PERKINS, A E (RAYLEIGH) LTD Nelson Street, Southend-on-Sea, Essex. 0702 48314	226	REES, W F LTD Westminster House, Old Woking, Surrey. Mayford 62221
208A	PERMABOND ADHESIVES & SEALANTS LTD 33 Clarence Street, Staines, Middlesex. 81 56041	227	RESADHESION LTD 29 Eve Road, Woking, Surrey. 04862 61719
209	PERTEC Griffin Lane, Aylesbury, Bucks. 0296 4201	228	RESINOUS CHEMICALS LTD Blaydon, Co Durham. 089425 2751
210	PHILLIPS PATENTS LTD Dantzic Street, Manchester. 061–834 5854	229	REVELL LTD Cranbourne Road, Potters Bar, Herts. 77 58261
211	PLANAX BINDING SYSTEMS LTD 15–17 Cheam Road, Sutton, Surrey. 01–643 4441	230	REVERTEX LTD 51/5 Strand, London WC2. 01–839 7888
212	PLOW PRODUCTS LTD 43A Portlock Road, Maidenhead, Berks. 0628 22614/20297	231	REX PASTE LTD 90 Lots Road, London SW10. 01–352 1456
213	PLUS PRODUCTS LTD Stella Works, Newburn Bridge Road, Blaydon-on-Tyne, Co Durham. 089425 2715		

continued

Table 10.3 continued

Code No.	Manufacturer	Code No.	Manufacturer
232	RICKARDS, L J CO LTD 76 Watling Street, London EC4M 9BL. 01–248 3761	251	SHEPPY GLUE & CHEMICAL WORKS LTD Bonehurst Road, Horley, Surrey. 02934 5151
233	ROBERTS SMOOTHEDGE LTD Victoria Road, Burgess Hill, Sussex. 04446 5454	252	SHELL CHEMICALS UK LTD Halifax House, 51–55 Strand, London WC2N 5PR. 01–839 9070
234	ROCOL LTD Rocol House, Swillton, Nr Leeds, Yorks. 09738 2261 also 01–892 6252	253	SHELL COMPOSITES LTD Trading Estate, Slough, Bucks SL1 4DL. 75 21261
235	ROMAC INDUSTRIES LTD Romac Works, The Hyde, London NW9. 01–205 6055/9	254	SIKA LTD 7 Buckingham Gate, London SW1. 01–834 3693
236	ROSS INSULATION PRODUCTS The Power House, Formby, Lancs.	255	SMITH & MCLAURIN LTD Wheatsheaf House, Carmelite Street, London EC4. 01–583 7481
237	RUBBER LATEX LTD Harling Road, Wythenshawe, Manchester M22 4TP. 061–998 3226	256	SMITH, H A LTD Braunston, Rugby, Warwickshire. 0788 89248
238	RUBERT & CO LTD Acru Works, Demmings Road, Cheadle, Cheshire. 061–428 5855	257	SMITH, HERBERT & CO (GRINDING) LTD 144 Leeds Road, Hull, Yorks. 0482 408915/6
		258	SONDHEIMER, M Sondal Works Ltd, Markfield Road, Tottenham, London N15. 01–808 1153
239	SAFETY TREAD LTD, THE Crown Wharf Ironworks, Dace Road, Old Ford, London E3. 01–985 4407	259	SPECTRA CHEMICALS LTD 5 Bridge Road, Haywards Heath, Sussex. 0444 2548
240	SANKEY, J H & SONS LTD Essex Works, Ripple Road, Barking, Essex. 01–592 6666	259A	STAIDENT PRODUCTS LTD 33 Clarence Street, Staines, Middlesex. 81 55454
241	S.B.D. CONSTRUCTION PRODUCTS LTD Denham Way, Maple Cross, Rickmansworth, Herts. 87 77311	260	STANCOURT SONS & MUIR Victoria House, Southampton Row, London WC1. 01–405 9959
242	SCHENECTADY-MIDLAND LTD Four Ashes, Wolverhampton, Staffs. 090–727 533/5	261	STARCH PRODUCTS (SCOTLAND) LTD Port Downie Works, Falkirk, Stirlingshire, Scotland. 0324 24254
243	SCHJELDAHL CO (UK) DIV Eastern Road, Bracknell, Berks. 0344 4331	262	STEPHENS LTD, HENRY C Britannia House, Drayton Park, London N5. 01–226 4455
244	SCOTT BADER & CO LTD Wollaston, Wellingborough, Northants. 09333 4881	262A	STERLING COATED MATERIALS LTD 7 Tower Mill, Park Road, Dukinfield, Cheshire. 061–330 2062
245	SCOTTISH ADHESIVES CO LTD 9/23 Farnell Street, Glasgow C4. 041–332 1736	263	STRUCTOPLAST LTD Ford Airfield, Ford, Nr Arundel, Sussex. 09064 6955
246	SEALOCRETE GROUP SALES LTD Head Office & Export, Atlantic Works, Hythe Road, London NW10 6RU. 01–969 9111	264	SWIFT CHEMICAL CO Spelthorne Lane, Ashford, Middlesex. 69 56333/7 also at Bridge Street, Chatteris, Cambs. 03543 2653
247	S. E. CHEMICAL CO LTD P O Box 73, Evington Vale Road, Leicester.		
248	SELLOTAPE PRODUCTS LTD Sellotape House, 54/8 High Street, Edgware, Middlesex HA8 7ER. 01–952 2345	265	SYNTHETIC & INDUSTRIAL FINISHES LTD Imperial Way, Balmoral Road, Watford, Herts. 92 28363/4
249	SERICOL GROUP LTD 24 Parsons Green Lane, London SW6. 01–736 8181		
250	SHAWINIGAN LTD 118 Southwark Street, London SE1. 01–928 2765		

continued

Table 10.3 continued

Code No.	Manufacturer
266	SYNTHETIC RESINS LTD Frodsham House, Edwards Lane, Speke, Liverpool L24 9HR. 051–486 1395 Kollercast Department, 103 High Street, Oxon OX9 3DZ. 084421 2282 and 2573
267	TEKNA PRODUCTS High Wycombe, Bucks.
268	TENNANT TRADING CO LTD 9 Harp Lane, Great Tower Street, London EC3. 01 626 4533
269	TIVOLI KAY ADHESIVES LTD P O Box 22, Elton Fold Mills, Bury, Lancs BL8 1NR. 061–764 4821 also at 7 Parker Drive, Leicester (Sales Office).
270	TOSELAND & SONS LTD Kettering, Northants. 0536 2175
271	TRANSDUCERS (C.E.L.) LTD 2 Trafford Road, Reading, Berks, RG1 8JH. 0734 580166
272	TREMCO LTD 27 St George's Road, London SW19 4DY. 01–947 3451
273	TRETOBOND LTD Tretol House, The Hyde, London NW9. 01–205 6191
274	TURNBRIDGES LTD 72 Longley Road, London SW17. 01–672 6581/3
275	TURNER BROS ASBESTOS CO LTD P O Box 40, Rochdale, Lancs. 0706 47422
276	UCAN PRODUCTS LTD 11–13 Old Esher Road, Hersham, Walton-on-Thames, Surrey KT12 5PT. 98 28921
277	UNBRAKO LTD Burnaby Road, Coventry CV6 4AE. 0203 88722
278	UNIBOND LTD Yorktown Industrial Estate, Glebeland Road, Camberley, Surrey. 0276 3135/7
279	UNIMATIC ENGRS LTD 16 Coverdale Road, London NW2. 01–459 2145/6
280	UNION CARBIDE UK LTD P O Box 2LR, 8 Grafton Street, London W1. 01–629 8100
281	UNION GLUE & GELATINE CO LTD Adhesives Division, P O Box 58, Manchester M17 1JD. 061–872 7607

Code No.	Manufacturer
282	UNITEX Knaresborough, Yorks. 042376 2455/6
283	VIK SUPPLIES LTD Stafford. 0785 2241
284	VINYL PRODUCTS LTD Mill Lane, Carshalton, Surrey. 01–669 4422
285	WALLER ADHESIVES LTD Greenpark Road, Bray, Co Wicklow, Ireland. Dublin 862931
286	WELWYN ELECTRIC LTD Bedlington, Northumberland. 067082 2181/9
287	WHITTAKER CORP Narmco Materials Division, 600 Victoria Street, Costa Mesa, California, USA.
288	WILLIAMS ADHESIVES LTD 179/180 Gresham Road, Trading Estate, Slough, Bucks. 75 24343 and 29481
289	WILSON & CO (ADHESIVES) LTD, HAROLD Brook Street Works, Hazel Grove, Stockport, Cheshire SK7 4PU. 061–480 9611
290	WITCO CHEMICAL CO LTD Droitwich, Worcs. 09057 2454
291	YORKSHIRE CHEMICALS LTD Adhesives Division, Kirkstall Road, Leeds LS3 1LL. 0532 443111

10.4. SOURCES OF FURTHER INFORMATION ON ADHESIVES

RESEARCH ASSOCIATIONS AND OTHER ORGANISATIONS

Adhesive Tape Manufacturers' Association,
18 Fleet Street,
London EC4Y 1AS
Telephone: 01-353 8894
Contact: Mr H Basil Shelby
Trade association to promote manufacture, development and sale of mechanical and industrial self-adhesive tape products in the U.K.

The British Adhesive Manufacturers' Association
(BAMA)
Secretarial Offices: 20 Pylewell Road,
                    Hythe,
                    Southampton,
                    Hants SO4 6YW
Telephone: Hythe 2765
Contact: Dr P Bosworth
*Association formed by leading manufacturers to
establish and provide quality standards for industries
concerned with adhesives usage. Activities include the
determination of a Code of Practice in connection with
safety and health safeguards to benefit users and
manufacturers of adhesives.*

The British Ceramic Research Association,
Queen's Road,
Penkhull,
Stoke-on-Trent ST4 7LQ
Telephone: 0782 45431
Contact: Mr D Murfin, Information Officer
*Evaluation of refractory cements and binders for
ceramic materials used in building construction, e.g.
tiles, pipes, bricks; environmental test facilities for
wall and floor tiling adhesives; studies on ceramic-to-
metal seals.*

The British Food Manufacturing Industries Research
Association,
Randalls Road,
Leatherhead,
Surrey KT22 7RY
Telephone: Leatherhead 76761
Contact: Dr A Courts
Information on glue and gelatin products.

British Standards Institution,
2 Park Street,
London W1A 2BS
Telephone: 01-629 9000
Contact: Mr M J Pater
*Publication of British Standards relating to adhesives
usage. BSI Yearbook lists current editions of standards
issued under the authority of the Adhesive Industry
Standards Committee. Revised editions or amend-
ment slips are reported in the monthly journal BSI
News.*

Cement and Concrete Association,
(Research and Development Division),
Wexham Springs,
Slough,
Bucks SL3 6PL
Telephone: Fulmer 2727
Contact: Mr A A Lilley
*Information on technology of adhesives usage with
concrete and allied materials in connection with road-
work, floors and precast concrete elements.*

The Furniture Industry Research Association
(FIRA),
Maxwell Road,
Stevenage,
Herts SG1 2EW
Telephone: Stevenage 3433
Contact: Mr A D Spillard
*Research, advisory and consultancy services for the
furniture industry; performance of adhesives in
furniture construction; advice and technical assistance
on materials selection, joint design and bonding
technology.*

Malaysian Rubber Producers' Research Association,
19 Buckingham Street,
London WC2N 6EJ
Telephone: 01-930 9314
Contact: Mr M G Rodway
also, MRPRA (Laboratories),
        Brickendonbury,
        Hertford SG13 8NP
Telephone: Hertford 4966
Information on natural rubber adhesives.

PIRA, Randalls Road,
Leatherhead,
Surrey KT22 7RU
Telephone: Leatherhead 76161
Contact: Mr C V Hawkes
*Research association for the paper and board, printing
and packaging industries. Research and development
work on adhesives for bookbinding, packaging,
laminating, corrugating, paper tapes and general paper
and board conversion; machine production and testing
of adhesive joints under service conditions; license the
manufacture of specific test equipment.*

The Plastics Institute,
11 Hobart Place,
London SW1W 0HL
Telephone: 01-245 9555
Contact: Mr M J Quinlan
*Objective of the Institute is the advancement of poly-
mer science and the technology of plastics. It organises
a wide range of conferences, symposia and other
technical meetings, some of which pertain to aspects of
adhesion to plastics materials. Publications include a
two-monthly journal Plastics and Polymers, and
technical monographs on polymer technology.*

Princes Risborough Laboratory,
Princes Risborough,
Aylesbury,
Bucks HP17 8BR
Telephone: Princes Risborough 3101
Contact: Mr J C Beech
*Timber and Technology Division: Research into
requirements for glued structures based on timber and
other construction materials. Current studies concern*

adhesives evaluation under laboratory and service conditions; bonding of preservative treated wood; efficiency of nail/glue techniques.

Production Engineering Research Association (PERA),
Melton Mowbray,
Leics LE13 0PB
Telephone: Melton Mowbray 4133
Contact: Mr D B Benton
Companies frequently experience difficulty in selecting the best joining or fastening technique for a particular application. Through PERA, companies can obtain impartial advice about all types of joining and fastening techniques and the related economic considerations. The PERA Joining Economic Evaluation and Process Selection system (JEEPS) provides a basis for analysing individual production joining applications so that the relative merits of adhesives, fasteners, welding, brazing and soldering processes, together with assorted production costs, can be evaluated. Companies can be assisted to apply the most appropriate techniques in production, preceded, if necessary, by practical proving trials.

Rubber and Plastics Research Association of Great Britain (RAPRA),
Shawbury,
Shrewsbury,
Shropshire SY4 4NR
Telephone: Shawbury 383
Contact: Mr R P Nelms, Information Officer
Information services in connection with technology of rubbers, plastics and textiles.

Science Research Council,
State House,
High Holborn,
London WC1R 4TA
Telephone: 01-242 1262
Contact: Miss R Lyster (Press Officer)
Information on SRC-supported research and post-graduate training in polymer science.

Shoe and Allied Trades Research Association (SATRA),
Satra House,
Rockingham Road,
Kettering,
Northants NN16 9JH
Telephone: Kettering 3151
Contact: Mr D Pettit
Technical service and research aspects of bonding to flexible substrates such as rubbers, leather, thermoplastics, for footwear and allied industries' membership; development work on surface pretreatments; surface studies; test facilities for bonded assemblies; manufacture of testing equipment for the shoe trade.

Sira Institute Ltd,
South Hill,
Chislehurst,
Kent BR7 5EH
Telephone: 01-467 2636
Contact: Mr J Shields
Adhesives advisory service offers information, technical assistance or research to industry; evaluation of new and established adhesives undertaken with respect to mechanical, optical, thermal, electrical and other properties; past research activities have included optical cements, strain gauge adhesives, structural bonding of metals, plastics bonding; optical and electron microscopy facilities are available for surface examination; test equipment for durability studies on bonded assemblies; extensive technical data bank on commercial adhesives.

Timber Research and Development Association,
Hughenden Valley,
High Wycombe,
Bucks HP14 4ND
Telephone: Naphill 3091
Contact: Mr Gavin Hall
Research, development and testing for the timber industry and other users. TRADA Wood Technology Laboratory undertakes variety of investigatory work which includes chemical analysis, microscopy and physical testing of adhesives and sealants.

UNIVERSITIES AND TECHNICAL COLLEGES

Cranfield Institute of Technology,
Department of Materials,
Cranfield,
Bedford MK43 0AL
Telephone: Bedford 750111
Contact: Dr R L Apps

Oxford Polytechnic,
Headington,
Oxford OX3 0BP
Telephone: Oxford 63434
Contact: Mr A Beevers, Principal Lecturer in Production Engineering
Zygology Centre offers consultancy and technical assistance to industry on all aspects of joining and fastening technology; facilities for research, evaluation and testing of bonded joints; educational courses on zygology.

University of Cambridge,
Department of Metallurgy and Materials Science,
Pembroke Street,
Cambridge CB2 3QZ
Telephone: Cambridge 65151
Contact: Dr T P Hoar
Information on adhesion of oxide and polymer layers to copper.

University of Manchester,
Department of Prosthetics,
Turner Dental School,
Bridgeford Street,
Manchester M15 6FH
Telephone: 061-273 5252
Contact: Dr E C Combe
Development of improved dental cements.

University of South Wales and Monmouthshire,
Cathays Park,
Cardiff CF1 3NS
Telephone: Cardiff 22656
Contact: Dr B H Williams
Structural adhesives research.

The City University,
St John Street,
London EC1V 4PB
Telephone: 01-253 4399
Contact: Mr K W Allen, Department of Chemistry
Organisation of an Annual Conference on 'Adhesion and Adhesives' for industrialists, research scientists and technologists. Scope—current developments in adhesion theory and adhesives technology. Conference proceedings are published by University of London Press Ltd. In addition to its research activity, the Department organises seminars on specific aspects of bonding technology, and offers educational and research facilities for post-graduate studies on adhesion.

GOVERNMENT ORGANISATIONS

Department of the Environment

Building Research Establishment,
Princes Risborough Laboratory,
Princes Risborough,
Aylesbury,
Bucks HP17 8BR
Telephone: Princes Risborough 3101
Contact: Mr C H Tack

The Transport and Road Research Laboratory,
Crowthorne,
Berkshire
Telephone: Crowthorne 3131

Department of Industry

National Engineering Laboratory,
East Kilbride,
Glasgow G75 0QU
Telephone: East Kilbride 20222
Contact: Dr W S Carswell or Mr W Paton
General advisory service on adhesives relating to engineering applications; testing facilities.

Department of Industry

Laboratory of the Government Chemist,
Cornwall House,
Stamford Street,
London SE1 9NQ
Telephone: 01-928 7900
Contact: Mr A D Wilson
Development work on Ionomer cements as adhesives for teeth and mineral substrates.

Procurement Executive, Ministry of Defence

Atomic Weapons Research Establishment,
Building SB24,
Aldermaston,
Reading,
Berks RG7 4PR
Telephone: Tadley 4111
Contact: Mr D Deverell, Chemical Technology Division
Development of special purpose adhesives for military applications; environmental testing; development of test methods.

Explosives Research and Development Establishment,
Waltham Abbey,
Essex EN9 1BP
Telephone: Lea Valley 713030
Contact: Mr W A Dukes
Adhesives, sealants and coated fabrics advisory service; research into test methods; effect of joint design on strength; fracture mechanics; durability and accelerated environmental testing; surface pretreatments and adhesion performance.

Royal Aircraft Establishment,
Farnborough,
Hants GU14 6TD
Telephone: Aldershot 24461
Contact: Mr L N Phillips, Materials Department
Evaluation and development of structural adhesive materials for aircraft assembly and related applications.

CHAPTER 11

Glossary of Terms Relating to Adhesives Technology

The rapid growth in the industrial use of adhesives has resulted in the development of an extensive technical vocabulary, descriptive of the various aspects of adhesion and adhesives, needed to communicate the technology of the adhesives industry.

The following pages present a selection from various authorities (see sources at the end) of 340 terms in current usage, which are covered by the broadly defined groups:

GENERAL, definitions, fundamental concepts and terms.

MATERIALS, terms embracing products, forms, additives and additional materials relating to adhesives and adherend materials and assemblies.

PROCESSES, terms covering preparative methods and procedures applicable to the manufacture of bonded assemblies; processing equipment.

PROPERTIES, descriptions of the physical and chemical properties of adhesives and joints; adhesion measurement and test equipment and methods; joint types.

Throughout the glossary, terms are to be found listed under the title of the main noun followed by an adjective (e.g. FAILURE, ADHEREND for ADHEREND FAILURE) unless this is inconvenient. This arrangement ensures that related terms are grouped together in a meaningful context and also simplifies the explanation required for each term. A perusal of the group terms is often suggestive of a more suitable alternative material or process which might otherwise not have readily occurred to the user. Thus, the classification extends the usefulness of the glossary to those people outside the adhesives industry whose knowledge is generalised.

The glossary includes the terms used throughout the text and also provides a ready reference to the terminology of the trade literature and other publications relating to adhesives technology. The meanings of the terms given are those pertaining to adhesives practice, although it is recognised that alternative meanings can apply in other connections. All definitions are believed to be correctly descriptive of modern usage but are not rigidly defined.

A-STAGE an early stage in the curing of thermosetting resins when the material is still fusible and soluble in certain solvents (cf. B-STAGE, C-STAGE).

ABHESIVE a material which is adhesive resistant and applicable as a non-sticking surface coating; release agent.

ABSORPTION the penetration of a liquid (adhesive) into an adherend by virtue of capillary action or interaction with the substrate surface.

ACCELERATOR a material that initiates or promotes chemical processes, such as condensation, vulcanisation and polymerisation, and finally forms an integral part of the product.

ACID ACCEPTOR an additive material to an adhesive formulation for the neutralisation of acid by-products resulting from the preparation, use or storage of the adhesive.

ADHERE fasten together two surfaces by adhesion.

ADHEREND a body which is attached to another body by an adhesive.

ADHEREND FAILURE see FAILURE, ADHEREND.

ADHESION the state of being held together by means of an interlayer of adhesive between adherend interfaces; the attachment of two surfaces by interfacial forces consisting of molecular forces, chemical bonding forces, interlocking action, or combinations of these.

ADHESION, MECHANICAL the adhesion between surfaces held together by the interlocking action of the adhesive (e.g. the bonding of porous materials with penetrant resins).

ADHESION, SPECIFIC the adhesion between surfaces held together by chemical bonding forces.

ADHESION TO BACKING the peeling force required to remove an adhesive tape applied to its own backing, under specified conditions.

ADHESIVE a material that binds other materials together by surface attachment.

ADHESIVE, ANAEROBIC a single component adhesive which cures only in the absence of air.

ADHESIVE, AQUEOUS an adhesive based on water as a solvent, e.g. dextrine, animal glues, or dispersing medium, e. g. latex suspensions.

ADHESIVE, ASSEMBLY an adhesive employed for bonding construction parts together, e.g. woodwork, model making, metal assembly work.

ADHESIVE, CLOSE CONTACT an adhesive intended for use with joint surfaces brought into close contact by maintained pressure application during curing; non-gap-filling adhesive providing glue-line thicknesses not exceeding 0·1 mm.

ADHESIVE, COLD SETTING an adhesive which sets at temperatures below 20°C.

ADHESIVE, CONTACT non-tacky adhesive with self adhesive properties. The conjunction of two surfaces coated with contact adhesive results in a joint not requiring a setting period or a sustained bonding pressure.

ADHESIVE, DISPERSION OR EMULSION a two phase system with one phase (the adhesive material) in a liquid suspension.

ADHESIVE, FOAMED an adhesive material containing a dispersion of gas bubbles to decrease its apparent density.

ADHESIVE, GAP-FILLING an adhesive subject to low shrinkage on setting employed as a sealant.

ADHESIVE, HEAT ACTIVATED a dry adhesive film activated by the application of heat, with or without pressure, to the joint assembly.

ADHESIVE, HEAT SEALING a thermoplastic film adhesive which is melted between the adherend surfaces by heat application to one or both of the adjacent adherend surfaces.

ADHESIVE, HOT MELT an adhesive material applied at a temperature above its melting point. Usually thermoplastic, wax or 100% solids adhesives applied between 150 and 200°C.

ADHESIVE, HOT SETTING an adhesive with a setting temperature of 100°C or more.

ADHESIVE, INTERMEDIATE TEMPERATURE SETTING an adhesive which sets in the temperature range 31–99°C.

ADHESIVE, LATEX an emulsion of rubber or thermoplastic resin in water.

ADHESIVE, LAYFLAT an adhesive with the property of non-warping in laminate structures. (cf. LAMINATE, CROSS, LAMINATE, PARALLEL).

ADHESIVE, MULTIPLE LAYER a film coated on each side with a different adhesive and intended to bond dissimilar materials.

ADHESIVE, ONE COMPONENT an adhesive material incorporating a latent hardener or catalyst activated by heat. Usually refers to thermosetting materials but also describes anaerobic, hot melt adhesives, or those dependent on solvent loss for adherence.

ADHESIVE, PRESSURE SENSITIVE an adhesive which adheres to a surface at room temperature by temporary application of pressure alone; permanently tacky adhesive capable of bonding a variety of surfaces.

ADHESIVE, PRESSURE SETTING an adhesive which sets permanently when under pressure.

ADHESIVE, ROOM TEMPERATURE SETTING an adhesive which sets in the temperature range 20–30°C.

ADHESIVE, SEPARATE APPLICATION a two component adhesive in which one part is applied to one adherend and the second part to the other adherend so that the subsequent union of the coated adherends forms the joint.

ADHESIVE, SOLVENT an adhesive containing a solvent as a fluid carrier, which depends on solvent loss or its absorption by adherends for bond formation.

ADHESIVE, SOLVENT ACTIVATED a dry adhesive film made tacky by solvent application before bonding.

ADHESIVE SPREAD the amount of adhesive applied to an adherend expressed in terms of grammes of liquid or solid adhesive per square metre of joint area. Application to one adherend surface is called 'single spread' whereas the term 'double spread' refers to adhesive application to both adherends.

ADHESIVE, STRUCTURAL a material employed to form high strength bonds in structural assemblies which perform load bearing functions, and which maybe used in extreme service conditions, e.g. high and low temperature exposure.

ADHESIVE, SYNTHETIC RESIN a synthetic resin(s) product, e.g. aminoplastic or phenolic resins or mixtures of these.

ADHESIVE, TWO COMPONENT an adhesive supplied in two parts which are mixed before application.

ADHESIVE, WARM SETTING an intermediate temperature setting adhesive (31–99°C).

ADSORPTION the interaction between a solid surface and a liquid without penetration by the latter.

AFFINITY the attraction between adhesive and adherend.

ANAEROBIC acting only in the absence of air (cf. ADHESIVE, ANAEROBIC).

APPLICATOR a means for applying adhesive to adherends.

AQUEOUS water based or containing water (e.g. ADHESIVE, AQUEOUS).

ASSEMBLY a completely bonded structure, or a process of applying an adhesive to adherends which are united to form a joint.

ASSEMBLY, DRY an assemblage of unglued adherends into a unit for the purpose of determining the joint positions, e.g. scarf joints in laminations.

ASSEMBLY TIME see TIME, ASSEMBLY

ASSEMBLY, WET an assemblage of glued adherends before application of bonding pressure.

AUTOCLAVE a closed container which provides controlled heat and pressure conditions.

B-STAGE an intermediate stage in the processing of certain thermosetting resins where the material softens with heat to a rubbery state but will not entirely melt or dissolve in some of the solvents which will dissolve resins in the A-stage. Uncured thermosetting resins in this stage are sometimes known as 'Resitols' (*cf.* RESITOL, A-STAGE, C-STAGE).

BACKING the flexible supporting material for an adhesive, e.g. pressure sensitive adhesives are commonly backed with paper, plastic films, fabric or metal foil; heat curing thermosetting adhesives are often supported on glass cloth backing.

BACTERICIDE an additive used to destroy bacteria occurring in adhesive formulations in order to prevent their attack on the adhesive. Natural products based on protein or carbohydrate are prone to biodeterioration.

BAG MOULDING a moulding or bonding process involving the application of pressure by means of air, water, steam or vacuum, to a flexible cover (bag) which completely encloses the material being bonded.

BATCH a production quantity derived from a manufacturing process or a mixture of these resulting from the same process conditions.

BINDER the component of an adhesive which is mainly responsible for adhesive properties; an adhesive material employed to adhere between particles, aggregates and powders.

BITE the penetration or dissolution of adherend surfaces by an adhesive.

BLANKET MOULDING an alternative term for bag moulding.

BLISTER a surface elevation on a flexible adherend resembling a skin blister, usually due to adhesive deficiency, entrapped air or solvent vapour.

BLOCKING undesired adhesion between touching layers of a material such as occurs under moderate pressure during storage or use.

BLOOM a visible exudation or efflorescence on the surface of materials; reference term for the 'gel strength' of animal glues.

BLOOM STRENGTH a measurement of 'gel strength' in Degrees Bloom made with a Bloom Gelometer.

BODY the consistency of an adhesive which is a function of viscosity, plasticity and rheological factors.

BOND (N) the union of two materials by adhesives.

BOND (V) the use of an adhesive to join materials.

BOND STRENGTH a term synonymous with adherence; the force required to break an adhesive assembly with failure occurring near, or at, the interface. The force may be applied in shear, compression, flexure, peel, impact or cleavage to disrupt the bond.

BRASHINESS the brittleness of an animal glue film as a result of drying or migration of plasticiser; loss of flexibility.

BRITISH GUM a special type of dextrine.

BUTT JOINT *see* JOINT, BUTT.

C-STAGE the final stage in the processing of certain thermosetting resins where the material becomes insoluble and infusible; sometimes called 'Resite' (*cf.* RESITE, A-STAGE, B-STAGE).

CATALYST a chemical substance which accelerates adhesive curing when added in small amounts to the larger quantities of the reactants; material which promotes cross linking in a polymer or accelerates adhesive drying.

CAULK the use of a plastic material to fill voids and cracks in the adherend surface to prevent moisture or solvent penetration by sealing.

CEMENT a synonym for adhesive; a mixture of water with finely powdered lime and clay which hardens and adheres to suitable aggregates to form concrete or mortar; material conforming to BS12, BS146, BS915, BS1370; an inorganic paste with adhesive properties.

CENTIPOISE one hundredth of a poise, unit of viscosity.

CLEAVAGE *see* STRESS, CLEAVAGE.

CLIMBING DRUM an instrument for measuring the peel strength of adhesive joints.

CLOSED ASSEMBLY TIME *see* TIME, ASSEMBLY.

COALESCENCE the process of unification or joining of liquid globules; the fusion of liquid films or tacky adhesives.

COBWEBBING the formation of emergent threads of adhesive during the operation of a spray-gun applicator.

COHESION internal adhesion; the ability to resist rupture within the bulk material; the state in which the particles of a material are held together by forces of adhesion.

COHESIVE FAILURE *see* FAILURE, COHESIVE.

COLD FLOW the deformation of a material at room temperature without applied load (*cf.* CREEP).

COLD PRESSING a bonding operation in which an assembly is subjected to pressure without the application of heat.

COLLAGEN the protein derived from bone and skin used to prepare animal glue and gelatine.

COLLOIDAL a state of suspension and dispersion of submicron particles in a liquid medium without their dissolution in the medium, e.g. animal glue adhesives.

COLOPHONY the resin gum obtained from various species of pine trees.

CONDENSATION the precipitation of moisture on adherend surfaces; chemical reaction of two or more molecules to form a new molecule and a by-product; poly-condensation refers to polymer formation.

CONDITIONING TIME *see* TIME, JOINT CONDITIONING.

CONSISTENCY the ability of a liquid adhesive to resist deformation or flow when shearing stresses are applied to it. The term is usually applied to materials whose deformations are not proportional to applied stresses.

CONTACT BONDING the deposition of cohesive materials on both adherend surfaces and their assembly under pressure.

CO-POLYMER a polymerisation product of two or more different monomers (*cf.* POLYMER).

CORE the honeycomb structure used in sandwich panel construction; innermost portion of a multi-layer adherend assembly.

CORROSION the chemical reaction between the adhesive or contamination and the adherend surfaces, due to reactive components in the adhesive film, leading to deterioration of the bond strength.

COTTONING the formation of web-like filaments of adhesive between the applicator and substrate surface.

COVERAGE the spreading power of an adhesive over the area of adherend.

CRAMPING PERIOD the time required for holding an adhesive joint under pressure during the curing of the adhesive.

CRAMPING PRESSURE the quotient of the force applied to adherend surfaces and the area of the surface over which the force acts.

CRAZING an extension of fine cracks on, within, or beneath, the surface of an adhesive layer; formation of fissures, voids in a film due to shrinkage or solvent action.

CREEP the dimensional change with time of a material under sustained load.

CRIMP to fold over and fasten under pressure; the indentation of the adherend surface to improve interfacial contact with the adhesive.

CROSS-LINKING the union of adjacent molecules of uncured adhesives (often existing as long polymer chains) by catalytic or curing agents.

CURE to alter the physical properties of an adhesive by chemical change, e.g. polymerisation, vulcanisation, brought about by the agency of heat, pressure, and catalysts, etc.

CURE TIME *see* TIME, CURING

CURING AGENT the term for accelerator, catalyst, or resin added to an adhesive composition to promote or control curing.

CURLING the distortion or warping of an adhesive layer due to solvent penetration into the adherend surfaces or inequalities of the expansion properties of the adherends.

CURTAIN COATER a liquid spreading machine which deposits a controlled thickness of coating liquid (adhesive) on a surface passing through it.

DECAY the decomposition of adhesives and adherends by fungi and micro-organisms, leading to loss of strength, weight and other properties.

DEGREASE to remove oil and grease from adherend surfaces.

DELAMINATION the separation of laminated layers due to failure of the adhesive.

DEMULSIFICATION the separation of an emulsion into its component phases (e.g. separation of oil-water emulsion on standing).

DEXTRINE a water based product derived from the acidification and/or roasting of starch.

DIELECTRIC CURING the use of a high frequency electric field through a joint to cure a synthetic thermosetting resin. Curing process for wood and other non-conductive joint materials. Curing results from the heat generated by the resonance of the molecules within the adhesive due to the imposed field.

DILATENT a liquid whose viscosity increases with shearing rate.

DILUENT an additive for the reduction of the concentration of bonding components; often used to reduce the viscosity of an adhesive formulation.

DISPERSION a two phase system with one phase suspended in the other.

DOCTOR BLADE OR BAR a scraper blade for regulating the amount of adhesive being deposited on a surface or on spreader rolls.

DOCTOR ROLL a regulating roller for controlling the amount of adhesive supplied to spreader rolls (*cf.* ROLLER COATER).

DOPE a cellulose ester used as a lacquer.

DOUBLE LAP JOINT *see* JOINT, LAP

DOUBLE SPREAD *see* ADHESIVE SPREAD

DRY to remove solvent constituents from an adhesive by evaporation or absorption, or both, and thus alter the physical state of the adhesive (*cf.* CURE, SET).

DRYING TEMPERATURE *see* TEMPERATURE, DRYING

DRYING TIME *see* TIME, DRY

DRY STRENGTH *see* STRENGTH, DRY

DRY TACK *see* TACK, DRY

DURABILITY the resistance to reduction in joint strength shown by adhesives to moisture, heat, chemicals and biodeterioration, etc.

ELASTICITY, MODULUS OF the ratio of stress to strain in an elastically deformed material.

ELASTIC MEDIUM a material obeying Hooke's Law (i.e. strain is proportional to the stress producing it).

ELASTOMER a polymeric rubbery material capable of deformation or distension under stress; a material which may be stretched repeatedly, at room temperature, to twice its original length or more and, upon release of the stress, recovers to its approximate original length rapidly. (BS 3588 describes many adhesives based on synthetic or natural rubbers).

EMBRITTLEMENT loss of plasticity resulting in the production of fissures, cracks and flaws, under low impact conditions, e.g. in an overdried adhesive; stratification of an adhesive by atmospheric oxidation, where plasticiser components migrate into the substrate, or by exposure to extreme service temperatures.

ENCAPSULANT a protective film enclosure of adhesive particles or liquids to prevent adhesive coalescence until pressure is applied; material used to separate an adhesive and its catalyst by coating each component individually.

EQUILIBRIUM MOISTURE CONTENT the moisture content at which timber or other material suffers no loss or gain in moisture when exposed to constant humidity and temperature conditions.

EXOTHERMIC REACTION a chemical reaction evolving heat as a by-product of the process.

EXTENDER a material added to an adhesive to reduce the amount of primary binder required per unit area of adherend surface; inert material added to an adhesive to improve void filling properties and reduce crazing (cf. BINDER, DILUENT, FILLER, THINNER).

EXTRUDE to expel material through an orifice; method of applying melted adhesives at the adhesive interface.

FAILURE, ADHEREND joint failure by cohesive failure of the adherend (cf. FAILURE, ADHESIVE and FAILURE, COHESIVE).

FAILURE, ADHESIVE joint failure such that the separation occurs at the surface of the adherend, e.g. the failure in adhesion of a pressure sensitive tape when peeled from an adherend.

FAILURE, COHESIVE joint failure by cohesive failure within the adhesive.

FAILURE, CONTACT the failure of an adhesive joint as a result of incomplete contact during assembly, between adherend and adhesive surfaces, or between adhesive surfaces.

FATIGUE a condition of stress from repeated flexing or impact force upon the adhesive—adherend interface; weakening of material caused by repetitive loading and unloading.

FATIGUE STRENGTH see STRENGTH, FATIGUE

FAYING SURFACE the surface of an adherend which makes contact with another adherend.

FEATHERING the tapering of an adherend on one side to form a wedge section, as used in a scarf joint.

FIBRE TEAR the disruption of fibres during separation of paper, textiles, etc., from the adhesive-adherend interface.

FILLER an adhesive additive intended to improve the strength and performance of the adhesive. Fillers are usually powdered inorganic or organic materials with non-adhesive properties by themselves (cf. BINDER, EXTENDER).

FILLER SHEET a deformable, resilient sheet material acting as a sandwich layer in assemblies; a layer between an assembly and a bonding pressure applicator used to ensure the uniform application of pressure over the bonding area.

FILLET that portion of an adhesive joint which fills the corner or angle formed by two adherends; the term for junction of outer skin and inner core in honeycomb assemblies.

FILM FORMING the ability of an adhesive to form a stable continuous film.

FILM, SUPPORTED ADHESIVE an adhesive material incorporating a carrier that remains in the bond when the adhesive is employed; carrier support material is usually composed of organic and/or inorganic fibres which may be in woven form.

FILM, UNSUPPORTED ADHESIVE an adhesive material in film form without a carrier support.

FINISHING the preparation of laminated timber to specification size by cutting, planing, sanding and pore sealing.

FLEXIBILITY the ability to sustain stretching, bending or compression.

FLEXURAL STRENGTH see STRENGTH, FLEXURAL

FLOW the movement of an adhesive during bond formation before it is set.

FUSION the formation of continuous adhesive film by melting; formation of homogeneous material, at an interface, from molten adhesive and adherend.

GEL a semi-solid system consisting of a network of solid aggregates in which liquid is held; a jelly; the semi-solid state of a colloidal system as opposed to sol, the liquid phase. The system is capable of deformation by heat or pressure, e.g. animal glues are strongly cohesive gels subject to deformation by low shearing forces.

GELATION the formation of a gel.

GELOMETER indentation instrument for measuring the deforming force on gels under specified conditions.

GEL POINT the stage at which a liquid begins to exhibit the properties of a gel.

GEL STRENGTH resistance of a gel to shear.

GEL TIME the time required for a liquid adhesive system to form a gel under specified conditions.

GLUE, (N) a term synonymous with 'adhesive' (cf. ADHESIVE, PASTE, GUM, MUCILAGE); a high molecular weight protein (gelatin) derived from skin, bone and connective tissues of animals and the basis of an aqueous adhesive.

GLUE, (V) to bond adherends.

GLUE-LINE STARVED a deficiency of adhesive between two adherends resulting in an unsatisfactory bond.

GLULAM glued laminated material.

GRAB see TACK

GREEN STRENGTH the strength of a joint on assembly with an unset adhesive.

GUM, (N) generic term describing a class of adhesive materials, synthetic and natural, which display good tack and flow properties. Natural resin from plant exudates, colloidal or water soluble organic acids and carbohydrates (cf. ADHESIVE, GLUE, RESIN).

GUM, (V) to bond adherends.

HARDENER a catalytic or cross-linking material used to promote the setting of adhesives; liquid or powder additive which accelerates curing (see BS 1203).

HEAT REACTIVATION the use of heat to effect adhesive

activity, e.g. hot-melt adhesive; completing the curing process of a B-staged resin.

HEAT SEAL the use of heat reactivation to prepare a joint with a thermoplastic material present, as a thin layer, on the adherends; bringing adherend surfaces to their melting point and bonding under pressure.

HIGH-FREQUENCY HEATING *see* DIELECTRIC CURING

HONEYCOMB CORE a sheet material, which may be metal, formed into cells (usually hexagonal) and used for sandwich construction in structural assemblies.

HOT PRESSING the curing of thermosets by heat and pressure application to a bonded assembly, e.g. plywood and laminate manufacture or production of multi-layer printed circuit boards.

HUMIDITY the water content of unit volume of air (*cf.* RELATIVE HUMIDITY).

HYGROSCOPIC describes materials which absorb and retain atmospheric moisture.

IMPACT STRENGTH *see* STRENGTH, IMPACT

IMPREGNATION the penetration of liquid into a porous or fibrous material.

INHIBITOR a material which retards a chemical reaction. Inhibitors are employed to extend the shelf-life and pot-life of certain adhesives.

INTERFACE the contact area between adherend and adhesive surfaces.

IRREVERSIBLE refers to chemical reactions which proceed in one direction and are not reversible. A term applied to thermosetting resins.

JIG a former used to hold a bonded assembly until the adhesive has cured; supporting frame for the production of laminate shapes under pressure.

JOINT the juncture of two adherends which are held together by an adhesive layer.

JOINT AGEING TIME *see* TIME, JOINT CONDITIONING

JOINT, BUTT the end joints formed by the conjunction of the squared ends of two adherends.

JOINT, CLOSE a test joint to check adhesive behaviour in a thin glue line.

JOINT CONDITIONING TIME *see* TIME, JOINT CONDITIONING

JOINT, EDGE a joint formed by bonding the edge faces of two adherends with an adhesive.

JOINT, GAP a test joint to check adhesive behaviour in a thick glue line.

JOINT, LAP a joint formed by bonding the overlapped portions of two adherends. Double lap joints involve the overlapping of opposing faces of one adherend.

JOINT, SCARF a joint formed by cutting away angular segments from the leading edges of two adherends and bonding the cut areas together.

JOINT, STARVED a joint with insufficient adhesive in the glue line to produce a satisfactory bond; glue deficiency in a joint due to gap formation between adherends resulting from thinly spread adhesive,

too short an assembly time, too much bonding pressure leading to adhesive exudation, excessive absorption of adhesive by adherend, or excessively rough adherend surfaces.

LAMINATE, (N) a product formed by bonding two or more layers of material(s) together (*cf.* LAMINATE, CROSS and LAMINATE, PARALLEL).

LAMINATE, (V) to bond layers of a material with an adhesive.

LAMINATE, CROSS a laminate in which the grain direction of some layers of material is oriented at right angles to the grain of the remaining layers (*cf.* LAMINATE, PARALLEL).

LAMINATED WOOD glued wood layers (*cf.* LAMINATE, CROSS and LAMINATE, PARALLEL).

LAMINATE, PARALLEL a laminate in which the grain of all layers of material are oriented approximately parallel to each other (*cf.* LAMINATE, CROSS).

LAY-FLAT *see* ADHESIVE, LAY-FLAT

LIQUIFIER a substance used to reduce the tendency to gel and lower the viscosity of protein or carbohydrate adhesive systems, e.g. urea.

MANUFACTURED UNIT a quantity of adhesive or adhesive component produced on a process run; a batch of material.

MASTIC a resinous cement derived from a resin exuding tree. A highly viscous adhesive material applied by trowel to give thick glue-lines with gap sealing properties.

MATERIAL FAILURE the rupture of an adherend within an assembly.

MECHANICAL ADHESION *see* ADHESION, MECHANICAL

MEMBER a structural part

MIGRATION the transfer of an adhesive plasticiser into the adherend.

MINERAL GLUE an adhesive based on inorganic materials, e.g. aqueous sodium silicate adhesive.

MODIFIER a chemically inert additive material which alters the properties of an adhesive (*cf.* FILLER, EXTENDER, PLASTICISER).

MODULUS OF ELASTICITY *see* ELASTICITY, MODULUS OF

MOISTURE METER an instrument for measuring the moisture content of timber, concrete, etc.

MOISTURE VAPOUR TRANSMISSION the ability of an adhesive film to allow moisture penetration into the adjacent adherend.

MONOMER a simple chemical compound which can react with itself, under catalytic action, to form a polymer (*cf.* POLYMER).

MUCILAGE an aqueous adhesive of low bond strength which is usually plasticised.

NEWTONIAN FLUID a fluid in which shearing rate is directly proportional to the applied torque (*cf.* VISCOSITY).

NON-COALESCENCE non-miscibility of two adhesive layers during bonding.

NON-IDEAL not conforming to the theoretical behaviour of an ideal system.

NON-WARP an adhesion system not subject to curling, shrinking, or other distortions.

NORMAL a geometrical term for a direction perpendicular to a plane surface.

NOVOLAK a phenolic-aldehyde thermoplastic resin.

ONE-WAY STICK the application of an adhesive to one adherend only.

OPEN ASSEMBLY TIME *see* TIME, ASSEMBLY

ORDINARY STORAGE indoor storage at temperatures not greater than 20°C.

OVERLAP the type of joint where one adherend surface extends beyond the leading edge of another.

OXIDATION a chemical reaction in which oxygen combines with a material to form an oxide; deterioration of an adhesive film as a result of atmospheric exposure.

PASTE a high viscosity adhesive composition having high yield value, e.g. the adhesive prepared by cooling a heated starch-water mixture (*cf.* ADHESIVE, GLUE, MUCILAGE).

PEEL STRENGTH *see* STRENGTH, PEEL

PENETRATION the passage of an adhesive into an adherend. The depth of adhesive layer within an adherend measures this property.

PERMANENCE the resistance of an adhesive joint to deterioration.

PH a measure of acidity or alkalinity on a scale from 1 to 14. The logarithm (to base 10) of the reciprocal of the hydrogen ion concentration in an aqueous solution determines the numerical value. Strong acids have a pH from 1 to 3; water is neutral at a pH of 7; strong alkalies have a pH range from 10 to 14 (*cf.* BS 1647).

PHASE a homogeneous and physically distinct part of a system separated from other parts of the system by definite bounding surfaces, e.g. ice, water and water vapour are three phases.

PHASING the separation of a multi-component adhesive into two or more layers.

PICK to experience tack; uneven transfer of an adhesive from a roller coater due to high surface tack.

PICK-UP ROLL a spreading roller which picks up adhesive from a reservoir.

PLASTIC capable of being moulded or otherwise deformed.

PLASTICISER a material added to an adhesive to increase its flexibility and plastic flow properties. High boiling point organic solvents are often used to promote molecular mobility of thermoplastic resins; humectants, e.g. calcium chloride, glycerol, are incorporated to impart flexibility to protein-adhesive films. Plasticisers may lower the elastic modulus of the cured adhesive or reduce its melt viscosity (*cf.* LIQUIFIER).

PLASTICITY the propensity for continuous and permanent deformation, of materials (adhesives), without rupture, upon the application of a force exceeding the yield value of the material.

PLASTICS a generic term for materials based on natural or synthetic polymers susceptible to shaping and moulding by virtue of their flow properties. Plastics processing is often assisted by heat and pressure application.

PLASTIC VISCOSITY a property of materials which flow only when a shear stress above a minimal value is applied.

POLARITY the surface activity of an adherend as determined by the chemical structure of the material surface, e.g. wood, paper, cloth are polar whereas rubber is non-polar.

POLYMER a compound formed by the reaction of identical simple molecules containing active functional groups which combine to produce high molecular weight material (*cf.* CO-POLYMER).

POLYMERISATION the reaction of monomers to form large, high molecular weight molecules (polymers). Co-polymerisation refers to the process involving two or more different monomers (*cf.* CONDENSATION).

POROSITY the ability of an adherend surface to absorb an adhesive.

POT-LIFE the effective working time for an adhesive after preparation; interval before the adhesive systems become unusable through an increase of viscosity or curing.

PRECISION the measure of technique reliability or reproducibility as indicated by the magnitude of the random errors (*cf.* REPRODUCIBILITY).

PRESSURE the application of force to adherends, normal to the adhesive layer, to increase spreading and wetting.

PRESSURE SENSITIVE a term describing adhesive materials which bond to adherend surfaces at room temperature immediately a low pressure is applied; a characteristic of sticky tapes coated with permanently tacky adhesives.

PRESSURE SETTING describes adhesives which require only pressure application to effect permanent adhesion to an adherend.

PRETREATMENT those treatments, mechanical, chemical or physical, which are applied to adherends to promote adhesive receptivity.

PRIMER an adherend surface coating applied before the adhesive to improve bond performance. A strippable primer is a protective surface coating which is removed prior to bonding.

PYROMETER an instrument for temperature measurement; equipment used to determine the glue-line temperature.

QUALIFICATION TEST the determination of conformity, or otherwise, of an adhesive product to the requirements of a relevant specification.

REACTIVE DILUENT a low viscosity liquid dilutant for solvent-free thermosetting resins of high viscosity. The diluent undergoes chemical reaction with the adhesive whilst curing proceeds.

RELATIVE HUMIDITY the ratio of the weight of water vapour in a given volume of air to the weight required to saturate it at the same temperature. High relative humidities favour the acquisition of water by materials.

RELEASE AGENT an abhesive material which prevents bond formation (*cf.* ABHESIVE).

RE-MOISTENING reactivation of an adhesive film system with water or solvent.

REPRODUCIBILITY the variation in results obtained by a standard test procedure (*cf.* PRECISION).

RESIN the general term for natural and synthetic polymers which are amorphous and have no fixed melting point. Resins tends to flow under stress and fracture conchoidally on impact.

RESIN, AMINO PLASTIC a synthetic thermosetting resin obtained from the reaction of ureas, melamines or mixtures of these compounds, with formaldehyde.

RESINOID a term describing synthetic thermosetting resins in a temporary fusible state (*cf.* THERMO-SET and NOVOLAK).

RESIN, PHENOLIC a synthetic resin; product from the reaction of a phenol with an aldehyde.

RESITE alternative term for a resin at the C-stage.

RESITOL alternative term for a resin at the B-stage.

RESOL alternative term for resin at the A-stage.

RETARDER an additive which slows the rate of a chemical reaction, e.g. the curing of an adhesive (*cf.* INHIBITOR).

RETROGRADATION a deterioration process involving reversion of a material to an original physical form, e.g. the change in consistency, from low to high, of certain vegetable adhesives on ageing.

RHEOLOGY the study of deformation and flow behaviour of materials under stress.

ROLLER COATER mechanical equipment for the application of adhesives to sheet materials.

ROSIN the resin remaining after distillation of pine tree turpentine derived from the sap (gum rosin), or wood stumps (wood rosin).

SAGGING run or flow-off of adhesive from an adherend surface due to application of excess or low viscosity material.

SANDWICH PANEL an assembly composed of metal skins (facings) bonded to both sides of a lightweight core.

SCARF JOINT *see* JOINT, SCARF

SEALANT a gap filling material to prevent excessive absorption of adhesive or penetration of liquid or gaseous substances.

SELF ADHESIVE a material which bonds to itself.

SELF VULCANISING the process by which an adhesive vulcanises without the aid of heat, i.e. self curing.

SERVICE CONDITIONS the environmental conditions to which a bonded structure is exposed, e.g. heat, cold, humidity, radiation, vibration, etc.

SET the conversion of an adhesive into a permanently cured state by chemical or physical means such as vulcanisation, polymerisation, dehydration, gelation, evaporation, etc. (*cf.* CURE and DRY).

SETTING TEMPERATURE *see* TEMPERATURE, SETTING

SETTING TIME *see* TIME, SETTING

SHEAR STRENGTH *see* STRENGTH, SHEAR

SHELF LIFE *see* STORAGE LIFE

SHORTNESS the non-formation of threads, strings or filaments during adhesive application.

SHRINKAGE the volume reduction occurring during adhesive curing. This is sometimes expressed as a percentage volume or linear shrinkage; size reduction of adhesive layer due to solvent loss or catalytic reaction.

SINGLE SPREAD *see* ADHESIVE SPREAD

SIZE a chemical substance for coating adherend surfaces to reduce the absorption of oil, water, or adhesive; protective primer coat to prevent scuffing, e.g. rosin or synthetic polymer.

SIZING the application of a material to a surface to seal pores and thus reduce adhesive absorption; to improve adhesion to a surface by changing the surface properties.

SKINNING the formation of a dry surface layer (skin) on an adhesive coating following too rapid evaporation of the solvent vehicle.

SLIP the ability of an adhesive to accommodate adherend movement or re-positioning after application to adherend surfaces.

SLIPPAGE the movement of adherends with respect to each other during adhesive bonding.

SLIP SHEET INTERLINER a sheet or film used to cover an adhesive whilst handling; protective film for a film adhesive.

SOFTENER a plasticising additive to reduce adhesive embrittlement; component of elastomeric films to increase their flexibility.

SOFTENING POINT the temperature at which an adhesive commences to flow or soften.

SOLIDS CONTENT the percentage weight of non volatile material in an adhesive formulation, as determined by a specified test method.

SOLVENT the liquid component of an adhesive which serves to reduce its viscosity and increase wetting of adherends; liquid which dissolves an adhesive.

SOLVENT CEMENT an adhesive utilising an organic solvent as the means of depositing the adhesive constituent.

SOLVENT REACTIVATION the application of solvent to a dry adhesive layer to regenerate its wetting and bonding properties (*cf.* RE-MOISTENING).

SPECIFIC ADHESION *see* ADHESION, SPECIFIC

SPREAD OF ADHESIVE *see* ADHESIVE SPREAD

SQUEEZE OUT the application of pressure to an assembly to expel excess adhesive from the glue-lines.

STABILISER an adhesive additive which prevents or minimises the change in properties, e.g. by adherend absorption, demulsification, rapid chemical reaction.

STAINING the discoloration of adherends by adhesives, curing agents or solvents.

STARVED JOINT *see* JOINT, STARVED

STORAGE LIFE the time period for which an adhesive remains usable when stored under specified temperature conditions. Adhesive refrigeration often extends the storage life.

STRENGTH, CLEAVAGE the tensile load expressed as force per unit width of bond required to cause cleavage separation of a test specimen of unit length.

STRENGTH, DRY the adhesive joint strength as determined immediately after drying or after a conditioning period under specified conditions (*cf*. STRENGTH, WET).

STRENGTH, FATIGUE the maximum load that a joint will sustain when subjected to repeated stress application under specified conditions, viz: range of stress and its mean value together with the frequency of stress application.

STRENGTH, FLEXURAL the maximum shear of a material when tested as a rectangular cross-section beam, loaded at midspan.

STRENGTH, IMPACT ability of material to resist shock by physical blow directed against it (e.g. adhesive) or sudden wrenching or shocks (e.g. packaging tapes). Impact shock is the transmission of stress to an adhesive interface by sudden vibration or jarring blow of the assembly, measured in work units per unit area.

STRENGTH, PEEL the resistance of an adhesive joint to peel stress, the force per unit width of the bond expresses the point of failure under peeling stress.

STRENGTH, SHEAR the resistance of an adhesive joint to shearing stresses. The force per unit area sheared, at failure.

STRENGTH, TENSILE the resistance of an adhesive joint to tensile stress; the force per unit area under tension at failure.

STRENGTH, WET the strength of an adhesive joint determined immediately after its removal from a liquid (at specified time, temperature and pressure) in which it has been immersed; the strength of a latex-adhesive joint determined whilst the adhesive remains wet (*cf*. GREEN STRENGTH).

STRESS the force per unit area resulting from an applied load.

STRESS, CLEAVAGE the stress resulting from the concentration of forces on a joint at the edge of a rigidly bonded area.

STRESS, PEEL the stress resulting when the forces exerted on a joint tend to strip off one adherend surface.

STRESS, SHEAR the stress resulting when the forces applied to a joint act in the same plane as the adhesive layer.

STRESS, TENSILE the stress resulting when the forces applied to a joint act perpendicular to the adhesive layer.

STRINGINESS the complete breakoff of adhesive film when it is divided between transfer rollers, stencils, picker plates, etc., uneven transfer of an adhesive to an adherend surface (*cf*. COTTONING).

STRIPPING TESTS a term referring to peeling tests (*cf*. STRENGTH, PEEL).

SUBSTRATE the material surface to which an adhesive material is applied for bonding or coating or other purposes.

SURFACE PREPARATION the physical and chemical methods employed to prepare an adherend surface for bonding, e.g. abrasion, solvent cleaning, etching, priming.

SUSTAINED LOAD TEST a test for the assessment of adhesive performance when placed under stress for an extended period.

SYNERESIS the exudation of liquid globules by gels on standing; the bleeding of plasticisers from polymer formulations.

SYNTHETIC RESIN any resin which is manufactured and not obtained from natural sources, e.g. polyester resins, formaldehyde resins, polyamides.

TACK the resistance offered by an adhesive film to detachment from the adherend surface; stickiness of an adhesive enabling it to form an instant bond when brought into low pressure contact with an adherend.

TACK, DRY the self-adhesion property of certain adhesives which are touch-dry (a stage in the evaporation of volatile constituents).

TACK-DRY the state of an adhesive which has lost sufficient volatiles (by evaporation or absorption into the adherend) to leave it in the required sticky condition.

TACKIFIER an additive intended to improve the stickiness of a cast adhesive film; usually a constituent of rubber-based and synthetic resin adhesives.

TACK, RANGE OR STAGE the interval of time during which an adhesive will remain in the dry tack state after its application to a surface, under standard conditions of humidity and temperature.

TAPE a film form of adhesive which may be supported on carrier material.

TAPE, GUMMED FABRIC a gum coated fabric which becomes adhesive on moistening.

TAPE, GUMMED PAPER a gum coated paper (usually Kraft) which becomes adhesive when moistened.

TAPE, HEAT SEALING ADHESIVE a strip of material (usually fabric, metal foil, paper or plastic film) coated with a heat sealing adhesive.

TAPE, PRESSURE SENSITIVE *see* PRESSURE SENSITIVE

TELEGRAPHING the visible transmission of faults, imperfections and patterned striations occurring in an inner layer of a laminate structure.

TEMPERATURE, CURING the temperature to which an adhesive or a bonded assembly is subjected to cure the adhesive. The temperature reached by the adhesive whilst curing is not necessarily the temperature of the atmosphere adjacent to the assembly (*cf.* TEMPERATURE, DRYING and TEMPERATURE, SETTING).

TEMPERATURE, DRYING the temperature to which an adhesive coated adherend, or bonded assembly is subjected to dry the adhesive. The temperature reached by the adhesive in drying usually differs from the temperature of the assembly environment (*cf.* TEMPERATURE, CURING and TEMPERATURE, SETTING).

TEMPERATURE, MATURING the temperature, for a given time and bonding procedure, which produces required characteristics in components bonded with ceramic adhesives.

TEMPERATURE, SETTING the temperature to which an adhesive or a bonded assembly is subjected to set the adhesive. The adhesive setting temperature may differ from the temperature of the assembly environment (*cf.* TEMPERATURE, CURING and TEMPERATURE, DRYING).

TENSILE SHEAR the apparent stress applied to an adhesive in a lap joint.

TESTS, ACCELERATED the testing of materials by exposure to intensified simulation of service conditions, e.g. weathering, radiation, etc.

TESTS, DESTRUCTIVE tests involving the destruction of assemblies in order to evaluate the maximum performance of the adhesive bond.

TESTS, NON-DESTRUCTIVE inspection tests for the evaluation of bond quality without damaging the assembly, e.g. ultrasonics, visual inspection, etc.

THERMOPLASTIC susceptible to repeated softening by heating and hardening by cooling.

THERMOSET a material which does not soften on heating, as a result of being formed from an irreversible chemical reaction initiated by catalysts, heat, light, radiation, etc.

THINNER a liquid additive intended to modify viscosity properties (*cf.* DILUENT and EXTENDER).

THIXOTROPY a property of materials which display a reduction in viscosity when a shearing action is applied. Some adhesive systems become thinner in consistency on agitation and thicken again when left undisturbed.

THROWING the transfer of adhesive globules from roller coater edges rotating at high speed. Also known as spitting.

TIME, ASSEMBLY the interval between adhesive application to the adherend and the application of heat and/or pressure to the assembly. Assembly time involves a time interval for assembly of parts after adhesive application (open assembly time) followed by a time interval between the assemblage of parts for bonding and the application of heat and/or pressure to the assembly closed assembly time).

TIME, CURING the period of time for which a bonded assembly is subjected to heat and/or pressure to cure the adhesive. N.B. curing may continue to take place on removal of the assembly from conditions of heat and/or pressure.

TIME, DRYING the period of time for which an adhesive coating is allowed to dry without the application of heat and/or pressure; time between adhesive application and adhesive particle coalescence (*cf.* TIME, CURING; TIME, JOINT CONDITIONING; and TIME, SETTING).

TIME, JOINT CONDITIONING the period between the removal of the joint from the bonding conditions, e.g. heat, pressure, radiation, and the attainment of maximum bond strength. Term is often described as Joint Ageing Time or Joint Conditioning Period.

TIME, SETTING the interval of time required for the adhesive to set in a bonded assembly subjected to heat and/or pressure (*cf.* TIME, CURING; TIME, DRYING; and TIME, JOINT CONDITIONING).

TOXICITY the effectiveness of a poisonous material related to its concentration.

TRANSFERENCE the retention of some adhesive material, by a substrate, on removal of an attached pressure sensitive adhesive.

TROWEL the use of a knife or spatula to spread high viscosity adhesive material.

TUNNELLING the formation of tunnel-like chambers in incompletely bonded laminates by the release and deformation of longitudinal sections of the adherend.

TWO-WAY STICK the application of an adhesive to both adherends before assembly.

UNROLLING OR UNWINDING ADHESION a term describing the bond which resists the peeling force needed to free tape from a roll.

VEHICLE the carrier medium (liquid) for an adhesive material which improves its ease of application to adherends; solvent component of an adhesive.

VISCOSITY a measure of the resistance to flow of a liquid. For Newtonian liquids, the shear rate is proportional to the shear stress between laminae of moving fluid; for non-Newtonian liquids it is not proportional.

VISCOSITY COEFFICIENT the viscous force is proportional to the velocity gradient between two parallel Newtonian fluid layers a short distance apart. The constant of proportionality is called the coefficient of viscosity of the fluid.

VULCANISATION the cross linking of an adhesive material by means of heat or catalysts, the chemical reaction of rubber with sulphur or other agents to alter its physical properties, e.g. to cause less tackiness, reduce plastic flow and increase tensile strength (*cf.* SELF-VULCANISING).

VULCANISE to subject to vulcanisation.

WARP a distortion of an adherend surface; a variation from the normal plane surface of a material.

WATERPROOF GLUE a cured adhesive which is resistant to water.

WATERPROOF MEDIUM reagents in solution or dispersion, used to improve the water resistance of a bond.

WEBBING the formation of threads or filaments on separation of adhesive transfer surfaces such as roller coaters or stencils (*cf.* STRINGINESS).

WET STRENGTH *see* STRENGTH, WET

WETTING a surface is said to be completely wet by a liquid if the contact angle is zero and incompletely wet if it is a finite angle. Surfaces are commonly regarded as unwettable if the angle exceeds 90°C.

WOOD FAILURE the cohesive failure of wood fibres in strength tests on bonded specimens expressed as a percentage of the total area displaying this type of failure.

WOOD, VENEER sheet wood, of thickness not greater than 7 mm, employed in laminate fabrication.

WORKING LIFE the period of time during which an adhesive remains suitable for use after it has been compounded with materials such as catalysts, solvents, plasticisers, etc. (*cf.* POT-LIFE).

YIELD STRESS the minimum stress required to deform a plastic material permanently.

REFERENCES

Plastics Engineering Handbook, 3rd Edn, Reinhold, New York, (1960).

GUTTMANN, W. H., *Concise Guide to Structural Adhesives* Reinhold, New York, (1961).

'Structural Sandwich Constructions; Wood; Adhesives', *ASTM Stand.,* Part 16, June, (1968).

CHAPTER 12 | # Bibliography

BOOKS

ALNER, D. J., (Ed.), *Aspects of Adhesion*, University of London (Series 1965 onwards).
Published papers from the annual international conferences of The City University, London. Fundamental theory and practical applications of the subject form the theme of these proceedings in which academic and industrial contributors participate. Series of eight volumes issued to date forms a valuable review of developments in the theory and practice of adhesive bonding over the last decade.

Adhesion—Fundamentals and Practice, by The Ministry of Technology (U.K.), Gordon and Breach, London, (1971).
The book contains authoritative papers originally presented at an International Conference on Adhesion, University of Nottingham (1969). The purpose of the work is to promote understanding of the whole field of adhesion by relating theory and practice. Section headings comprise the following: The Adherend Surface; The Preparation of the Adherend Surface; The Adhesive; Joint Design; Methods of Test. A useful reference book on the principles and practice of adhesion.

Adhesives in Surgery, published by B. Braun Foundation, Melsungen, W. Germany, (1968).

Adhésion des Elastomères (Adhesion of Elastomers), Association Française des Ingenieurs du Caoutchoue et des Plastiques, Paris, (1972).
Proceedings of the International Rubber Conference organised by the French Association of Rubber and Plastics Engineers in Paris (1970).
Twenty-one papers (16 in English) covering scientific and technical aspects of adhesion of elastomers presented by outstanding experts in the field.

BODNAR, M. J., (Ed.), 'Structural Adhesives Bonding', *Polymer Symposia, No. 3*, Interscience, (1966).
The book comprises all the papers presented at the Symposium on Structural Adhesives Bonding held at the Stevens Institute of Technology in September 1965. Emphasis of text is on military and industrial applications for structural adhesives in metal bonding.

BRUNO, E. J., (Ed.), *Adhesives in Modern Manufacturing*, Palmerton Pub. Co. Inc., New York, (1973).
Written by manufacturing managers and field engineers with experience in the industrial usage of adhesive bonding.

Emphasis is on practical low-cost applications and discussion of assembly operations and quality control procedures.

BUCHAN, S., *Rubber to Metal Bonding*, Crosby, Lockwood & Son, London, (1960).

CAGLE, C. V., *Adhesive Bonding, Techniques and Applications*, McGraw-Hill, New York, (1968).
Case history studies of industrial companies concerned with adhesive bonding as a production technique. Recent bonding techniques are discussed and detailed consideration given to non-destructive methods of bond testing. Includes a glossary of terms and lists military and other specifications for structural adhesives.

CHUGG, W. A., *Glulam: The Manufacture of Glued Laminated Structures*, Benn, (1964).
Properties of adhesive materials and discussion of bonding technology and production of laminated wood and timber structures.

COOK, J. P., *Construction Sealants and Adhesives*, Wiley-Interscience, New York, (1970).
Comprehensive and practical guide to selection and application of adhesives and sealants used in construction industries. Emphasis is on the performance of the materials rather than their chemistry and formulation.

DELMONTE, J., *The Technology of Adhesives*, Hafner, (1965).
Discusses the formulation of adhesives and aspects of industrial bonding technology.

DELOLLIS, N. J., *Adhesives for Metals; Theory and Technology*, Industrial Press Inc., New York, (1970).
Survey of the factors involved with adhesion, materials, and bonding technology. Suitable for students, and materials and production engineers at the practical applications level, as an introduction to adhesives usage. Several federal and military specifications are listed in an appendix.

DIETZ, A. G. H., (Ed.), *Composite Engineering Laminates*, MIT Press, (1969).

GUTTMAN, W. H., *Concise Guide to Structural Adhesives*, Reinhold, (1961).
Text covers the principles of adhesion, selection of adhesives, precautions in handling resins and curing agents, preparation of adherend surfaces, and bonding technology. Main content of the book concerns a compilation of structural

adhesives and their properties, and a cross reference index to enable the user to select adhesives for metal and non-metal bonds. Official federal and military specifications for structural adhesives are described in detail. Essentially, a reference work for industrial designers and engineers engaged with production research and development.

HOUWINK, R., and SALOMON, G., *Adhesion and Adhesives*, Vol. 1, *Adhesives*, Elsevier, (1965).

Revised edition of a standard reference book compiled by authorities in the field and first published in 1951. Properties of natural and synthetic, organic, inorganic, and metallic adhesives are concisely surveyed with reference to the mechanical properties of adhesives and adherends. The bonding of metals and wood for industrial applications is described in detail, and consideration given to testing procedures and stress analysis of bonded joints. Emphasis of the book is on adhesives as engineering materials.

Vol. 2, *Applications*, Elsevier, (1967).

Major developments in the industrial usage of adhesives are discussed in this revised edition to which 16 specialists have contributed. Joint design, the selection of surface treatment, and the choice of adhesive and processing method are discussed in detail. Other sections consider the performance of adhesives in composite structures, metal joints, rubber-textile structures, wood structures, and pressure sensitive tape technology. Includes an account of joint testing by destructive and non-destructive methods. Appended Author Index covers the contents of both Volumes 1 and 2 of the work.

KATZ, I., *Adhesive materials—their properties and usage*, revised edition (C. V. Cagle), Foster Publishing Co., Long Beach, California, (1973).

Reference work containing government, industrial, and international specifications for adhesives and allied materials. Provides chemical, physical, and performance properties of bonding materials together with specification guidelines for adhesives use. Surface preparation for bonding is dealt with in a separate section.

KNIGHT, R. A. G., *Adhesives for Wood*, Chapman and Hall, (1952).

MCGUIRE, E. P., *Adhesive Raw Materials Handbook*, Padric Publishing Co., New Jersey, (1964).

PARKER, R. S. R., and TAYLOR, P., *Adhesion and Adhesives*, Pergamon, (1966).

Introductory text to the chemistry and usage of adhesive materials.

LEE, L. H., (Ed.), *Recent Advances in Adhesion*, Gordon and Breach, London, (1973).

Proceedings of the Symposium on Recent Advances in Adhesion held in Washington 1972. Important reference book for scientists interested in applying theories to the solution of practical adhesion problems.

PATRICK, R. L., (Ed.), *Treatise on Adhesion and Adhesives*, Vol 1, *Theory*, (1967), Vol 2, *Materials*, (1969), Marcel Dekker, New York.

Existing theories, hypotheses, and postulates on the phenomenon of adhesion are discussed in the text by outstanding specialists in the field. Of major importance to those interested in theoretical aspects of adhesion and the interpretation of fundamental knowledge.

PERRY, H. A., *Adhesive Bonding of Reinforced Plastics*, McGraw-Hill, New York, (1959).

Standard reference work on the adhesives technology of reinforced plastics which deals with properties and usage of adhesives, joint design, surface preparation and other aspects of composite manufacture.

SEMERDJIEV, S., *Metal-to-Metal Adhesive Bonding*, Business Books, London, (1970).

The text is primarily concerned with metal-to-metal bonding practice in the various branches of industrial engineering (electrical, mechanical, civil, etc.). A separate section deals with honeycomb sandwich constructions. Bonding techniques are discussed and consideration given to the design and testing of joints. Principles of adhesion and the chemical composition of adhesives and curing reactions are also described.

SERAFINI, T. T., and KOENIG, J. L., *Cryogenic Properties of Polymers*, Marcel Dekker, New York, (1968).

SHIELDS, J., *Adhesive Bonding*, Oxford University Press, (1974).

One of a series of monographs on fastening technology outlining main features of existing adhesives materials and their industrial usage. Suitable for students, technicians and designers seeking an introduction to bonding technology and interested in a choice of fastening method.

SKEIST, I., (Ed.), *Handbook of Adhesives*, Reinhold, (1962).

Reference work on adhesive materials and bonding technology. Fundamental aspects of bonding are considered but most of the emphasis in the text is on properties of the basic adhesive materials and industrial applications areas. Suitable for the non-specialist.

SNOGREN, R. C., *Handbook of Surface Preparation*, Palmerton Publishing Co., New York, (1974).

Comprehensive text on practical techniques of surface preparation for adhesion. Deals with selection and optimisation of surface treatments and implementation of process control procedures, for metallic and non-metallic engineering materials.

WEISS, P., *Adhesion and Cohesion*, Elsevier, (1962).

Proceedings of a symposium on 'Adhesion and Cohesion' sponsored by General Motors Research Laboratories, Michigan, 1961. Text deals with discussion and interpretation of fundamental knowledge and its relation to adhesion and cohesion phenomena. Of value to the specialist and research worker concerned with theoretical aspects of adhesion rather than practical applications of bonding.

'Adhesion and Bonding', reprint of a section in *Encyclopaedia of Polymer Science and Technology*, Wiley-Interscience, (1971).

Concise practical guide for students and designers. Text covers surface properties, composition and applications of adhesives. Joint design and evaluation.

REVIEWS

General:

WAKE, W. C., 'Adhesives', *Royal Institute of Chemistry Lecture Series 1966*, No. 4, Heffer, (1968).

WICK, C. H., 'Adhesive Bonding' (in four parts), *Machinery and Production Engineering*, 112 (2895), Part 1, 732, April, (1968); 112 (2896), Part 2, 836, May, (1968); 112 (2897), Part 3, 932, (1968); 112 (2898), Part 4, 1028, (1968).

ANDERSON, C. C., 'Adhesives', (annual reviews), *Industrial and Engineering Chemistry*, **60**(12), No. 80, August, (1968) and **61**(12), No. 48, August, (1969).

PERKINS, R. B., and GLARUM, S. N., *Adhesives, sealants and gaskets, a survey*, U.S.N.A.S.A. Tech. Utilization Div NASA SP 5066.60, (1967).

Engineering applications:

NEWELL, G. S., 'Adhesives and engineering fabrications', *Light Prod. Engineering*, **4**(7), 4, (1966).

EPSTEIN, G., 'Adhesive bonds for sandwich constructions', *Adhes. Age*, **6**(3), 30, (1963).

HESS, E. F., 'Adhesives for fabricating high strength structures', *Metal Progress*, **10**, 97, June, (1963).

BUTZLAFF, H. R., and CHARTER, K. F., 'Adhesives for the bonding of metals', *Mech. Engineering, N.Y.*, **83** No. 5, 52–55, (1961).

BURGMAN, H. A., 'Selecting structural adhesive materials', *Electro-Tech. N.Y.*, **75**, No. 6, 69–75, (1965).

TWISS, S. B., 'Structural adhesive bonding', *Adhes. Age*, Part 1, 'Adhesive characteristics', **7**(12), 26, (1964), Part 2, 'Adhesive classification', **8**(1), 30, (1965).

HOLLAND, T., 'Adhesives', *Engineering*, **213**(9), 632–637, (1973).

REINHARDT, T. J., 'Engineering properties of adhesives', *Adhes. Age*, **16**(7), 35–41, (1973).

ASTROP, A. W., 'Adhesives and steel replace solder and brass', *Mach. Prod. E.*, **124** (3192), 106–112, (1974).

MINFORD, J. D., 'Evaluating adhesives for joining aluminium', *Metals Eng. Q.*, **12**, 48, (1972).

ROLF, R. L., JOMBOCK, J. R., and PETERS, L. K., 'Adhesives-bonded structural joints in aluminium', *Adhes. Age*, **14**(7), 23–27, (1971).

HARRISS, R. W., 'Role of adhesives and sealants in the automotive industry', *Adhes. Age*, **13**(9), 45–50, (1970).

DRISKO, R. W., 'Underwater marine applications of coatings and adhesives', *J. Paint Tech.*, **47**(600), 40–42, (1975).

EPSTEIN, G., 'Adhesives for space systems', *Adhes. Age*, Part 1, **15**(6), 18–23, (1972); Part 2, **15**(7), 27–31, (1972).

REINHART, T. J., and SCARDINO, W. M., 'Composite to metal bonding using structural thermosetting adhesives', *Adhes. Age*, **18**(2), 23–28, (1975).

MINFORD, J. D., and VADEE, E. M., 'Aluminium-faced sandwich panels and laminates', *Adhes. Age*, **18**(2), 30–35, (1975).

WEBER, C. D., and GROSS, M. E., 'Modified epoxy adhesives speed honeycomb bonding', *Mater. Eng.*, **79**(5), 92–93, (1974).

DUKE, A. J., 'Structural Adhesives—to use or neglect?', *Engng. Mat. Des.*, **11**(7), 937, (1968).

BOCHKARE, V. P., and GLEVITSKI, T., 'Adhesive bonded and welded joints in shipbuilding', *Weld. Prod. R.*, **17**, 43, (1970).

Building Industry:

SLACK, T. K., 'Trends in the use of adhesives in construction', *Adhes. Age*, **17**(3), 17–21, (1974).

LOWE, G. B., 'Structural adhesives and compounds for use in the construction industry', *Adhes. Age*, **16**(12), 41–42, (1974).

HUGENSCH, F., 'Epoxy adhesives in precast prestressed concrete construction', *J. Pre. Concs.*, **19**(2), 112–124, (1974).

ARENS, G., 'Modern adhesives for the construction industry', *Chem. Tech.*, **23**, 615–620, (1971).

Electrical/Electronics:

DE LOLLIS, N. J., 'The use of adhesives and sealants', IEEE, *Electronics*, Vol. PMP **1**(3), 4, (1965), 'Transactions on Parts, Materials and Packaging'.

RIDER, D. K., 'Adhesives in printed circuit applications', *Symposium on Adhesives for Structural Applications*, (Picatinny Arsenal, 1961), 49, Interscience, (1962).

HESS, E. F., 'The outlook for adhesives in electronics', *Electron. Ind.*, **22** No. 6, G10–G13, (1963).

HOLLAND, D. L., 'Adhesives for flexible printed circuits', *Adhes. Age*, **16**(4), 16–25, (1973).

ALBERS, H. J., 'Cyanoacrylate adhesives for fast bonding in the electronics industry', *Adhes. Age*, **14**(6), 23–26, (1971).

MURPHY, T. S., 'Polyester adhesives—vital links in electrical film laminates', *Electr. Prod.*, **13**, 79–82, (1971).

KATZ, I., 'Adhesives for optical and electro-optical applications', *Electro-Tech.*, **71** No. 4, 161–168, (1963).

SHIELDS, J., 'Strain gauge adhesives', *Metron*, **2**(5), 163, May, (1970).

ANZALONE, B., 'Understanding and applying strain gauges, carriers, adhesives and coatings', *Instr. Contr.*, **46**(1), 60–62, (1973).

High and low temperature applications:

KENDALL, B. R. F., and ZABIELSKI, M. F., 'High temperature insulating adhesives for vacuum applications', *J. Vacuum Sci. Technol.*, **3** No. 3, 114–119, (1966).

KAUSEN, R. C., 'Adhesives for high and low temperature', *Mat. Design Engng.*, Part 2, **60**(3), 120, (1964), also Part 1, **60**(2), 108, (1964).

KANTNER, R., and LITVAK, S., 'Adhesives for bonding large high temperature sandwich structures', *Adhes. Age*, **12**(11), 24, (1969).

SMITH, M. B., and SUSMAN, S. E., 'Adhesives for low temperatures', *Mach. Design*, **34** No. 25, (1962).

HERTZ, J., 'Adhesives for extreme-low-temperature applications', *Electrotechnology*, **17**(3), 93, (1962).

CONWAY, D. J., 'Metal-to-metal adhesives for structural applications at elevated temperatures', *Adhes. Age*, **11** No. 5, 30–34, (1968).

TANNER, W. C., 'Adhesives in ordnance applications', *Symposium on Adhesives for Structural Applications*, (Picatinny Arsenal, 1961), 89, Interscience, (1962).

MEINERTZHAGEN, M., 'Using adhesives in thermal insulation', *Adhes. Age*, **16**(6), 31–34, (1973).

Wood technology:

SELBO, M. L., 'Selecting adhesives for wood products', *Adhes. Age*, **16**(10), 36–41, (1973).

PAGE, W. D., 'Using adhesives in plywood components', *Adhes. Age*, **16**(3), 27–29, (1973).

Adhesives for bonding wood to metal, Publication N65–18544, FPL–082, of United States Forest Products Laboratory, (1964).

GRAY, V. R., 'Adhesives in the timber trades', *Aspects of Adhesion*, 1 (Almer, Ed.), University of London Press, (1967).

Paper and Packaging applications:

Adhesives for Packaging, BS 1133, 'Packaging Code', Section 16, (1953) revised (1967).

DEBNAR, D., 'Adhesives used in packaging and converting industries', *Adhes. Age*, **9** No. 2, 30–31, (1966).

EGLI, G. A., 'How to select adhesives for foil/paper laminations', *Adhes. Age*, **14**(9), 39–40, (1971).

BARTLETT, F. P., 'Adhesives for films and foils', *Adhes. Age*, **9** No. 2, 34–35, (1966).

LAMBERT, P. L., 'Recent developments in adhesives for flexible packaging', *Adhes. Age*, **16**(7), 22–25, (1973).

'Hot Melts—review issue of materials and technology', *Adhes. Age*, **15**(8), 21–50, (1972).

348

Plastics and rubber:

TOY, L. E., 'Plastics-Metals: Can they be united?', *Adhes. Age*, **17**(10), 19–24, (1974).

DUNCAN, T. F., 'Adhesives and methods for bonding metal and plastics to porous substrates', *Adhes. Age*, **12**(4), 24, (1969).

MOORE, D. F., and GEYER, W., 'Review of adhesion theories for elastomers', *Wear*, **22**, 113–116, (1972).

MCLACHLAN, I., 'Rubber to metal bonding', *Light Prod. Engng.*, 4 Sept., (1968).

HUDSON, R. E., 'Adhesive systems for fabric-elastomer applications', *Adhes. Age*, **11**(10), 21, (1968).

ROWAND, E., and MACGRANDLE, T. D., 'The age of film adhesives', *Adhes. Age*, **14**(9), 34–38, (1971).

MACGRANDLE, T. D., 'Film adhesives', *Mach. Design*, **43**, 76–80, (1971).

FAHRENDORF, P. M., 'Dry laminating processes', *Adhes. Age*, **14**(3), 32–36, (1971).

ADHESIVES JOURNALS

Adhesives Age, Palmerton Publishing Co. Inc., New York.

Sub-Assembly Components Fastening, Morgan-Grampian Publishing Co., London.

Journal of Adhesion (Ed. L. H. Sharpe), Technomic Publishing Co., Stanford, Conn.

Adhäsion, Ullstein GmbH., Berlin 42, Germany.

ABSTRACTS

A current awareness information service is provided by the Scientific Documentation Centre Ltd., Halbeath, Dunfermline, U.K., for readers requiring more recent abstracts on specific areas of adhesives technology.

The Adhesives Documentum (1968 onwards):

Information retrieval system based on punched card data which cover the areas of applications, bonding technology, production, and raw materials. Comprehensive coverage of trade journals, book reviews and patent literature relating to adhesives and allied products (e.g. sealants, putties, cements, coatings). Originally prepared in German by Hinterwaldner-Verlag, the system is now available as an English version. Updating documentation cards are issued every 1–2 months to subscribers.

Hinterwaldner-Verlag, D. 8000 Munich 90/P O, Box 900–425 Kastanienstr. 13, Germany.